A Failed Eldorado

Colonial Capitalism, Rural Industrialization, African Land Rights in Kenya, and the Kakamega Gold Rush, 1930–1952

Priscilla M. Shilaro

UNIVERSITY PRESS OF AMERICA,® INC.
Lanham • Boulder • New York • Toronto • Plymouth, UK

Copyright © 2008 by
University Press of America,® Inc.
4501 Forbes Boulevard
Suite 200
Lanham, Maryland 20706
UPA Acquisitions Department (301) 459-3366

Estover Road
Plymouth PL6 7PY
United Kingdom

All rights reserved
Printed in the United States of America
British Library Cataloging in Publication Information Available

Library of Congress Control Number: 2007938473
ISBN-13: 978-0-7618-3606-3 (paperback : alk. paper)
ISBN-10: 0-7618-3606-3 (paperback : alk. paper)

Library
University of Texas
at San Antonio

∞™ The paper used in this publication meets the minimum
requirements of American National Standard for Information
Sciences—Permanence of Paper for Printed Library Materials,
ANSI Z39.48—1984

Contents

Maps and Tables		xi-xii
Abbreviation		vii
Preface		xiii
1	Prelude to the Gold Rush	1
	Introduction	1
	Luyia Indigenous Economy	5
	Notes	10
2	The Kakamega Gold Rush and the Kenya Land Commission, 1932–34	13
	Introduction	13
	The Kenya Land Commission: Terms of Reference	14
	The KLC at Kakamega	16
	The Luyia Grievances	18
	The Official View	22
	"Inviolable" African Reserves: The Native Lands Trust Ordinance	23
	Negotiating the NLTO of 1930	24
	Amery Exits and Passfield Imposes the Bill	25
	Changing Fortunes: The Gold Rush and the NLT(A)O of 1932	27
	The NLT(A)O of 1932: A Storm of Protest	32
	Protest in Kenya	32
	Protest in England	34
	The CO Defends the NLT(A)O of 1932	37
	The KLC on the NLT(O) of 1930	39
	The Recommendations of the KLC	42

	Conclusion	44
	Notes	45
3	Rural Industrialization, 1931–52	59
	Introduction	59
	Prospecting, 1931–34	61
	Development by Large Companies, 1934–44	71
	Trends in Mining, 1934–44	76
	"A Failed Eldorado," 1945–52	82
	Conclusion	90
	Notes	90
4	Politics of Land, 1931–52	98
	Introduction	98
	Early Forms of Protest	98
	Organized Protest: The North Kavirondo Central Association (NKCA)	104
	The Colonial State's Response	109
	The Anti-Soil Conservation Campaign	114
	Conclusion	131
	Notes	132
5	Rural Industrialization: The Economic Balance Sheet	143
	Land Alienation, Mining Leases, and Compensation	144
	Stifled African Entrepreneurship	149
	Gold Mining and Labor	157
	World War II and Labor Conscription	163
	Working Conditions	166
	Industrial Labor Protest	169
	Gold Mines as Death Traps	174
	Kakamega: "The Boom Town"	179
	Industrial Penetration and the Market Economy	182
	Conclusion	186
	Notes	187
6	At the Crossroads: Social-Cultural Transformation	204
	Gold Mining and Health	204
	Gold Mining and Rural Socio-Cultural Transformation	211
	Conclusion	216
	Notes	216
7	"A Failed Eldorado"	220

Appendix 1: The NLT(A) O of 1932: Letters of Protest
in England — 231

Appendix II: Kimingini Gold Mine Ltd., Commuted Rent — 233

Appendix III: List of Owners, Property and Value on the
Kimingini Lease — 237

Appendix IV: List of Those Affected by the Edzawa Ridge
Mine Lease (Maragoli) and the Amount of Land Alienated — 239

Appendix V: List of Those Affected by the Edzawa Ridge Mine
Lease (Tintax, Kisa Location) and the Amount of Land Alienated — 241

Bibliography — 243

Index — 257

WITHDRAWN
UTSA Libraries

Abbreviations

AAAO	African Assistant Agricultural Officer
AAO	Assistant Agricultural Officer
ACA	Abaluhya Central Association
ADC	African District Council
AO	Agricultural Officer
ASAPS	Anti-Slavery and Aborigines Protection Society
CARS	Colonial Annual Reports
CK	Central Kavirondo
CMIR	Confidential Monthly Intelligence Report
CMS	Church Missionary Society
CNC	Chief Native Commissioner
CO	Colonial Office
CS	Colonial Secretary
DAO	District Agricultural Officer
DofAAR	Department of Agriculture Annual Report
DC	District Commissioner
DO	District Officer
DYM	Dini Ya Msambwa
EACM	East African Chamber of Mines
EAD	East Africa Department
EAEP	East African Educational Publishers
EAEB	East African Educational Bureau
EAPH	East African Publishing House
EPL	Exclusive Prospecting Licence
EMS	Eldoret Mining Syndicate
FAM	Friends African Mission

GAS	Government African School
HC	House of Commons
HL	House of Lords
HMSO	His Majesty's Stationery Office
HOR	Handing Over Report
ILAC	Isukha Locational Advisory Council
KA	Kikuyu Association
KCA	Kikuyu Central Association
KLC	Kenya Land Commission
KMC	Kenya Missionary Council
KNA	Kenya National Archives
KTWA	Kavirondo Taxpayers Welfare Association
LAC	Locational Advisory Council
LegCo	Legislative Council
LMSDAR	Land, Mines & Surveys Department Annual Report
LNC	Local Native Council
MGDAR	Mining and Geological Department Annual Report
MOH	Medical Officer of Health
MP	Member of Parliament
NADAR	Native Affairs Department Annual Report
NK	North Kavirondo
NKCA	North Kavirondo Central Association
NKCofC	North Kavirondo Chamber of Commerce
NKDAR	North Kavirondo District Annual Report
NKDIR	North Kavirondo District Intelligence Report
NKLLB	North Kavirondo Local Land Board
NKLNC	North Kavirondo Local Native Council
NKAgARs	North Kavirondo Agricultural Annual Reports
NKAgMR	North Kavirondo Agricultural Monthly Report
NKCR	North Kavirondo Crop Report
NKTWA	North Kavirondo Taxpayers' Welfare Association
NLT(A)O	Native Land Trust (Amendment) Ordinance
NLTB	Native Land Trust Bill
NLTO	Native Land Trust Ordinance
NNADC	North Nyanza African District Council
NN	North Nyanza
NNAgAR	North Nyanza Agricultural Annual Report
NNDAR	North Nyanza District Annual Report
NNLNC	North Nyanza Local Native Council
NPAR	Nyanza Province Annual Report
NPIR	Nyanza Province Intelligence Report

OAG	Officer Administering Government
OI	Oral Interview
OUP	Oxford University Press
PC	Provincial Commissioner
POWs	Prisoners of War
P & P	Private and Personal
PQ	Parliamentary Question
PRO	Public Record Office
PWD	Public Works Department
RCNLTNKR	*Report of the Committee on Native Land Tenure in North Kavirondo Reserve*
RHO	Rhodes House Oxford
SAO	Senior Agricultural Officer
SDA	Seventh Day Adventist
SK	South Kavirondo
S of S	Secretary of State
TLGOAA	The London Group on African Affairs
UMA	Ukamba Members Association
WEA	Workers' Educational Association
WWII	World War II

Maps and Tables

Map 1	Nyanza Province Districts and Ethnic Groups in 1962	xiv
Map 2	Luyia Chieftaincies in the 1930s	xvi
Map 3	Nyanza Province District Administrative Units in relation to European Settled Areas	xvii
Map 4	Eldoret Mining Syndicate (EMS) Concession	66
Map 5	Nyanza Province Mining Area Zones	68
Map 6	Tanganyika Concession (1,550 Square Miles)	69
Table 3.1.	Number of Claims Registered Between 1931 and 1934	65
Table 3.2.	Percentage Increase in Gold Production, 1933–38	78
Table 3.3.	The Value of Gold Exports Visa-Viz Value of the Four Major Agricultural Exports of Kenya Colony, 1936–45	82
Table 3.4.	Government Revenue from Gold Mining Fees, 1933–52	83
Table 3.5.	European Mining Population in Kakamega, 1932–47	85
Table 3.6.	Output (milled, fine gold value) and Cost per Ton at Rosterman Gold Mine at Kakamega, 1936–51	87
Table 3.7.	Refined Gold Output for Kenya Colony, 1935–53	88
Table 3.8.	Local Costs of Mining Requisites Supplied by John Riddoch, Esq., M. L.C	89

Table 5.1.	Numbers Dispossessed and Compensation Values for Rosterman Lease	147
Table 5.2.	Twenty-One Year Surface Mining Leases Granted in NK District	149
Table 5.3.	Licensed African Producers, Localities and Gold Production Figures, 1949	156
Table 5.4.	Mining (African) Labor Force—October 1935: Tribal (Ethnic) Composition	159
Table 5.5.	Occupations, Total Earnings for Each Occupation, and Monthly Wages, 1935	168
Table 5.6.	Labor Categories, Types of Labor and Wages, Rosterman Gold Mining Ltd., 1941	171
Table 5.7.	Mining Accidents, Numbers Killed and Injured, 1933–52	176
Table 5.8.	Types and Quantities of Vegetable Seed Distributed by the NKLNC, 1933	182
Table 5.9.	Prices of Vegetables at Kisumu Market, 1951	185

Preface

This study focuses on the impact of rural industrial transformation in a portion of Western Kenya between 1930 and 1952. During the period, colonial industrial capitalism based on gold mining engulfed the region following the discovery of gold at Kakamega in colonial North Kavirondo (NK) district in 1931. NK (after 1948, North Nyanza (NN)) district was a large administrative unit inhabited by the Luyia sub-ethnic groups. The district was part of Nyanza Province, the administrative entity created by the colonial state that encompassed the southwestern portion of colonial Kenya.[1]

Until 1952 when the gold mining industry ended in Kakamega, this colonial rural industrial project proved a most controversial discourse. In that year, the last gold mining company, Rosterman Gold Mining Ltd., closed down, thus ending a dramatic period in the colonial economy of Western Kenya. Due to the undercapitalization of the companies, the enormous costs of production, and the paucity of large gold deposits, miners decamped, and Kakamega never became another Johannesburg. Nonetheless, the "Failed Eldorado" demonstrates how the colonial state and the imperial government negated Luyia land rights in the service of European industrial capitalist interests.

This study was motivated by several factors. First, while monumental work has been done on the mining industry in southern Africa, very few studies have focused on the mining industry in eastern Africa in general and Kenya in particular. Second, scholarship on Kenya's colonial economy has intensely favored the agrarian crises that characterized the encounter between Africans and European settler farmers in Central and Rift Valley Provinces of Kenya. This means that the experience of rural agricultural communities that were significantly affected by colonial industrial capitalism such as the Luyia, has received little academic attention. This study explores how the incorporation of the Luyia agrarian economy into the colonial industrial capitalist economy based on gold mining shaped and transformed the Luyia people. By conducting interviews in Buluyia, the study captures the largely undocumented histories of "the common people". In addition to addressing some of these issues, this study shifts the focus to the conflict arising out of the ambiguous laws relating to mineral and land rights in colonial Kenya.

Although colonial impositions such as taxes had appreciably affected the peoples of Western Kenya, the dawn of the gold mining industry proved a most transforming experience for the Luyia community. The colonial state had introduced the hut tax in 1901, followed by the poll tax in 1910. Both

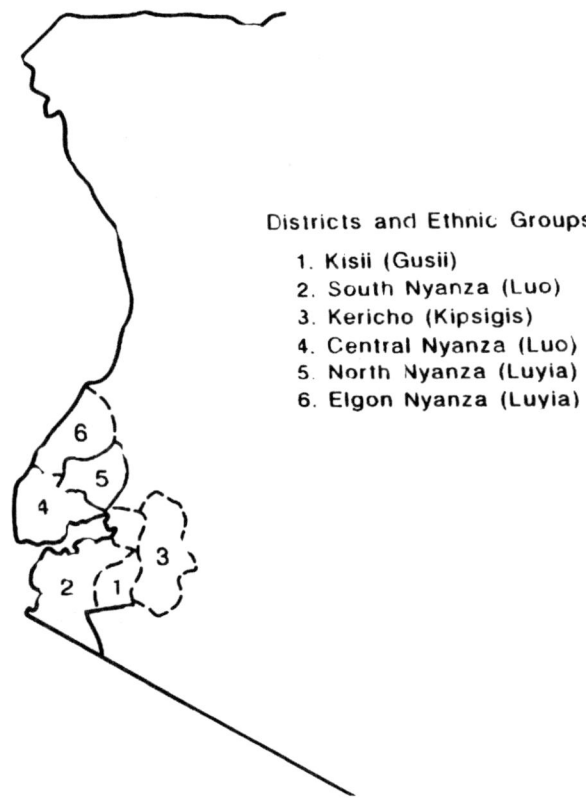

Map 1. Nyanza Province Districts and Ethnic Groups in 1962

Source: Robert M. Maxon. *Conflict and Accommodation in Western Kenya* (Rutherford, Madison, Teaneck: Fairleigh Dickinson University Press, 1989), 159.

taxes aimed at generating revenue for the administration of the colony, and quickly became fiscal coercive apparatuses geared toward compelling Africans to engage in the colonial capitalist economy either as wage laborers or peasant producers.[2] This study argues that the colonial state's attempt to superimpose a capitalist industrial economy on the Luyia of Western Kenya failed to efface the significance of land and peasant production in the Luyia political economy.

The inauguration of the gold mining industry in NK reserve, and the threat this colonial rural industrialization program posed to Luyia landowners produced diverse responses among the local peasants. The Luyia acted and reacted first as individuals, and later the resistance crystallized into a militant organized protest by the North Kavirondo Central Association (NKCA). While initial resistance entailed obstructionist activities and physical assaults against prospectors, organized protest resorted to deputations, memoranda, petitions, and other official ways of highlighting Luyia grievances against the gold mining industry and the colonial land policies. With time, NKCA fastened its protest around colonial soil conservation measures that led to widespread failure of colonial agrarian initiatives. Although the question of population pressure loomed large in Luyia opposition to soil conservation measures, the fear of losing reconditioned land to the colonial state and European settlers, largely shaped and continued to steer Luyia resistance to these measures.

Kakamega, today the headquarters of Western Province of Kenya, lies north of Kisumu, which sits on the shores of Lake Victoria. It was a town of little significance even by 1920 when it became the government headquarters of the then NK district. A generally fertile country of rolling plains interspersed with numerous perennial streams; the area receives sufficient rainfall and has rich soils that sustain a variety of agricultural activities. The town borders the largely uninhabited Kakamega Forest to the east, and sections of the Luyia people including the Kabras to the north, the Isukha to the south, and the Tsotso to the west.[3] Administratively, NK district was one of the five districts that made up the then Nyanza Province. The others were Central (CK) and South Kavirondo (SK), South Lumbwa (Kericho) and Kisumu-Londiani districts (See Map 2). These districts were large and diverse. In the 1930s, NK was home to the Bantu-speaking Luyia[4] with minority non-Luyia such as the Luo, Teso, and Kalenjin. The district borders the Trans Nzoia and Uasin Gishu settled areas to the north, the Nandi reserve to the east and Uganda to the west.[5] (See Map 3). Central Kavirondo was home to the Nilotic Luo, although the colonialists included the Samia subgroup of the Luyia in the district. The Luo, Gusii, and Kuria inhabited SK, while South Lumbwa was home to the Kipsigis as well as European settlers, as the case was with Kisumu-Londiani. Luo and Asian farmers also resided in Kisumu-Londiani.[6]

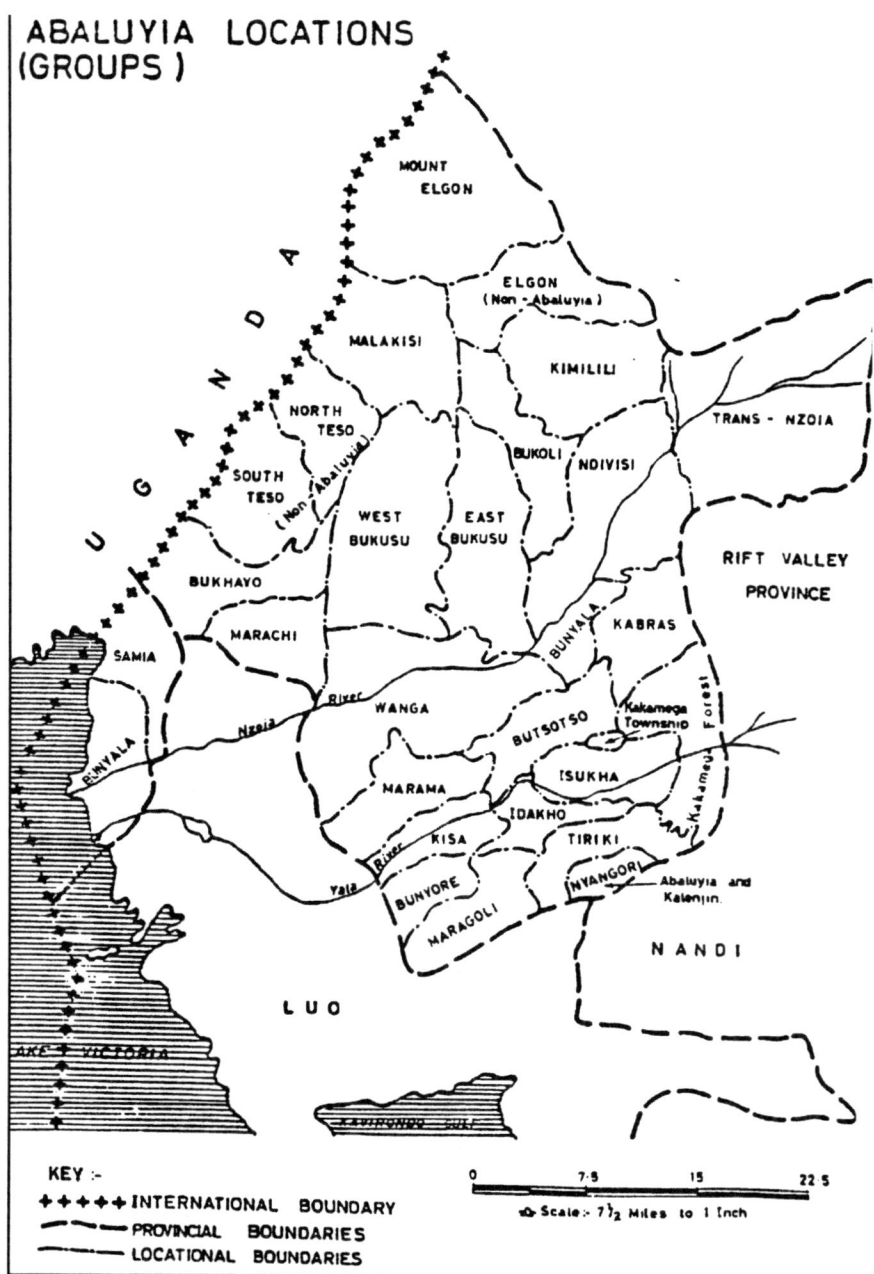

Map 2.   Luyia Chieftaincies in the 1930s
Source: Gideon S. Were Were. *A History of the Abaluyia of Western Kenya. c. 1500-1930* (Nairobi: EAPH, 1967), 28.

Map 3. Nyanza Province District Administrative Units in Relation to European Settled Areas
Source: Hugh Fearn, *An African Economy* (London: OUP, 1961), 19.

This study is divided into seven chapters. Chapter 1 provides a background to the work examining the context and conceptual framework, aspects of the pre-colonial Luyia land tenure system and the epistemological, historiographical, and methodological contribution of this study to Kenyan historiography. Chapter 2 examines the role of the imperial government, the colonial state and the Kenya Land Commission (KLC) in the execution, or failure to execute, trusteeship with regard to Luyia land rights. The chapter details the factors responsible for the reversal of land policy following the discovery of gold in the NK reserve in October 1931.

Chapter 3 zeroes in on the major trajectories that characterized the gold mining industry in NK district from the initial discovery in 1931 to the closure of the last mining company in 1952. The rush to Kakamega, the development of alluvial mining, the emergence of large mining companies and the role of both local and foreign capital, constitute the major thrust of the narrative. The study emphasizes the importance of gold in the colonial economy and the myriad complex factors, both domestic and international, that led to the failed Eldorado in 1952.

Chapter 4 examines the diverse forms of Luyia responses toward the development of gold mining in their locations. After World War II (WWII), Luyia politics of land transformed itself from the anti-mining stance to the more dramatic anti-soil conservation movement.

Chapter 5 is an account of the economic ramifications of the gold mining industry for the Luyia community. By examining indices such as land alienation and cash compensation rates, African entrepreneurship in the gold mining industry, labor opportunities, mining accidents, urban growth and market opportunities, the study demonstrates the largely negative impact of the gold mining industry in Buluyia.

Chapter 6 focuses on the social and cultural effects of the gold mining industry in Western Kenya. By examining the health and environmental challenges resulting from mining activities and the remarkable social-cultural changes that accompanied the gold mining industry, this study highlights how the forces of colonial industrial capitalism affected the Luyia society.

Chapter 7 provides a synopsis of the major arguments and conclusions.

This study is based on published, archival and oral sources obtained from the United States of America (USA), the United Kingdom (UK), and Kenya. Various funding opportunities made this research possible: numerous grants from the Department of History and the Office of Academic Affairs at West Virginia University (WVU), facilitated travel to the Kenya National Archives (KNA) collections at Syracuse University and Ohio University. Similarly, grants from the WVU Office of International Programs, funded my research travel to the Public Record Office (PRO) in London, and Rhodes House Library (RHO) in Oxford in the UK. These sites provided important information that illuminated the imperial view, the views and ideas of leading British humanitarians as well as former colonial officials who kept meticulous records.

The Rockefeller African Dissertation Internship Fellowship actualized my one-year fieldwork in Kenya during 1998. I collected data through in-depth interviews with Luyia men and women aged over fifty years, a majority of whom were from the communities in the former goldfield and areas of dense population pressure such as Maragoli and Bunyore. These included families or relatives who lost land to the gold mining industry, those who worked or knew people who served as laborers in the mines, surviving or relatives of colonial functionaries such as chiefs, village headmen, compensation assessors, health workers, members or supporters of the NKCA, traders, and members of the NKLNC. The extensive oral interviews unearthed the largely undocumented voices of Luyia men and women whose lives were significantly impacted by the gold mining industry.

I am equally grateful to The American Association of University Women International Fellowship for funding my dissertation writing time. My appreciation goes to the East African Educational Publishers in Kenya for granting me permission to use portions of my chapter which appeared in William R. Ochieng' ed. *Historical Studies and Social Change in Western Kenya: Essays in Memory of Professor S. Gideon Were* (2002).

I also wish to extend my appreciation to the staffs of the libraries and archives where my research was done: WVU Libraries; Bird Library, Syracuse University; Ohio University; and the PRO and RHO in the UK. Staff at Jomo Kenyatta Memorial Library at the University of Nairobi (UON); the Moi Library at Kenyatta University and the Margaret Thatcher Library at Moi University assisted my research in Kenya. Also helpful were, staff at the Provincial Public Record Center in Kakamega, the KNA, and the Mines and Geological Department Library in Nairobi.

Special thanks to my mentor Professor Robert M. Maxon, for his guidance and continued interest in this work. Dr. Amos Beyan and Dr. George Ndege have always encouraged me and offered valuable advice. Professor Bethwell Ogot, and Professor William R. Ochieng', have supported my work with interest. Professors Tabitha Kanogo, Toyin Falola, and Wangari Mwai; Drs. Isaiah Kulecho, Kenda Mutungi, Okia Opolot, and Nameeta Mathur; Mr. John Hagan, Mr. Philip Mukonyi, and Mrs. Margaret Kalo, have supported and inspired me. I am grateful to Donna MacIssac for formatting the manuscript, Benson Njoroge and Randy Crowe, for valuable assistance with maps, and Dr. Phyllis Belt-Beyan and Tracy Hall for your generous help with indexing.

I want to thank my family, Dr. Dorothy Smith-Akubue-Brice, my sister Esther, late brother Luta Tom Mboya, Kawa, Peter and Jacob, and my father and mother, Justmore Shilaro and Ruth Wekaya for supporting me in my endless search for knowledge. You have boldly endured my long absence from home and suffered immense inconveniences on my behalf. My daughter Anne Mutashi, I thank you for "allowing" me to deny you the carefree life of a child. I dedicate this work to my late brother Luta Tom Mboya, who never lived to see the product of my intellectual endeavors.

NOTES

1. Robert M. Maxon, *Going Their Separate Ways: Agrarian Transformation in Kenya, 1930–1950* (Madison. Teaneck: Fairleigh Dicknson University Press, 2003), 11.

2. Hugh Fearn, *An African Economy: A Study of the Economic Development of Nyanza Province of Kenya, 1903-1953* (London: Oxford University Press (OUP), 1961), 109. In this study, peasant households refer to rural households involved in the production of agricultural commodities for both consumption and sale using family labor. For details see F. Ellis, *Peasant Economics: Farm Households and Agrarian Development* (Cambridge: Cambridge University Press, 1988), 9–13.

3. A. Huddleston, *Geology of the Kakamega District* (Nairobi: Government Printer, 1952), 1–3, 5; Norman Humphrey, *The Liguru and the Land: Sociological Aspects of Some Agricultural Problems of North Kavirondo* (Nairobi: Government Printer, 1947), 1. The district consisted of several locations that often coincided with the colonial ethnic boundaries under the leadership of colonial chiefs. Map 1 shows Luyia chieftaincies in the 1930s.

4. Before the 1940s, the Luyia were referred to as the "Bantu-Kavirondo". The origin and etymology of the term "Kavirondo" is obscure. However, it is believed to have been employed and spread by the Swahili-Arab traders and later adopted by Europeans to refer to the Bantu-speaking groups of Western Kenya comprising the Abanyore, Abalogoli, Abatiriki, Abesukha, Abedakho, Abamarama, Abashisa, Abawanga, Abatsotso, Abatachoni, Ababukusu, Abakabras, Abasamia, Abanyala, Abamarachi, Abakhayo, and Abasonga. Gunter Wagner, *The Bantu of North Kavirondo*. Vol. I. (New York: OUP, 1949), 19, 20; See also J. D. Otiende, *Habari Za Abaluyia* (Nairobi: The Eagle Press, 1949),1; Gideon S. Were, *History of the Abaluyia of Western Kenya c. 1500–1930* (Nairobi: East African Publishing House (EAPH), 1967), 29–31. A 1930 inquiry by the editor of the Swahili newspaper, *Habari,* revealed that its Luyia readership disliked "Kavirondo", the English name derived from "caravan, "which later became "Kaviron" and then "Kavirondo". Many considered it derogatory. See *Habari. Augusti* (Swahili for August) 1930; Gunter Wagner, *The Bantu of North Kavirondo: Economic Life*. Vol. II. (New York: OUP, 1956), 38; Erasto Ayiekha, Oral Interview (OI), February 15, 1998.

Similarly, the origin of the name Abaluyia is a matter of speculation. Archival and oral sources show that it was coined in 1939 and adopted in 1942 as a vehicle for realizing Luyia unity. North Kavirondo District Annual Report (NKDAR) 1942, Kenya National Archives (KNA): DC/NN/1/24; J. D. Otiende and Jairo Eloga Akibaya, OI, August 20, 1998; Otiende and E. M. Aseka concur that "Luyhia" refers to the evening bonfire around which elders warmed themselves, while transmitting the history of their communities to the youth orally. Abaluyia, therefore, denoted a people who shared a common bonfire and ethnic affiliation. E. M. Aseka, "A Political Economy of Buluyia: 1900–1964," (Ph. D. dissertation, Kenyatta University, 1989). In this study, the term Luyia is used to refer to the corporate Luyia body politic and, Buluyia for the geographical area occupied by the Luyia peoples. Individual Luyia ethnicities are referred to as Nyore, Logoli, Tiriki, Isukha, Idakho, Marama, Kisa, Wanga, Tsotso, Tachoni, Bukusu, Kabras, Samia, Nyala, Marachi, Khayo, and Songa by eliminating the prefix "Aba/Ava" which dominate local vernacular references to the various Luyia ethnic groups.

5. Humphrey, *The Liguru*, 1

6. Peter Odhiambo Ndege, "Struggles for the Market: The Political Economy of Commodity Production and Trade in Western Kenya, 1929–1939," (Ph D. dissertation, West Virginia University, 1993), viii.

Chapter One

Prelude to the Gold Rush

INTRODUCTION

The period between 1930 and 1952 witnessed a dramatic transformation of a portion of Western Kenya into a colonial industrial enclave based on the gold mining industry. This study traces the process of colonial rural industrial capitalist transformation and its impact on the Bantu-Luyia peoples of Western Kenya. The discovery of gold in 1931 at Kakamega in colonial NK district launched a new form of colonial capitalism based on gold mining in a virtually rural agrarian terrain. The establishment of an industrial capitalist economy in Western Kenya had significant political, economic, social and cultural ramifications for the Luyia community. Prior to the gold discovery, the colonial government had received the Report of the Committee appointed to investigate the land tenure systems in NK reserve in 1930. Similarly, in the same year, the Colonial Office (CO) imposed the NLTO of 1930 on the colonial state and the powerful European settler community in Kenya. This ordinance aimed at protecting African reserves from further encroachment by immigrant races. However, following the discovery of gold in NK in October 1931, the colonial state amended this ordinance in 1932 to pave the way for lease grants to European industrial interests. The amending ordinance sought to facilitate the alienation of land for gold mining in Buluyia. Until 1952, when the last gold mining company in the Kakamega goldfield closed down, land remained the central rallying point for Luyia opposition to the gold mining industry.

This study explains why the colonial state altered its land policy in 1932 thereby violating the public pledges entailed in the NLTO of 1930. Moreover, the imperial government approved the amendment of the NLTO of 1930 to pave way for the gold mining leases. The resultant NLT(A)O of 1932, allowed the colonial state to alienate land from NK reserve for gold mining interests. By elucidating how and why the CO came to approve the implementation of the amending ordinance in spite of great opposition from critics of British colonial policies in Kenya and the House of Commons and the House of Lords in Britain, this work places economic considerations at the center of this controversy. Did economic motives outweigh the concept of trusteeship at the time? Were the imperial officials no longer keen to stand up for African land rights in Kenya? Answers to these questions demonstrate the anticipated "Kakamega Eldorado" to both the colonial and imperial economies. This also illustrates the incongruence between colonial policies relating to security of tenure for African landholders and the overriding rights of the state over all mineral resources in Kenya.

This study draws significantly from the liberal tradition of analyzing imperial history and the political economy paradigms. This paradigm places both the policy formulation process and economic exploitation within the wider context of colonial domination. Colonial land policy in Kenya was a product of the larger British imperial economic design. The Kakamega gold mining industry clearly illustrates the direct link between politics and economics. The gold discovery occurred against the background of a depressed economy in Kenya. The colonial state thus noted its potential as the harbinger of economic recovery. Hence, European settlers, prospectors, and the colonial state emphasized the economic imperative of exploiting gold above all else. Consequently, the colonial state shelved Luyia claims to land and its security in NK reserve as profit making gained ascendancy.

Fundamentally, this study examines the modes of the Luyia peoples' response to the gold mining industry. How did the Luyia people respond to the new industrial enterprise that had in fact disinherited them? Although this question little troubled the colonial state, it points to the importance of the issue of land security for Africans in colonial and post-colonial Kenya. Luyia men and women, robbed of their only means of livelihood, sought alternative survival strategies such as paid labor, hawking around the mining compounds and trade. In South Africa, for example, gold mining afforded women an opportunity to engage in beer brewing from which they obtained additional income to support their families. The myriad survival strategies the Luyia adopted reveals the gendered consequences of the superimposition of a colonial industrial economy on the Luyia people. In addition, the analysis historicizes the actions and reactions of Luyia men and women who struggled against the crippling economic, social and cultural transformation brought on by the gold mining industry. Through spontaneous individual action and organized protest, the Luyia took initiative in challenging the oppressive British colonial land policies.

Indeed, as the noted University of Minnesota (Twin Cities) Social Historian Allen F. Isaacman aptly observes, it is only by listening to such voices that researchers are able to penetrate their inner world and thereby reconstruct their notions of justice, property and community.[1] Thus, the life histories and the popular culture of the women and men in Kakamega in their every day resistance toward colonial land policies relating to gold mining illustrate how these localized movements crystallized into a long tradition of resistance to authority.

This study also evaluates the economic and social benefits of the gold mining industry to the Luyia of Western Kenya. Did the gold mining industry usher in two decades of economic prosperity for the Luyia? Did Africans directly benefit as entrepreneurs in the gold mining industry? In what ways did the gold mining industry affect the general welfare and lifestyle of the local population in terms of land loss, employment opportunities, working conditions, wages, and welfare and health facilities? How did the presence of European miners and mine labor affect Luyia social relations? Answers to these questions reveal that the gold mining industry provided short-lived benefits to the Luyia people.

Land is the most coveted and jealously guarded resource for economic development among rural Kenyan households. Throughout the colonial period, land remained a major contested issue, pitting the CO against the colonial state,[2] Kenya's European settlers, and the African men landholders and women who enjoyed usufruct rights to land by virtue of their relations to such men.[3] In 1931, gold was discovered near Kakamega in Western Kenya. Initially, prospectors extracted gold from alluvial deposits; however, from 1932 onwards, reef mining surpassed alluvial mining.[4] This development led to growing demands by European prospectors and mining companies for the alienation of land from the Luyia reserve. The colonial states' decision to expropriate land from the NK reserve for the gold mining industry, prompted significant opposition from the Luyia and intensified fear of security of tenure that found expression in diverse forms of protest against the gold mining industry, land alienation and colonial agrarian policies.

This study examines how the injection of a colonial industrial capitalist economy centered on the extraction of gold affected the Luyia agrarian economy. At the center of the debate is the way in which the Luyia men and women responded to the loss of land to European gold mining interests and how this experience continued to shape Luyia reaction to colonial agrarian policies.

British colonial policy in Kenya revolved around the concept of trusteeship. This meant, *inter alia*, the pledge that Africans would have adequate land with security of tenure forever. Trusteeship also implied that the British were not controlling Kenya for their own benefit, but also for the benefit of the original inhabitants of the country. Therefore, Britain had an obligation, at least in theory, to pursue policies that benefited both Britain and her subject peoples.

While the imperial government in London represented the ultimate trustee for African lands in Kenya, the colonial state had some latitude with regard to policy formulation in the colony. Sometimes, the CO found itself at loggerheads when dealing with pro-settler governors. For the purpose of this study, the classic case is the struggle between the CO and the colonial state in the passage and adoption of the NLTO of 1930. Although the colonial state had demarcated and gazetted African reserves in 1926, no legal guarantees of security of tenure of these reserves existed. Indeed, it was not until the Labour Secretary of States (S of S) for the Colonies, Sydney Webb (later Lord Passifield) took over from the Conservative S of S for the Colonies, L. Emery, that the CO imposed the NLTO of 1930 on the pro-settler governor, Sir Edward Grigg and the Kenya settler community. Until 1930, the concept of trusteeship remained a theoretical postulation with little applicability.

The NLTO of 1930 provided legal protection of African reserves from intrusion by immigrant races without the consent of the concerned Africans and the Local Native Councils (LNCs). The colonial state established LNCs in 1924 as official avenues for articulating and expressing African concerns. In addition, the Ordinance endorsed the policy that in the event of appropriating land from African reserves, the state had to add land of equal size and value to the affected reserve. Nevertheless, with the discovery of gold in NK in 1931, the colonial state passed the NLT(A)O of 1932 at a remarkable speed, with the consent of the CO and the KLC, which was then investigating the perennial land question in Kenya. The NLT(A)O of 1932 sought to meet the sudden demand for land in NK reserve for European gold miners and mining companies.

The NLT(A)O of 1932 eliminated two core principles of the NLTO of 1930. First, the amending ordinance legalized cash compensation for land alienated to the gold mining industry. Second, the ordinance eliminated the provision requiring the colonial state to consult the relevant LNC before alienating land for gold mining. This policy reversal constituted a major setback to African land rights in Western Kenya. Hitherto, colonial land policies legally excluded Africans from purchasing land outside their designated reserve boundaries. The policy change and the subsequent appropriation of the Luyia reserve significantly shaped the Luyia response to the gold mining enterprise, and impacted their perception of both the colonial and post-colonial land policies. The alienation of land in Kakamega for gold mining affords an interesting case where the colonial state, (with the approval of the CO) flagrantly violated Luyia land rights. The repressive colonial policies led to mistrust for state initiatives including rural industrialization schemes, state-driven agrarian policies such as soil conservation programs, and the regulation of on-farm activities and the marketing of produce. This work illuminates the bitter memories of colonial land policies in Western Kenya and the impact it had on peasant response to state industrial and agrarian initiatives, thereby highlighting Kenya's development dilemma. This is important because while agriculture undeniably remains the backbone of most African economies, the peoples' struggles for independent production versus government emphasis on close control have led to agricultural decline in many countries.[5]

The creation of colonial boundaries, European settlement and the alienation of land for gold diggers in NK reserve in the 1930s and 40s significantly affected the pre-colonial Luyia land tenure system. To understand and contextualize these effects, therefore, a brief discussion of the major aspects of the pre-conquest Luyia land tenure system pertinent to this study is in order.

LUYIA INDIGENOUS ECONOMY

Gunter Wagner's authoritative work, *The Bantu of North Kavirondo*,[6] based on research among the Bukusu and Logoli subgroups of the Luyia, ably details the Luyia land use patterns. Wagner's work provides major common characteristics applicable to the rest of the Luyia subgroups. According to Wagner, the Luyia land tenure system remained relatively undisturbed until 1926 when the colonial state altered the role of clans by replacing the indigenous and, therefore, the legitimate clan leaders, *amaguru,* with the newly created paid *milango* headmen.[7]

Land among the Luyia was owned on clan basis. Clan elders surveyed and selected areas suitable for cultivation. The choice of an area was dependent on the presence of water, salt licks, absence of tsetse fly and the fertility of the soil. Land so selected became the home of the clan, which was the proprietary unit in respect of the whole area of land within its boundary. In due course, the group and its land became one and indissoluble.[8] Although the clan owned the land, each family had exclusive occupational and usufruct rights over its own holding. These rights were inalienable and automatically passed from father to son. The absence of an individual from the clan lands for an extended period did not invalidate these land rights.[9]

The Luyia customary unit of land administration was the *olugongo*, literally a ridge or clan settlement. Each *olugongo* was under the control of an elder known as the *liguru*. The customary functions of the *liguru* included keeping the peace, protecting the land and its occupants, and settling disputes. The clan leader among the Luyia was not an autocrat. In all cases, he acted in consultation with the elders of the families living on the *olugongo*. He derived his authority from direct descent from the original founder of the clan. In addition, the *liguru* derived his political authority as the representative of the clan. Members of the clan, therefore, had rights in the *olugongo* as members of a political community that had asserted its authority over lands controlled by it.[10]

Land was acquired either by inheritance of ancestral lands cultivated by one's grandparents, or by right of first occupation or cultivation of a portion of virgin land known variously as *omutsuru, eshitsimi,* or *oluanjerekha*. In some cases, clans or ethnic groups acquired land through outright military conquest. In general, there is no evidence to show that Luyia purchased or paid for land on previously existing rights such as was the case in part of Central Province, where the Kikuyu *ithaka* purchased hunting grounds from the Ndorobo.[11]

Due to the centrality of land in Luyia social, political and economic life, several distinct categories of land existed. These categories referred to the economic value of land and the specific use of such lands. *Obulimu* (bush land) comprised uncultivated virgin lands; *ovutsaka (ofutsakha)*, included uncultivated land on the fringes of streams and valleys; *eligyinga (elisino)*, wild bush country; *omulimu*, individual strip of cultivation; *omukunda, a* piece of fallow or reclaimed fallow land (second year and subsequent years); *elilimilo*, a first year cultivation on freshly cleared land; *eshayo*, pasture or grazing land; *eshiangalangwe (egyangalangwi)*, waste land; *mwigendanyama (mwichendanyama)*, no man's land; and *elidala*, homestead.[12] The Luyia had special names for water holes (*eshitikho, ekidaho)*, salt licks (*eshilongo)*, grazing land (*eshayo, obwayiro)*, ceremonial land (*eshitavita)*, cultivated land (*lilimilo)*, and unused bush land (*lisino)*.[13]

There were several ways of holding land in Buluyia: the most important being the owner (*Omwene*) and *omumenya*: `a host resident on a clan land. One could gain access to clan land through inheritance or by staking off and cultivating a strip of virgin bush land (*oluanjelekha, oluangereka*). In addition, one established ownership by cultivating bush land (*obulimu*) in the uninhabited border-zone between two ethnic groups (*mwigendanyama*).[14] Any such land occupied by a single clan and its host clients constituted the *olugongo*, containing holdings of the various families, known as *limenya*. These consisted of individual homesteads, *amadala* or *tsingo*, and their respective cultivated plots. Each family enjoyed exclusive occupation and usufruct rights over its cultivated plot *(omugunda*, or *Omukunda)*. Prolonged absence or allowing such land to revert to bush, did not affect these inalienable rights. The only exception was when an individual was excommunicated from clan land for extreme misconduct.[15]

There was no concept of individual land ownership among the Luyia. All rights to land emanated from clan membership and inheritance. No one could hold land apart from his family, and no one could hold land otherwise than under the tribal authority represented by the *liguru* and the lineage elders.[16] Custom bound individuals not to alienate clan land without the consent of clan elders.

Arguably, although the pre-colonial Luyia land tenure intricately bound the clan and its land, the community recognized a class of strangers who often resided on land outside of their clans. Such "foreigners", known as *abamenya* (singular *omumenya*), acquired land rights in clans other than their own through clientilist relations. In most cases, however, the *abamenya* had relationships with their host clan by either blood or marriage.[17] The hosting clan head and elders approved such rights after a thorough investigation into the history and behavior of the *abamenya*. Although the abamenya had no rental obligations to their hosts, continued rights in such lands depended on their good behavior, often crowned by hospitality and token presents to their hosts.[18] In the early period of settlement of such clients, only grievous offense to the host clan mandated their expulsion. Such an offense included quarrelsomeness and unpopularity resulting from witchcraft accusations. Sometimes, the needs of the clan for land settled by the *abamenya* necessitated such evictions. In such cases, the clan often squeezed out the *abamenya* by extending their holdings into land previously used by the former. In any case, it was difficult to get rid of the *abamenya* related to the clan by blood. Those who retained rights of residence, however, had no rights to alienate land granted to them by host clan. Typically, the *abamenya* submitted to the authority and control of the head of the host clan and their children inherited their father's position as tenants-at-will. After two or three generations, they acquired the status of *abamilikha*, meaning the "absorbed" or "merged."[19]

These complex modes of land ownership and the attendant intricate land use forms represented an equally complex economic and social structure among the Luyia. For instance, livestock, such as cattle, sheep and goats, formed a significant facet of the Luyia economy. Livestock constituted a means of exchange, besides serving a variety of social transactions and ceremonial functions. They were also a source of meat, milk and blood, which supplemented the people's diet.[20] Due to the centrality of livestock, particularly cattle, in the Luyia political economy, each clan had common grazing lands for the collective use of its members. The clan head protected the *eshayo* or *obwayiro* (grazing/pastureland) against encroachment by cultivators.[21] Likewise, salt licks (*shilongo*) belonged to the clan, and, therefore, all its members. Like grazing lands, salt licks were free from cultivation and recognized routes existed to these areas. This general principle applied to water holes, springs and wells.[22]

Apart from livestock keeping, land resources played a significant role in the evolution and survival of numerous specializations among the Luyia. The Logoli were and are renowned pot-makers in Western Kenya. The Batovo and Bamachina clans among the Kabras also specialized in pottery making. Pots played a very important role in daily life in Buluyia: they prepared delicacies such as chicken and beef, storing water and foodstuffs, and served in mythological or religious ritual. These ranged from ceremonial beer drinking to rituals surrounding rainmaking. Due to the pervading use of pots in Luyia political economy, deposits of gummy clay suitable for pottery were highly valued; together with ochre, which served a variety of functions, they were strictly communal property. Normally, the clan head had control over land containing such minerals, and strangers had no access without the consent of the respective clan head.[23]

With regard to trees and forest, where supply was abundant there was no restriction on the right of members of a clan to cut trees anywhere on the clan land. Natural trees on an individual *omugunda* (*omukunda*), however, belonged to the owner. Trees found in bush land were common clan property. Among the Isukha and the Idakho of East and West Kakamega, forests and the hunting rights therein belonged to the senior elder of the ethnic group who controlled them with the help of other elders. The elders had exclusive rights to game and honey found within the forest. Among the Kabras and the Banyala of Navakholo, all members of the clan had right to hunt for game in forests.[24] This would drastically change in the 1930s and 1940s because of the introduction of the gold mining industry.

The issue of land and the attendant population problem, particularly in Kakamega area, was a point of focus for the colonial administration throughout the 1930s and 1940s. Due to high population densities in the southern portion of NK, particularly in Maragoli and Bunyore locations, the colonial state proposed several different alternatives in succession as possible solutions to attaining sustainable economic development in the region. At first, the administration considered the possibility of a state sponsored population resettlement program. This option quickly fizzed out because government officials could not control the reproduction of the remaining population. A second option sought to create a landless class, deriving a portion of its income from wages earned outside the district. This option also proved an inadequate measure in regulating the increasing population. Consequently, colonial officials contemplated the development of secondary industries and "native" cities, as well as the introduction of high value cash crops in areas such as North Kavirondo.[25] As Robert Maxon aptly notes, by the end of the 1940s, however, the government had not implemented any of these policies.[26] For example, the colonial state did not initiate any new rural industries in Western Kenya, and even in the Kakamega gold mining industry, few Africans benefited directly as prospectors and miners. In addition, while the colonial state supported cash crop production in Gusiiland in SK, this did not occur for NK. Nevertheless, a rural landless "class" emerged and once again, most Luyia fell back on labor export[27] and the cultivation of "low priced" maize, legumes and vegetables for local consumption. Instead, the colonial state emphasized soil conservation and reclamation, a movement that aroused opposition among disillusioned Luyia households. The issue of population and sustainable rural economic development continues to haunt independent Western Kenya. While there are no legal obstacles to cash crop production, many rural households in areas impacted most by gold mining continue to shun cash crops. This negative response to a cash crop economy has its roots in unfavorable colonial land alienation policies associated with the Kakamega gold mining industry.

This study focuses on the Kakamega gold rush and NK reserve for several reasons. The history of the interaction between land ownership and rural industrialization in Western Kenya is still largely uncharted. Most studies on the Kenya land question tend to assume that the Luyia of Western Kenya had no significant land grievances throughout the colonial period. Against this background, this genre of scholarship has concentrated on the Kikuyu, Maasai, the Kipsigis and the Nandi communities that lost large tracts of land to European settlers. Thus, land grievances among these ethnic groups were a direct consequence of settler demands for prime farmland in the Kenya highlands. The loss of land in NK is unique because it resulted from the demands of the gold mining industry. A critical examination of how colonial industrial capitalism based on gold mining interacted with the established Luyia rural agrarian economy, contributes significantly to Kenya's colonial economic historiography.

Land remains a major factor of production among the Luyia. Both in colonial and post-colonial Kenya, while ownership of the land by individual tillers was and is recognized, mineral resources were/are reserved for the government.[28] The colonial land and mining laws reserved all mineral rights to the British Crown, while acknowledging African ownership of the surface.[29] As if to reiterate the fact that gold in Kakamega was not the property of the Kenyan "natives", the British Parliament passed the Petroleum Bill in 1934, declaring petroleum the property of the Crown. By examining the disharmony, resulting from the dichotomy between the Luyia ownership of land and the government's rights to gold in Kakamega, the incongruence of such legislation becomes apparent.

This work contributes to the historiography of Kenya in epistemological terms. It unearths new information from sources hitherto not adequately utilized. Moreover, the gold mining industry in Kakamega is important when placed within the global context of the Great Depression of the 1930s. Coming at a time of economic hopelessness, the discovery of gold was for a while thought to be the panacea to the Kenyan European settler economic difficulties, and probably a boost to the ailing British economy. Such economic considerations determined the radical change in land policy in Kenya in the 1930s to facilitate gold mining. Hence, the gold rush, the loss of land, and the European invasion of NK, a Luyia gazetted reserve, aroused significant protest from both the Luyia of Western Kenya and the British humanitarians, although this did not serve to move the colonial state and the CO. Indeed, the continuing politicization of the land question, which has had a negative impact on agricultural development in the Kakamega area, is the most consequential legacy of the colonial land policies surrounding the gold mining industry. Nevertheless, as the research shows, the gold mining industry did not really provide long-term economic benefits to the Kenyan European settlers, Kenyan exports or the British economy. More significantly, it provided few long-term answers to the problem of economic viability in areas of heavy rural population density such as NK reserve. In fact, the colonial rural industrial undertaking in Western Kenya only served to exacerbate the land problem in Buluyia.

NOTES

1. Allen F. Isaacman, "Peasants and Rural Social Protest in Africa," in *Confronting Historical Paradigms: Peasants, Labor and the Capitalist World System in Africa and Latin America*, eds. Frederick Cooper, Allen F. Isaacman *et. al.* (Madison: University of Wisconsin Press, 1993).

2. The colonial state in Kenya has exercised the minds of scholars for some time. Bruce Berman and John Lonsdale maintain that although the colonial state was an external creation, and, therefore, a dependent structure of the metropole, it was never a blind and loyal replica of the metropole. Rather, they assert that the colonial state was shaped, modeled, and influenced by both local conditions and interests in the colony, as well as the metropolitan demands that dictated its inception. This means the colonial state was not an all-powerful ruler presiding over servile and malleable subjects. Indeed, though the colonial state was imposed by force, it was dependent on local col-

laborators for survival. With time, it evolved into a set of institutions and practices that both reflected and shaped the contradictions involved in articulating the myriad conflicting interests in the metropole and in the colony. See John Lonsdale, "States and Social Processes in Africa: A Historiographical Survey," *African Studies Review* XXIV, nos. 2/3 (1981): 139–225; Bruce Berman and John Lonsdale, "Coping With the Contradictions: The Development of the Colonial State in Kenya, 1895–1914," *Journal of African History* 20, no. 4 (1979): 487–505; Lonsdale, "The Conquest State, 1895–1904," in *A Modern History of Kenya, 1895–1980*, ed. W. R. Ochieng', (Nairobi: Evans Brothers, 1989), 6–34. Throughout this study, an attempt is made to highlight the dynamics, both foreign and local, European and African, personal and institutional, that characterized policy formulation with regard to the Kakamega gold mining industry.

3. See Bruce Berman, *Control & Crisis in Colonial Kenya: The Dialectics of Domination* (Athens: Ohio University Press, 1990); Tabitha Kanogo, *Squatters & the Roots of Mau Mau 1905–63* (Athens: Ohio University Press, 1987); Berman and Lonsdale, *Unhappy Valley Conflict in Kenya & Africa*. Vol. One and Two (Athens: Ohio University Press), 1992.

4. C. G. B. DuBois, *Geological Survey of Kenya. Minerals of Kenya* (Nairobi: Government Printer, 1966), 26.

5. Robert H. Bates, *Markets and States in Tropical Africa: The Political Basis of Agricultural Policies* (Berkeley, Los Angeles: University of California Press, 1981).

6. Wagner, *The Bantu,* Vol. II, 75–100.

7. Ibid. 75–76.

8. *Report of the Committee on Native Land Tenure in the North Kavirondo Reserve (RCNLTNKR)* (Nairobi: Government Printer, 1931), 4; Priscilla M. Shilaro, "Kabras Culture under Colonial Rule: A Study of the Impact of Christianity and Western Education," (M. A Thesis, Kenyatta University, 1991), 75; Joseph Imbali, Oral Interview (OI), (OI), February 18, 1998; Andrea Mukabwa, OI, February 12, 1998; Clement Muhati OI, February 13, 1998; Gabriel Musalimwa, OI, February 17, 1998; Christopher Mavia, OI, March 2, 1998; Francis Mutsotso, OI, March 3, 1998.

9. Wagner, *The Bantu,* Vol. II, 11; Shilaro, "Kabras Culture," 75–76; Abaluyia Land Law and Custom,1930, KNA: DC/EN/3/3/2; Ayiekha, Oral Interview(OI), February 15, 1998; Musalimwa, OI, February 17, 1998; Beti Joseph Isenjia, OI, February 24, 1998; Peter Likhaya Mbwabi, OI, February 26, 1998; Zacharia Muganzi Andove, OI, February 28, 1998; Mavia, OI, March 2, 1998; Josphat Amagadu and Zablon Shikali, OI, March 3, 1998, Adriano Vihembo and Andrew Ludenyo, OI, March 26, 1998.

10. *RCNLTNKR*, 4; Wagner, *The Bantu,* Vol. II, 87–88.

11. *RCNLTNKR*, 3; Wagner, *The Bantu,* Vol. II, 90; Shilaro, "Kabras Culture," 76–77; Francis Mutsotso, OI, March 3, 1998; Imbali, OI, February 18, 1998; Mukabwa, OI, February 12, 1998; Muhati, OI, February 13, 1998; Musalimwa, OI, February 17, 1998; Mavia, OI, March 2, 1998.

12. Wagner, *The Bantu,* 75–76; Mukabwa, OI, February 12, 1998; Muhati, OI, February 13, 1998; Mavia, OI, March 2, 1998.

13. Sakwa Mwela, OI, January 18, 1990; Laban Ikhale Teka, OI, January 26, 1990; Eugene C. Burt, "Toward an Art History of the Baluyia of Western Kenya," (Ph. D. dissertation, University of Washington, 1980), 328.

14. Burt, "Toward an Art History," 77.

15. *RCNLTNKR*, 4–5; Wagner, *The Bantu*, Vol. II, 78, 79; Amagadu and Shikali, March 3, 1998; Mavia, OI, March 2, 1998.

16. *RCNLTNKR*, 5.

17. Wagner, *The Bantu*, Vol. 1, 56; Kaitani Mung'alo, OI, February 18, 1998; Muhati, OI, February 13, 1998; Musalimwa, OI, February 17, 1998; Isenjia, OI, February 24, 1998; Mavia, OI, March 2, 1998; Amagadu and Shikali, OI, March 3, 1998; Francis Mutsotso, OI, March 4, 1998; Litavakha, OI, March 9, 1998; Frederick Mutembei, OI, March 17, 1998.

18. Wagner, *The Bantu*, Vol. II, 78–79.

19. *RCNLTNKR*, 11–13; Shilaro, "Kabras Culture," 76; Francis Mutsotso, OI, March 4, 1998.

20. Shilaro, "Kabras Culture," 78.

21. Wagner, *The Bantu, Vol. II*, 80; *RCNLTNKR*, 10; Mwela, OI, January 23, 1990.

22. *RCNLTNKR*, 10–11; Wagner, *The Bantu*, Vol. II, 80, 84; Shilaro, "Kabras Culture," 78; Msalimwa, February 17, 1998; Isenjia, OI, February 24, 1998; Mbwabi, OI, February 26, 1998; Andove, OI, February 28, 1998; Amagadu and Shikali, OI, March 3, 1998; Francis Mutsotso, OI, March 4, 1998; Litavakha, OI, March 9, 1998.

23. *RCNLTNKR*, 11; Wagner, *The Bantu*, Vol. II, 84; Shilaro, "Kabras Culture," 79; Mukabwa, OI, February 12, 1998; Musalimwa, OI, February 17, 1998; Imbali and Mung'alo, OI, February 18, 1998; Isenjia, OI, February 24,1998; Andove, OI, February 28, 1998; Mavia, OI, March 2, 1998; Litavakha, OI, March 9, 1998; Mutembei, OI, March 17, 1998.

24. *RCNLTNKR*, 11.

25. Minutes of the Agricultural Officers (AOs) Meeting, June 20, 1943, KNA: AK/6/8; Senior Agricultural Officer (SAO) Nyanza to Director of Agriculture, June 28, 1943, KNA: AK/4/1.

26. Maxon, *Going Their Separate Ways*, 121–35; North Nyanza District Annual Report (NNDAR) 1948, KNA: DC/NN/1/30.

27. Joyce Lewinger Moock, "The Migration Process and Differential Economic Behavior in South Maragoli, Western Kenya," (Ph. D. dissertation, Columbia University, 1975).

28. Republic of Kenya, *Laws of Kenya: The Mining Act Chapter 306* (Nairobi: Government Printer, 1987), 7.

29. See The Crown Land Ordinances, 1902, 1915; Mining Ordinance 1925, and the Mining Amendment Ordinance 1931, *East African Standard*, 22 August 1931.

Chapter Two

The Kakamega Gold Rush and the Kenya Land Commission, 1932–34

INTRODUCTION

This chapter examines the impact of the discovery of gold in NK district in 1931 on colonial land policy in Kenya. The narrative describes the Kakamega gold rush, the new complex issues surrounding Luyia land rights, and the conflicts engendered by the thrust of a virtual rural agrarian terrain into a capitalist industrial economy based on gold mining. Shortly after the gold discovery, the imperial government had appointed the KLC (also known as the Carter Land Commission), to examine and reach a lasting solution to African grievances borne of the colonial land alienation policies in Kenya. By examining the actions of the colonial state, the imperial government and the recommendations of the KLC, this study sheds much light on the primacy of economic imperatives in the violation of Luyia land rights. The colonial state moved swiftly to amend the NLTO of 1930, which guaranteed the security of tenure for African reserves. Both the imperial government and the KLC approved the NLT(A)O of 1932, which allowed the colonial state to alienate land from NK for gold mining.[1]

In addition, the chapter re-examines the Kenya land question by addressing a unique case in which conflict over land emanated from the needs of rural industrial capitalist expansion as opposed to the perennial European settler demands for prime farmland. Hitherto, scholarship on this theme provides a biased historical account, which treats the land question in colonial Kenya under the rubric of the Kikuyu stolen lands. By shifting the analysis to the clash between European industrial capitalism and the Luyia agrarian economy, this chapter provides significant insights into the complex nature of such interaction. The chapter demonstrates that by 1932, the Luyia and their

Nilotic Luo neighbors had significant land grievances accentuated by the discovery of gold in Kakamega in 1931. The discovery sparked an influx of European gold prospectors into the Luyia reserve. Moreover, the pressure for land for gold mining in Kakamega coincided with the investigations of the KLC, whose charge included finding a lasting resolution to the daunting Kenya land question. An examination of the Commission's terms of reference, the Luyia land grievances presented before the Commission, and its recommendations, show that the Commission failed to protect African land rights in general and the Luyia in particular.[2]

Land was the central factor in Kenya's politics in the 1920s and 1930s. African grievances surrounding the alienation of land for European settlers had reached great heights by 1931, prompting the British government to appoint the KLC on April 1, 1932.[3] The appointment of the KLC was a result of the recommendations of the Joint Select Committee of Parliament of 1931. The primary objective of the Joint Select Committee was to study economic and political issues relating to the Closer Union of East Africa. Nonetheless, the evidence presented by the Kenya African representatives before it, namely Chief Koinange wa Mbiu from Central Province, Headman James Mutua from Ukambani, and Ezekiel Apindi, a Church Missionary Society (CMS) head teacher at Maseno in Nyanza Province, demonstrated the aggravated nature of African land grievances in Kenya.[4]

Based on the evidence of the African representatives, the committee recommended a full and authoritative inquiry into the land needs of Kenya's African population, both present and prospective. The committee further urged the colonial state to halt the alienation of Crown Land to immigrant races without the approval of the Secretary of State (S of S) for the Colonies.[5] Accordingly, the imperial government appointed the KLC in April 1932, and it commenced work in Kenya four months later.[6]

KENYA LAND COMMISSION: TERMS OF REFERENCE

The Commission's terms of reference were to:

1. consider land requirements of African population, present and prospective;
2. look into the desirability and practicality of setting aside more land for present or future occupation of African populations;
3. determine the nature and extent of African claims to land already alienated to non-Africans;
4. Examine African claims to land not yet alienated;

5. Consider the nature and extent of African land rights under the Crown Lands Ordinance of 1915;
6. Define the Kenya Highlands, within which Europeans were to enjoy a privileged position;
7. Review the working of the NLTO of 1930;
8. Examine the working of the Crown Lands Ordinance of 1915.[7]

The appointment of the KLC aroused much enthusiasm among the Africans in Kenya. Kenya Africans viewed the Commission as a permanent solution to the vexed land question in the colony. This optimism was manifest in the evidence that the Luyia presented to the commissioners. That the Africans felt a sense of betrayal at the recommendations of the Commission vividly demonstrates their great expectations of a favorable settlement of the land issue.[8]

The Commission's ability to reach a resolution acceptable to Africans in the colony was slim. The Commission was bedeviled with contradictions that immensely restricted its ability to reach an agreeable resolution to the evasive and hotly contested land question. First, the Commission's Chairman, Sir William Morris Carter had recommended racial segregation in land ownership when he chaired the Southern Rhodesian Land Commission of 1925. During his tenure as Chief Justice in Uganda and Tanganyika (now Tanzania), and later, as acting Governor of Uganda, Carter's policies remained consistently pro-settler. Carter's critics in East Africa considered his resolution of the Nkore/Toro land problem in Uganda as "hopelessly pro-European." The other two commissioners, Mr. Rupert W. Hemsted, a retired Kenya PC, and Mr. Frank O'Brien Wilson, were settlers with stakes in the Kenya Highlands. The Kenya Missionary Council considered Wilson as a "purely and solely the settlers' man."[9] The third commissioner, Mr. Rupert W. Hemstead had initially distinguished himself as the most pro-African colonial government official had became a settler farmer in Kenya after his retirement. This transition tilted his views in favor of the European settler community.[10] The Commission's Secretary, Mr. S. H. Fazan, was at the time the DC of Kiambu District, in Central Kenya.[11] In his address to the House of Commons, S of S for the Colonies, Sir Philip Cunliffe-Lister strongly defended the composition of the KLC. Cunliffe-Lister argued that Carter had extensive knowledge of African Affairs, that Sir Hemsted was a model "successful PC in Kenya," and that Mr. Wilson was a "practical farmer and a model employer."[12]

Second, an all-European Commission constituted another element of criticism and distrust. In Britain, critics of colonial land policies in Kenya in both Houses of Parliament demanded that the CO appoint one or two Europeans with no land, financial, or political interests in Kenya, and one or two African representatives.[13] The CO ignored such demands.

THE KLC AT KAKAMEGA

From the time of arrival in August 1932, the Commission interviewed 736 witnesses, 487 of whom were Africans in forty-two places throughout Kenya. The Commission received indirect evidence including 400 letters, 212 statements and 507 memoranda. The Commissioners arrived in Nyanza Province in September 1932, where they interviewed colonial officials, administrators, settlers, missionaries, Asians, various associations, and members of various clans. Most of these groups and individuals also submitted written memoranda.[14]

The evidence and demands made before the Commission in Nyanza were indicative of the deep-seated land grievances among the Luyia and their Nilotic Luo neighbors. African witnesses presented impressive testimony before the Commission illustrative of the broad knowledge and avid understanding of the task before the Commission. The passionate appeals also perhaps epitomized the heightened optimism among the Luyia for a final resolution to the land question that had hitherto dodged the colonial state.

The Luyia made several demands before the KLC. They demanded the return of, and compensation for, land already alienated from their reserve. They wanted the colonial state to repeal the Crown Lands Ordinance of 1915, and to clarify the core of the NLTO of 1930, which provided for inviolable African reserves. The Luyia called for the expansion of the reserve to accommodate the rapidly increasing population. In addition, the Luyia articulated their desire to return land control matters in the reserve to local tribunals.

Except for the mission-educated elites, the majority of Luyia witnesses opposed land registration on individual basis, and demanded that all African reserves be renamed Native Lands. The Luyia also called for the appointment of Paramount chiefs.[15] For the Luyia and other groups in Kenya colony, paramount chiefs reflected the sentiment that such chiefs guaranteed the security of their land as was the case in Buganda.

The colonial state had alienated land in Nyanza Province. Some communities had lost land when the colonial state alienated land for individual settlers, Asians, private companies like the British East Africa Corporation, the Swahili, Nubians, missions and government facilities such as aerodromes, townships, agricultural experimental farms and veterinary stations, railways, roads, and golf courses, among others.[16] The Luyia specifically claimed land alienated for European settlers in the Kaimosi, Kamukuywa, and the Kipkarren areas.¹⁷ The colonial state excised Kamukuywa and Kipkarren in Trans Nzoia District from NK to reward demobilized British World War I (WWI) veterans. Wa-Githumo notes that the colonial government demarcated 129½ square miles of land in Kipkarren for the Soldier Settlement Scheme.¹⁸ In his evidence to the Commission, Rev. Monsignor Brandsma, the Prefect Apostolic of Kavirondo, corroborated the Luyia claim to Kaimosi settled area. However, Brandsma argued that the alienated land was solely a Luyia expansion frontier for surplus population.¹⁹ Indeed, the Commission acknowledged that in 1913, the colonial government had alienated 10,726 acres of land in Kaimosi for European settlers on 999-year leases.²⁰

Apart from the question of a closed expansion frontier, the rapidly growing population, coupled with the fear of further encroachment from European settlers, private companies, Asians, and other alien groups like the Swahili, Nubians, and the Ganda, remained a major threat to the Luyia.²¹ In the precolonial era, the frontier offered a vent for surplus population. The Luyia claimed the Kaimosi areas on the border with Nandi, the Uasin Gishu and the Trans Nzoia areas as their open frontier.²² Similarly, in the pre-conquest era, the Nilotic Luo frequently engaged in skirmishes with the Nandi over the Asian settled regions like Kibos, Muhoroni, Chemilil, and Songhor near Kisumu.

Colonial policies, including land alienation, the fixing of reserve boundaries, and the closure of forests to Africans, eliminated the benefits of the frontier. These policies intensified intra-and inter-community conflicts over land as both the Luyia and Luo communities grappled with resettling surplus population within and between their restricted reserves. For instance, both groups claimed ownership of land on the border between Maragoli and Kisumu.²³ Moreover, CK, then reeling under increased population pressure, witnessed immigration into South Nyanza, NK, particularly the Wanga and Kabras areas; the Trans Nzoia settled areas, and the Kenya-Uganda border.²⁴

Except for a few areas like Kabras in NK district, most parts of Nyanza Province were experiencing land shortage by 1932. The Luyia and the neighboring Luo populations were undoubtedly increasing rapidly. Sections of NK reserve such as Bunyore, North and South Maragoli, had population densities of between 1,000 and 1,100 persons per square mile. These regions, together with the Bukusu inhabited northern part of the reserve, were crowded locations.²⁵

Apart from dense population, the Luyia practiced intensive crop cultivation. As early as 1894, when Charles William Hobley, the first British officer to administer the region, arrived in Kavirondo, the entire stretch of land between the northeastern borders from Kabras to the western part of Mumias constituted "fairly continuous cultivation."[26] In addition, pressure on land in Buluyia resulted from livestock keeping, a practice that wielded significant economic and social value. Walter Edwin Owen, the Church Missionary Society (CMS) Archdeacon of Kavirondo,[27] vividly demonstrated the difficulty involved in eking out a living in the African areas when he wrote:

> In very many areas . . . the industry and patience of the Africans in cultivating the most rocky and seemingly hopeless hillsides speaks in striking terms of the pressure of the population on the soil. These difficult sides are cultivated only because better places are not available.[28]

Negley Farson, a British journalist, reiterated this when he wrote more than a decade later:

> In the huge Kavirondo Reserve—which holds over 1,500,000 Africans, where the population rises to 800–900 to the square mile in some of the worst districts, I have seen four families, twenty-five people, living and trying to make a living out of 2¼ acres. . . . It is on a hillside called Lotego, near Majengo, in the South Maragoli District: four families, four huts, twenty-five people, on 2¼ acres. All farmers! It was a Salvador Dali nightmare of granite boulders among which the natives were huddled like rabbits.[29]

While Farson's Lotego might be an exaggerated case, other factors minimized usable land in Nyanza Province. Any available empty land was largely uninhabitable because of the prevalence of both human and animal diseases such as sleeping sickness and trypanosomiasis, respectively. In areas such as Kabras and Samia locations of NK reserve, swampy conditions, insufficient rainfall and lack of water, poor soils and the menace posed by wild animals compounded the problem.[30] Indeed, in 1932, the Agricultural Officer's (AO) surveys of land use in NK district show that out of a total of 1,539,200 acres, 545,687 acres were not suitable for cultivation.[31]

THE LUYIA GRIEVANCES

These factors indicate that the Luyia had legitimate land grievances. An increasing population, a closed frontier, land alienation and the hostile environmental factors, collectively conspired to thwart agriculture and animal husbandry in Buluyia. Accordingly, the Luyia and their Luo neighbors demanded ameliorative measures from the Commission.

First, the Luyia wanted the colonial government to curtail further alienation of land for European settlers, and called for the return of previously alienated lands. They challenged the practice of European settlers accumulating vast tracts of lands, which remained unutilized, and demanded that the colonial state open up the Kenya highlands for sale to Africans as well as Asians.[32] This demand received support from some Europeans and Asians in the colony.[33] Archdeacon Owen favored the redistribution of African population by allowing immigration between NK and CK, as well as the exclusion of Kaimosi Settlement from the Highlands.[34] In cases where the colonial state could not return alienated land to African reserves, Africans demanded adequate compensation to dispossessed groups.[35]

With regard to the development of gold mining, which had begun to have an impact on the NK reserve, the Luyia called for prospecting by the Crown, compensation directly to those displaced by the gold miners,[36] and the payment of a percentage of the royalties to the Local Native Council (LNC).[37] The Luyia vividly expressed uncertainty over the long-term impact of the gold mining industry on their reserve. They stated: "[We] are unable to estimate, . . . [or] tell, to what extreme the development will extend. Some Europeans say that most probably it will be like Johannesburg in South Africa where Africans have now been rendered landless. We do not know what lies in future for us."[38]

The Luyia distrusted the colonial government. The concretization of this mistrust drew heavily from stories of the African experience in the goldfields of Southern Africa. They submitted, "No one know[s] what the Kenya Government will do in ten, twenty, or thirty years hence. . . . We presume more land will be taken in the years to come, and compensated in money therefore, but money cannot compensate fully for the land so taken from us."[39]

Luyia protestations to the KLC embodied the central question of the future of African reserves in Kenya. The Kavirondo Taxpayers Welfare Association (KTWA), composed mainly of CMS Maseno School alumni, sent a protest memorandum to the S of S, opposing the European invasion of Kakamega.[40] KTWA started as a radical political association. However, at the urging of the colonial government, its patron Archdeacon Owen transformed the association into a welfare-oriented body. KTWA called for the registration of land on individual basis, the payment of mining lease rents directly to dispossessed African households, and requested that Africans be granted sufficient time to review the Commission's report before its adoption.[41]

Contrary to the KTWA's demand, a majority of the Luyia objected to land registration. The Native Chamber of Commerce, for example, desired the old communal landholding system. Members maintained that individual tenure was complex, expensive, and unnecessarily legalistic. Besides, the system had the potential of legalizing the dispossession of the poor by the rich. Moreover, members stated that the system had the proclivity for facilitating the breakdown of entrenched social relations. Consequently, they derided KTWA's decision as reflecting the views of Archdeacon Owen, which were unsuitable for Africans.[42] The pervasive sense of suspicion that surrounded the colonial state's attempt at registering African lands is vividly captured in Chief Arean's argument, "One thing that I hear which is wrong, . . . [is that] the land is not ours. Before we heard [the] Government say [that] the land was ours. In these days, we hear talk about a register. . . . If the land is ours, who is going to take it from us?"[43]

The Luyia, government officials and administrators, objected to land registration for different reasons. For the Luyia, land registration did not guarantee security of tenure. It also entailed extra financial obligations, and threatened societal cohesion.[44] Later, the NKCA, a Luyia political protest movement against the alienation of land in Kakamega for the gold mining industry, succinctly articulated this sentiment.[45]

Colonial land policies in Kenya reflected the contradictions that characterized the conquest state. These contradictions revolved around the question of African land rights. In a most perceptive assertion, missionary Monsignor Brandsma noted, ". . . [O]n many occasions natives have been told: 'Now this is your reserve and nobody will touch it. It is your land in perpetuity, etc.', and at the same time it is not. It is Crown Land and if the Crown requires it, it will take it."[46]

Luyia distrust for the colonial state emanated from its failure at consulting Africans on matters relating to land. In many cases, when the colonial state alienated land in the reserve for a cause that was beneficial to Africans, such excisions involved vast acres of land. Alienations of this magnitude often displaced numerous people. The advent of gold mining in Kakamega demonstrated the impotence of the LNC chaired by the DC who gravely compromised its independence.[47]

Although Archdeacon Owen corroborated Brandsma's evidence, he contended that Africans did not oppose land registration *per se*, but the unfavorable regulations associated with *the Report of the Committee on Native Land Tenure in North Kavirondo Reserve (RCNLTNK) (1930)*. The *RCNLTNK* had recommended in favor of individual land registration. Owen thus agreed that this recommendation represented the wishes of the minority-educated members of the KTWA.[48]

The Luyia also opposed individual registration of land due to the cost of the title. A majority of Luyia landholders considered the land registration fee as a purchase of one's own land. Cash payments obviated the fact of prior right to the land. In addition, landowners feared the possibility of increased registration fees, which posed an economic burden to Luyia peasants. In his evidence, Owen also referred to the impact of the *Kipande* (identification certificate) on the Luyia community. The Luyia were averse to the term "registration", as it evoked the untold suffering the *Kipande* laws had inflicted on Africans in Kenya.[49]

Both the Luyia and the missionaries contended that individual land tenure, involving free land exchanges (sales), would precipitate unfavorable social-economic stratification pitting the accumulators and their tenants against each other. Equally, Luyia peasants who owned land opposed registration because they risked forfeiting their land to tenants. The latter objected to land registration because such a system virtually turned them into permanent tenants with no hope of eventual absorption into their host clans. Using these arguments as a basis, Owen turned to the characteristic European attitude that individual land tenure among the Luyia was premature. Instead, Owen called for the gradual modification of the indigenous land tenure system.[50]

The evidence of E. B. Hosking, the government officer charged with implementing land registration in NK district, was revealing with regard to the myopic nature of the colonial state's perception of African land tenure systems. When Hosking visited the reserve, the Luyia treated him with suspicion. He reported:

> I could find no demand whatsoever for the registration of individual titles to land, and I speedily came to the conclusion that any such system as was advocated in the report was premature and unworkable. I . . . Submit that seventeen days is [sic] totally insufficient for the investigation of a system of land tenure and for the formulation of definite recommendations.[51]

Hosking contended that the notables interviewed by the Committee had used the opportunity to unduly glorify their own families. This means their demands represented the views of a tiny unrepresentative, partially educated group, which was allegedly highly influenced by similar demands in Central Province.[52] He ended his scathing criticism of the investigation procedure with a quotation from the renowned anthropologist, Bronislaw K. Malinowski: ". . . [I]t is futile to summon, as political Committees usually do, a number of witnesses and just ask them simply what is their form of ownership, or, worse, what in their opinion ownership should be."[53] To an untrained European like Hosking, African land tenure systems were utterly incomprehensible.[54]

It was to this kind of pressure that the DC NK, C. B. Thompson, capitulated in his oral evidence before the Commission. Thompson conceded that the administration could only institutionalize individual land tenure when the people themselves desired it.[55] In a separate memorandum, the DC noted the rejection of state appointed tribunals in settling land matters, and the need to restrict migration into the reserve by outsiders.[56]

Notwithstanding, the North Kavirondo Land Tenure Committee had clearly predicted the inevitability of alienating land in Kakamega for gold mining.[57] It is notable that the impetus for postponing land registration in NK lay in the demands and needs of the emerging rural industrial enclave centered on gold mining. If land registration in NK proceeded, the colonial state had no legal right to settle households displaced by gold mining leases on land registered as belonging to other clans or families. Since the administration anticipated these forms of expropriation, it was eager to halt a process that limited its ability to use compulsion in creating room for those evicted from the goldfield.[58]

In addition to the anxiety the Luyia expressed over imminent land alienation for the nascent gold mining industry, the Luyia also challenged the colonial state to sanction the cultivation of coffee by Africans and the encouragement of cotton growing. Citing the success of African coffee production in Uganda, the Luyia argued, and rightly so, that European demand for labor, and not poor disease management skills among African producers, was the major factor behind the ban.[59] Moreover, they demanded that two African representatives accompany the Commission when presenting the report to the imperial government.[60] Once again, the Luyia challenged the colonial state to take a solemn indigenous oath, pledging the security of African lands. In his evidence before the Commission, Chief Sore of Isukha declared, "Let us take a mutual oath with [the] Government by cutting a dog in half, and let it be sworn whether this is our land or not." [61] Predictably, this plea did not alter the official view.

Apart from land related problems and the cultivation of high value cash crops; the Luyia feared the perceived detrimental effects of a mining industrial economy on their society. The Luyia contended that the influx of miners into their reserve would facilitate the breakdown of societal norms, which held the community together. Moreover, the Luyia also opposed mining because of the potential for both the colonial state and European mine owners resorting to the strict enforcement of the *Kipande* or pass laws to regulate African labor.[62]

THE OFFICIAL VIEW

The PC's response to the Africans' claim to the alienated land is instructive of the colonial administration's disregard for African land rights. The PC Nyanza, H. R. Montgomery, dismissed African claims to lands excised from the reserve. Montgomery claimed that with only one exception, the NK LNC had consented to land leases for the establishment of mission stations and trading centers. The PC also dismissed the Logoli claim to land on the border with the Luo, arguing that the area had been "a battlefield" for the two groups. In his view, Luyia settlement in that area was a recent phenomenon. Notwithstanding, the PC contradicted his position by asserting that the disputed area was in fact, a Luo location under Luyia encroachment.[63] Similarly, the PC contended that the Kaimosi settler farms and the Asian owned sugar estates between Kisumu and Muhoroni served as war zones for the Nandi and their Luo neighbors, as well as warring Luo clans.[64]

PC Montgomery also quashed Luyia claims to land in Trans Nzoia and Kipkarren. Citing a 1912 report, Montgomery argued that Kamukuywa River was the legitimate boundary between NK reserve and the Trans Nzoia region, occupied by European settlers. To augment his argument, Montgomery alleged that the colonial government had evicted only two recently established Kabras villages from the Kipkarren farms. Since the evicted Kabras had paid taxes only twice on their Kipkarren residence, the colonial government could not honor their claims.[65] An entirely new colonial apparatus of oppression had become an instrument of gauging African rights to land. In fact, the PC declared that claims of land shortage in Nyanza Province were a myth. He wrote:

> In both these districts several witnesses stated that the reserves are too small for the needs of the people. . . . I do not agree. . . . Claims were made before the commission which had never, . . . been raised before, and one is led to infer that witnesses thought they were in duty bound to claim something . . . as natives are NO. 1 of the Terms of Reference.[66]

Montgomery was not alone in dismissing Luyia and Luo claims of land shortage. The Agricultural Officer (AO) for CK, A. S. Hartley, argued, "there is a lot of good land not being used."[67] He advised the people of Nyanza Province to adopt improved methods of cultivation including the adoption of new varieties of crops, versatile technology, irrigation, and the elimination of sleeping sickness. He urged peasants to embrace new industries, agricultural instruction and export crops, besides increasing the acreage of existing crops such as oil grasses, other oil seeds, cotton, groundnuts (peanuts) and wattle.[68] Boosted by the confident "expertise" of the AO, Montgomery insisted that owing to the lack of forest conservation skills among Africans, those seeking access to forest could do so by permit only. He rejected Owen's proposal of relieving population pressure through immigration between the two African reserves.[69]

"INVIOLABLE" AFRICAN RESERVES: THE NATIVE LANDS TRUST ORDINANCE

Although land alienation was a grievance in Nyanza Province, the central factor in African concern was the fear of insecurity of land not yet alienated. This insecurity reached a fever pitch in NK reserve with the discovery of gold in Kakamega in 1931.[70] The advent of gold mining led to an influx of European prospectors into the area, so that by October, there were seventy-five prospectors in Kakamega and by the end of the year, over 400 prospecting permits had been issued.[71] The convergence of prospectors on Kakamega and the discovery of gold reefs necessitated the amendment of the NLTO of 1930 with lightning speed in December 1932. The newly promulgated NLT(A)O of 1932 heightened the sense of land insecurity among the Luyia.

The Crown Lands Amendment Ordinance of 1926 provided for the inviolability of African reserves in Kenya.[72] This promulgation created between twenty-four and twenty-seven African reserves in the colony.[73] The inviolability of the reserves so constituted remained a theoretical postulation as land alienation persisted.

The passage and adoption of the NLTO of 1930 was a daunting task for the imperial government. The debate embroiled the colonial state, the European settler community in Kenya and the imperial government, in a major contest that lasted for four years. As noted above, the creation of African reserves in 1926 necessitated the promulgation of a legislation to secure these lands for African occupation for posterity. Pressure from critics of the British imperial edifice in Britain, the representatives of African interests in Kenya and most notably, the appointment of Labour's S of S for the Colonies, Sidney Webb (Lord Passfield) in 1929, played a fundamental role in forcing the CO to stand up against the colonial state and the European settler community in Kenya. As trustee of its African subjects, the imperial government had an obligation to safeguard African lands against the competing and contradictory settler interests. Between 1924 and 1929, neither the imperial government nor the colonial state enforced the doctrine of African paramountcy.[74] The period also coincided with the tenure of the Conservative Leopold S. Amery at the CO as S of S for the Colonies.

NEGOTIATING THE NLTO OF 1930

The drama that characterized the contest between these polarized interests began in earnest in May 1926. The major points of contention revolved around the following questions: Was the bill to pass as an ordinary legislation by the LegCo or an Order-in-Council? Were African reserves to be vested in the colonial state or an independent board? What was the nature of African and European settler representation on the envisaged board? What was the maximum lease term for land alienated from African reserves? Under what conditions were African reserves to be alienated for Europeans? What was the nature of compensation for land so alienated?

AMERY EXITS AND PASSFIELD IMPOSES THE BILL

In the General Election of June 2, 1929, the Labour Party swept the Conservatives from power in Britain. Sir Sidney James Webb, Lord Passfield (1859–1947) replaced the Conservative S of S Amery at the COs.[75] Ultimately it took the administrative rigor of the Labour Party and the personal will and humanitarian commitment of Lord Passfield to force the bill through the Kenya LegCo.

Following intense consultations with permanent CO officials, particularly Bottomley, Parkinson, and Wilson, on the Native Lands Trust Bill (NLTB) saga, Passfield was eager and committed to ensuring the security of African lands. His major points of concern, namely the lease terms and the nature of compensation payable to dispossessed Africans, became the most complex arena in the contest between the CO and the pro-settler governor, Grigg and the elected members of the LegCo in Kenya colony. Passfield reduced the maximum term of lease to thirty-three years. Notably, he strongly pushed for the policy of land for land, which required the colonial state to provide equal, suitable, and contiguous areas of land for land alienated from African reserves for public purposes.[76]

When discussion on the Bill resumed in the LegCo on October 20, the European elected members, led by Conway Harvey, the member for Nyanza, unanimously condemned Passfield's amendments. Harvey's declaration during the LegCo proceedings vividly captures the general feeling among European elected members:

> Why Sir, the natives of Africa should be put in a class apart and put above the law in this respect passes my comprehension. . . . This, to my mind Sir, represents class legislation with a vengence [sic], and in conclusion, Sir, I should like to warn you that if the Government persists in moving the proposed amendment . . . , Elected Members will feel it their duty to leave the House.[77]

With this declaration, the colonial government led by the strong pro-settler Governor Grigg, the European settlers and the CO had reached an impasse. Caught between an uncompromising S of S for the Colonies and a recalcitrant LegCo, the colonial state bowed to local pressure and adjourned the LegCo *sine die*.[78]

In spite of this, Passfield and the permanent officials at the CO were ready to take on governor Grigg, his Executive Council and the settler faction in

Kenya. The fundamental differences between Grigg and Passfield's views, guaranteed a permanent rupture between Nairobi and London.[79] Shortly after, Passfield offered to relieve the governor and the Kenya legislature of their lawmaking duties and sought to effect the legislation by Order-in-Council.[80] Grigg capitulated and informed Passfield that the LegCo would pass the Bill in accordance with his instructions.[81]

When the CNC, Maxwell recommitted the Bill to the LegCo, European elected members were uncompromising. Captain E. M. V. Kenealy, in particular, gained notoriety for his ignominious contempt for the NLTB. He castigated Passfield's views as "utterly illogical and irreconcilable," and called the Bill "a rotten . . . and . . . silly Bill" which "commonsense could stop."[82] The only members supporting the Bill were the elected Asian representative, Malik and the nominated representative of African interests, Rev. Leakey.[83] Indeed, when the Bill went through the third reading on April 11, 1930, of the eleven elected European members, three voted against it and eight abstained. Nevertheless, the Bill passed with the assistance of official majority, supported by the unofficial representative of African interests and the Asian elected member with twenty votes.[84] Thus, pressure from London dealt a deadly blow to the alleged "limit of [settler] endurance." Consequently, on April 13, Grigg assented to the Bill, and submitted it to the CO on May 23, 1930.[85] Concisely, Passfield imposed the NLTO of 1930 on a recalcitrant and uncompromising pro-settler governor and the European settlers in Kenya. The negotiating process proved a *longue durée* for the British Government.

Although Grigg and the settlers dismissed such criticisms, they dismally failed to carry the CO with them. Hostile press publicity, immense Parliamentary criticism, opposition from the Hilton Young Commission and the British humanitarians compelled the CO to take action. Moreover, the permanent officials at the East African Department (EAD) of the CO were under pressure to maintain CO control over Kenyan affairs. Indeed, the change of Government in Britain and the ascendancy of Lord Passfield to the CO marked the end of cooperation between London and Nairobi. Within a year at the CO, Passfield took Governor Grigg and the Kenya settlers head on, despite the radicalization of settler politics. Ultimately, it took the combined force of Lord Passfield, the representatives of African interests in Kenya, and the ever-present scrutiny of critics of the British imperial edifice at home, to force the CO to stand up against Governor Grigg and the Kenya settlers.[86] This was, however, a partial success. While the gestation period of the NLTO of 1930 lasted a long and arduous four year period, the discovery of gold in Kakamega in October 1931, led to a speedy amendment of the ordinance to allow the colonial state to alienate land in NK for European industrial interests in the gold mining industry.

Although not quite a blueprint of Passfield's ideal, the NLTO of 1930 provided several safeguards for African lands. First, the ordinance adopted the policy of land for land. Second, the ordinance required the colonial state to consult both the Local Land Board and the LNC before alienating land from an African reserve even on temporary basis. Third, leases for land excised from African reserves were limited to thirty-three years.[87] By decreeing compensation in land, and emphasizing the consent of the LNCs, the NLTO of 1930 gave a semblance of security for African lands.

Passfield reiterated his pro-African policy in two White Papers of June 1930, which became the "Passfield pledge."[88] In the envisaged policy document most pertinent to this study, *Memorandum on Native Policy in East Africa,* Passfield reaffirmed the doctrine of African paramountcy; accepted and reserved African trusteeship to the British Crown; and pledged to eliminate the sense of insecurity of land among Africans besides ensuring the adequacy of reserves as provided for in the NLTO of 1930.[89] Labeled "Black Papers" by the Kenya settlers and roundly condemned in pro-settler circles in Britain, the Labor Party's position as laid out in the White Paper on Native Policy was simply ignored.[90] Thus, the NLTO of 1930 remained the only safeguard for African lands. Yet, the ordinance suffered serious handicaps from its inception.

For instance, settler farms were surrounded by African reserves. How feasible was it to find land in the midst of European estates for African settlement? Then, European settlers and members of the Executive Council dominated the Central Native Lands Trust Board. Was such a body capable of articulating and upholding African interests? In addition, as Chairmen of LNCs, DCs had the power of veto against LNC resolutions. Were they, therefore, free institutions for securing African interests? The answers to these questions are simply negative as the Kakamega case reveals.

CHANGING FORTUNES: THE GOLD RUSH AND THE NLT(A)O OF 1932

The discovery of gold in Kakamega in 1931 constituted the testing ground for the efficacy of the NLTO of 1930 in securing African lands in Kenya colony. Many settlers viewed the gold discovery as salvation from the economic crises that came in the wake of the global Economic Depression of the 1930s. Devastation from locust infestations which inflicted Nyanza Province and the White Highlands between 1929 and 1932, served to worsen the hopeless economic situation.[91] European settlers facing declining economic fortunes flocked to Kakamega with the hope of making quick profit from gold prospecting.The mining venture laid bare the hollowness of the NLTO of 1930 and heightened African distrust for British legislation.

The gold discovery and the changing fortunes in the security of African lands in Kenya coincided with significant governmental changes in Britain. In August 1931, the Coalition or "National" Government replaced the Labour Government. The General election of October 27, reaffirmed the new government and effectively routed the Labour Party out of power.[92] Rather than join the National Government, the pro-African S of S, Lord Passfield retired, leaving the CO to the Conservative Sir Philip Cunliffe-Lister. With the collapse of the second Labour Government and a change of guard at the CO, the imperial government dominated by Conservatives embarked on a mission of reversing British policy in East Africa to what it had been before 1923.[93] The controversial NLTO of 1930 was among the earliest casualties of this decidedly anti-African and pro-European policy. As European prospectors descended on the Luyia reserve, the glitter of gold quickly overshadowed the burden of trusteeship.

Due to increasing fear of loss of land to miners among the Luyia, the Governor, Brigadier General Sir Joseph A. Byrne, visited Kakamega on July 25, 1932, in order to allay these fears. Byrne succeeded Grigg as governor of Kenya in February 1931. Byrne had a distinguished military and civil service record. He had served in the South African Anglo-Boer War and the First World War. From 1916 to 1922, Byrne rose to the position of Inspector General of the Royal Irish Constabulary and two years later, he was elevated to the position of governor of the Seychelles, a position he held for five years, before taking up the governorship of Sierra Leone, West Africa, in 1927. In the letter of appointment, S of S Passfield expressed satisfaction that Byrne was "well qualified for the appointment." Byrne subsequently assumed governorship of Kenya on February 13, 1931.[94]

During his visit to Kakamega, Byrne verbally assured a large crowd of Africans that the colonial state was committed to securing their land rights. In his pledge, Byrne emphasized the sufficiency of his "word," asserting that the government had no intention of depriving them of their land. Despite this, the Luyia expressed extreme suspicion and distrust for the colonial state's pledge to secure their rights to land. Accordingly, they challenged the governor to take an indigenous Luyia solemn oath of honesty, by cutting a dog into half.[95] Byrne declined to take up a challenge he perceived as a "savage" and "barbaric" act.[96]

The Luyia attempt to make the colonial establishment captive of their customary laws failed dismally to capture the governor's imagination. Nevertheless, it was a powerful statement indicative of their immense distrust for the British government. Indeed, Bryne conveniently shelved his pledge as the colonial state and gold mining interests became embroiled in a protracted struggle with Luyia landholders over the exploitation of gold.

The incongruence between the colonial land and mining laws in Kenya was evident in the Kakamega gold rush. While the former recognized African surface rights, the latter reserved all minerals to the Crown, and thereby, empowered the colonial state to develop all the mineral resources of the colony. Owing to the dichotomy between these two diametrically opposed laws, the discovery of gold, the most economically valuable mineral in an inviolable African reserve provided an interesting arena for the contest between the ideals of British trusteeship and economic pragmatism. It soon became clear to the colonial state, the Kenya settlers and the imperial government that the NLTO of 1930 was an obstacle to the development of the gold mining industry.

It is perhaps conceivable that both the colonial state and the imperial government took the pledge of securing African reserves very lightly. If this was not the case, then it is plausible that both London and Nairobi cared little about the Luyia concept of land ownership and justice. Byrne's failure to partake of the Luyia oath of honesty quickly came to represent his untrustworthiness. As governor Byrne pledged the security of tenure for NK reserve, the colonial state drafted and passed the NLT(A)O, which received favorable approval of the S of S for the Colonies in July 1932. The government subsequently published the amending ordinance on December 6, 1932.[97]

The brevity of the debate and the absence of dissent in the LegCo over the NLT(A)O of 1932, the consensus between the colonial state and the metropolitan government, and the lightning speed that characterized the passage of the amending ordinance are indicative of the hollowness of the principle of British trusteeship. The Labour Government had imposed the NLTO of 1930 specifically to safeguard African reserves from the insatiable settler demands for arable land. The ordinance, therefore, bound the metropolitan government to secure these lands for African use and occupation forever. While the British government had previously issued such pledges, the burden of trusteeship would not survive the ultimate litmus test: economics.

On December 19, 1932, the CNC Wade, (the supposed guardian of African interests) introduced the amending ordinance in the LegCo.The gist of his argument for amending the NLTO of 1930 revolved around the economic significance of developing the gold mining industry in Kakamega. The 1931 Mining Ordinance gave prospectors unrestricted rights to exploit the mineral wealth of the colony. While alluvial extraction did not require leases, pit/reef mining did. Wade contended that any obstacles put in the way of gold mining would be "unreasonable and detrimental" to the colony's economy.[98]

The CNC recognized the inherent conflict between Crown mineral rights and African surface land rights in the reserves. Colonial mining laws restricted the exploitation of mineral resources in the colony, to land not cultivated by Africans. However, NK reserve was both intensely cultivated and thickly populated. Therefore, it was impossible for the colonial state to grant mining leases to prospectors. Although the colonial state could alienate land in any reserve for mineral exploitation, it was incumbent upon the state to compensate the reserve in land. Wade considered this requirement "cumbersome and impractical business", due to the temporary nature of the leases. Against this background, the CNC urged the LegCo to amend the NLTO of 1930 to facilitate cash compensation for gold mining leases. The sum of cash paid as compensation depended on the value of private land outside the reserve. In the words of the CNC, "such exclusions would not make one single African one penny the worse for the gold having been found in his own country."[99]

Despite his resolve to have the land laws amended, Wade had no mistaken faith in the popularity of his proposals among the Africans in general, and the Luyia in particular. In the same vein, the CNC demonstrated an admirable grasp of the dialectic opposition between Luyia customary laws pertaining to land proprietorship and the colonial land and mining laws. Put into perspective, it was clear that cash compensation was no magical solution to this dilemma. As he aptly put it:

> No amount of compensation will probably induce them [Luyia] to agree voluntarily to the leasing of any land. I am afraid that we have got to hurt their feelings, . . . wound their susceptibilities and in some cases . . . we may even have to violate some of their most cherished and possibly even sacred traditions if we have to move natives from land on which according to their customary law they have an inalienable right to live, and settle them on land from which the owner has under that same customary law an indisputable right to eject them. Nevertheless, we have to face these difficulties—I can think of no other alternative, unless we are prepared to allow to lie underground and useless wealth which will bring prosperity to this country, in which every single community and every individual will share.[100]

Wade knew that the amending ordinance would elicit open hostility from Luyia landholders. To consult them and alienate the land against their wish would have provided a blood transfusion to the already heightened sense of grievance against the colonial state. In his own words, the Luyia would have argued, "we do not want to do it" to which the colonial state would have answered, "it doesn't matter whether you want to do it or not, it has got to be done."[101] To circumvent such an eventuality, Wade recommended the exclusion of the LNC and the Luyia landholders from consultations for lease grants to gold prospectors.

The debate on the amending ordinance demonstrated a strange cord of unanimity hitherto unknown in Kenya's politics among the colonial officials and the unofficial European settler faction in the LegCo. The burden of trusteeship could not hamper the march of European industrial capitalism. Indeed, the representative of African interests in the LegCo, Rev. Canon Burns, agreed that whatever objections Africans harbored against gold mining, the development of the industry was inevitable. Burns correctly noted that land was the lifeblood of the Luyia economy. Hence, the idea of consensual alienation of land in Buluyia was farcical.[102]

The CNC justified the amending ordinance as serving the interests of the Luyia. Contrary to the pessimistic view, which painted the Luyia as the "losers" in the new economic configuration based on the gold industry, Wade asserted that Luyia households stood to benefit immensely from the development of the goldfields. Wade equated the gold mining industry to untold opportunities of wealth and prosperity for the Luyia. Opportunities for wealth included the "handsome" monetary compensation for their agricultural holdings and common grazing lands.[103]

Against this background, the LegCo passed the NLT(A)O on December 21, 1932 after two days of debate. Ten days later, the Governor assented to the ordinance thereby making it law.[104] On January 17, 1933, the Governor submitted the ordinance to the imperial government. While this controversial piece of legislation was a brief two-page document, the amendments were nevertheless significant. The NLT(A)O of 1932 eliminated the two central provisions of the NLTO of 1930, namely compensation in land and consulting the LNCs. By this Ordinance, the colonial state could alienate land from African reserves either on temporary or permanent basis without consulting the LNCs and the affected Africans. More important, the ordinance provided for cash compensation. By a stroke of a pen, the colonial state and the imperial government nullified the consent of LNCs and legalized monetary compensation.[105]

In his communication to the CO, Governor Byrne emphasized the temporariness of the measure, the possibility of minimal evictions and the sparseness of population density in NK reserve. The colony's legal advisor, the Solicitor General T. D. H. Bruce, reiterated Byrne's views affirming the right for the Crown to exploit its gold resources, the temporary nature of mining leases and sufficiency of cash compensation.[106]

Officials at the CO received Byrne's correspondence with enthusiasm. L. B. Freeston of the EAD agreed with the pre-suppositions and conjectures of the London officials regarding numbers displaced by the gold mining industry and the support of the representatives of African interests in the colony. While Freeston advised against a premature response to Kenya, he keenly observed that the governor would have known from the S of S's personal communications that his "policy is fully approved and supported." In fact, the official suggested that the CO present Parliament with a *fait accompli*.[107]

Sir William Cecil Bottomley, an Assistant Under-Secretary of State in the CO, disagreed with Freeston. Following a discussion with the S of S, Sir Philip Cunliffe-Lister, Bottomley contended that the LegCo proceedings confirmed the consensus both in London and Nairobi that Africans would be distressed at all stages of the development of the gold mining industry. Notwithstanding, he recommended that the CO delay its approval of the ordinance for three weeks or one month, purely as an exercise in public relations. Bottomley contended that critics at home would have perceived a hasty approval of the ordinance and the proposal to present Parliament with a *fait accompli* as acts of hostility.[108] Concisely, it was the fear of criticism at home rather than disagreement with the colonial state's policies in Kenya that delayed official endorsement of the amended bill until February 13, 1933.[109] That Sir Philip Cunliffe-Lister approved the ordinance amidst protest from the Africans of Kenya, British humanitarians, and other critics of the British Empire, speaks to the economic supremacy of gold relative to Luyia land rights.

THE NLT(A)O OF 1932: A STORM OF PROTEST

PROTEST IN KENYA

The NLT(A)O of 1932 raised a storm of protest in both Kenya and England. This protest occurred at various levels of both the imperial and colonial administrative hierarchies. In Kenya, Africans expressed their disapproval in *barazas* (public meetings), delegations to both the district and provincial administration, and, on several occasions, they addressed their grievances to their friends in England or directly to the House of Commons. In England, the press, Houses of Parliament, letters and delegations to the S of S and the Prime Minister occupied center stage. Interestingly, this avalanche of protest moved neither the colonial administration in Nairobi nor the imperial government in London.

Both the Africans and critics of British land policies in Kenya considered the NLT(A)O of 1932 a breach of faith. To the Kenyan Africans, the amendment epitomized the British government's breach of an imperial pledge made to them. Africans could not trust the British justice system. Consequently, Nyanza Province became a beehive of political activity.

Ten days after the passage of the amending ordinance, the KTWA compiled an impressive list of grievances against the colonial state in Kenya. With regard to the NLT(A)O, the association challenged the S of S's allegations in the House of Commons that Africans had been informed of the changes to be associated with the mining venture in Kakamega. Based on a telegraphic message from Byrne, Sir Philip Cunliffe-Lister had informed the House of Commons that the CNC had circulated a memorandum "in clear and simple vernacular" explaining the process of prospecting and mining and the necessity for leases. He also alleged that the pamphlet specified the security of African lands.[110] Shortly after, it became public that the colonial administration had not circulated a notice of the amending bill among Africans. The S of S had thus made a false statement.[111]

It is instructive to note that the CNC's memorandum was in English and circulated to missions operating in Nyanza Province. From the CNC's instructions, it is evident that the colonial state abdicated its administrative duties, leaving the responsibility of translating, printing and circulating a public policy document to the African population in Nyanza to the discretion of the men and women of the pulpit. The instructions read thus, "A memorandum from the Chief Native Commissioner on the subject of mining is forwarded herewith. It is requested that you will bring this to the notice of your adherents. *Should you desire to issue copies in a translation into the vernacular please do so.*"[112] Owen's revelation unveiled the sloppy operation of the colonial administration in Kenya, besides confirming London's inability to monitor with accuracy the political pulse rate in Kenya.

The CNC's memorandum is nevertheless instructive of the colonial state's position on mining in Kakamega. The colonial state upheld the view that African unrest over the loss of land to gold miners was baseless. In addition, the CNC reiterated the supremacy of developing the goldfield asserting that the colonial administration was prepared not to "allow this gold to remain hidden and idle, but [we] must allow people to look for it and take it out when they find it."[113]

Wade was at pains to explain the leasing process. Leases would be required for reef mining and not alluvial mining; pegging an area did not constitute a claim to the land in *lieu* of a permit. In any case, Wade argued that a permit was not a land purchasing fee. Prospecting by permit left room for the affected Africans to remain on the land, while receiving cash compensation for alienated or derelict land, trees, and crop damages. The sinking of deep shafts and the installation of machinery, as well as the erection of permanent buildings, was dependent on the size of the gold deposit found on the pegged land. Only an economically paying gold discovery could justify the issuance of a lease, and, therefore, full compensation to the owners. At the expiration of the lease, such land could revert to the reserve.[114] In conclusion, the CNC emphasized the economic benefits the Luyia stood to reap from the development of the gold mining industry. These included full compensation, employment opportunities, good wages, and a lucrative market for their produce. Once again, Wade implored literate Africans to explain the conditions to the old men and the illiterate members of the community.[115]

Despite such eloquence, Nyanza Africans, mainly CMS elites, addressed a petition to the British House of Commons protesting the unilateral decision to amend the NLTO of 1930. They decried the inconsistency of the new ordinance with the common justice and moral fairness associated with the British Empire, and condemned the decreeing of monetary compensation, when Africans could not purchase land outside their reserves. They expressed . . . "surprise that His Majesty's Government should go back on its word given to the natives of Kenya . . ."[116]

The petitioners also reiterated their call for prospecting by a government utility, payment of rents and compensation to dispossessed individuals, expending of a percentage of the gold royalties on Luyia welfare, and demanded that the colonial state look for alternative land to settle Luyia families displaced by the gold mining industry. In addition, the petitioners expressed fear over the potential decay of their social life, and the threat of miners enforcing the oppressive and much hated *Kipande* laws in the reserve.[117] Although nine of the signatories were from NK reserve, the S of S rejected the petition on the pretext than none of them resided in either Isukha or Idakho locations, where most of the mining operations were concentrated.[118]

PROTEST IN ENGLAND

Protest in England passed through a well-coordinated channel of communication with missionaries in charge of specific missions in Kenya and the KMC in particular. The men of the pulpit on the spot and other critics of British policies in Kenya provided information appealing to their friends in England to put pressure on the government at home.

As early as November 1932, The London Group on African Affairs (TLGOAA) had forwarded suggestions to the S of S regarding the administration of the Kakamega goldfield. In his response, Sir Philip Cunliffe-Lister emphasized the Crown's right to the mineral wealth in Kenya and the importance of exploiting the goldfield for the benefit of the colony, vowing to safeguard Luyia interest.[119]

On December 16, 1932, Archdeacon Owen informed the Secretary of the Africa Section of the Missionary Council in London of the move to amend the NLTO of 1930. Owen had led a delegation of the KMC to the governor and the CNC to discuss the possibility of a state enterprise exploiting the Kakamega goldfield. Although the governor rejected the suggestion, he submitted that the final decision rested with the recommendations of the government geologist, Sir Albert Kitson who was then in the field. Byrne also warned Owen of the seriousness of telegraphing his friends in London the terms of the impending NLT(A)O. This was repeated in an extraordinary step in which the CNC categorically asked Owen to refrain from doing so until the Bill had passed its second reading in the LegCo. Owen, however, ignored this order and communicated to England.[120]

A week later, Rev. Arthur Pitt Pitts of the CMS, Nairobi, informed his superiors in London of the hurried passage of the NLT(A)O in the Kenya LegCo. Pitts termed it "a black week" and asked the missionaries in London to publicize the plight of the Africans in Kenya. Pitts graphically contrasted the treatment the Luyia had received with that of the Africans of Freretown (Sierra Leone) whose land was alienated for the CMS station for freed slaves. He wrote:

> They were after all squatters at will with no claim on the land at all and for the sake of 80 households, the Society gave 50 acres, plus £3000 in cash to the Africans, plus £500 to spend on the village wells and camps. It makes me wild to think the way they are behaving in Kavirondo.[121]

Such desperate appeals for intervention to critics at home led to swift action from humanitarians in England. Lord Lugard (1858–1945) was at the forefront in organizing and co-coordinating protest against the NLT(A)O. Lugard had extensive knowledge on the British imperial edifice, and was the foremost authority of British administration in Africa. After serving as High Commissioner and Commander-in-Chief of Northern Nigeria (1906), Lugard served as Governor of Hong-Kong (1906–1911), and subsequently, Governor-General of Nigeria between 1912 and 1919. In 1922, he issued his famed work, *The Dual Mandate in Tropical Africa*, which quickly became the Bible for British administrators in Africa. Later, Lugard distinguished himself as an ardent humanitarian.[122]

According to Lugard, the NLT(A)O of 1932 made the British pledge of trusteeship to the Africans of Kenya a mere "scrap of paper." In his view, any charge of lack of good faith on the part of the British Government risked arousing strong criticism in Africa, India and in international circles. A progressive imperialist, ardent believer and model practitioner of British trusteeship in Africa, Lugard sought the support of leading groups and personalities in England in an attempt to enlist public support against the NLT(A)O of 1932. Following his letter to *The Times* on January 4, 1933, the British public experienced a profuse outpouring of condemnation of the imperial government's support for the NLT(A)O of 1932.[123] Appendix I is a partial list of protest letters.

At the time of the passage of the NLT(A)O of 1932 the British Parliament was on recess. This meant critics of the ordinance operated outside the corridors of the British Parliament. Similarly, humanitarians needed to gather more information from their friends in Kenya, and could not recommend a commission of inquiry as the KLC was already in Kenya. Owing to this state of affairs, Lugard sought to mobilize public opinion through the British press.[124] In his scheme of work, Lugard was to utilize TLGOAA, the ASAPS, as well as other humanitarian fora in England. Moreover, he was especially eager to enlist the support of the Archbishop of Canterbury; Dr. Drummond Shiels and Sir Robert Hamilton, both former Under-Secretaries for the Colonies, as well as members of both Houses of Parliament with humanitarian leanings.[125]

For instance, at Lugard's direct appeal, the Archbishop of Canterbury challenged the British government to defer opening further areas in Kakamega to prospectors until public opinion on the issue was resolved. The Archbishop employed moral and economic arguments asserting that both the imperial pledge to secure African lands and the influx of a European mining population into an African reserve were moral and economic issues of gravest importance.[126]

Former Parliamentary Under-Secretary of State Drummond Shiels termed the NLT(A)O of 1932 "short-sighted and foolish, as well as wrong."[127] Most of the critics emphasized the centrality of land to the Luyia households and condemned monetary compensation. Since the colonial legal apparatus in Kenya prohibited Africans from purchasing land outside the reserves, critics demanded the reversal of the ban if the colonial state sought to institutionalize cash compensation.[128]

Other critics including Humphrey Leggett, former Director of Mining Companies in South Africa,[129] Louis Leakey, an authority on Kenya and official interpreter to the KLC, and Bishop James W. C. Dougall of Uganda, believed strongly in keeping the imperial pledge. In their view, the colonial state could repossess the Kisumu-Kaimosi, Muhoroni and Kipkarren farms to provide alternative land for displaced Luyia families.[130] In spite of this, the colonial officials and the settlers in Kenya maintained that the breach of faith was justified because the CO made the pledge against the advice of the colonial state.[131]

Once Parliament resumed, Lugard put down a motion in the House of Lords to discuss the Kakamega land question.[132] The debate focused on the breach of faith and the possible remedies. Although Lugard, Passfield and the

Marquess of Reading, put up an admirable argument in support of the NLTO of 1930, their protestations came to naught.

Replying on behalf of the S of S, the Secretary of War, Viscount Hailsham, regurgitated the imperial government's position. According to Hailsham, criticism labeled against the NLT(A)O of 1932, was based on misconception of policy. Hailsham invoked the British law that reserved all mineral wealth to the state, mandating it to develop such wealth. In addition, Hailsham argued that NK reserve was large, and pointed out that the Mining Ordinance of 1931, which had passed under the Labour regime provided for lease grants anywhere in Kenya. Furthermore, Hailsham claimed that the amending ordinance was not hurried, adding that the exclusions of land for mining were temporary. Hailsham claimed that Africans received adequate compensations. In addition, Hailsham emphasized the economic significance of developing the goldfield, the role of the KLC in resolving the Luyia land issue, the strong support the imperial government received from the men on the spot, whom the S of S Sir Philip Cunliffe-Lister asserted, "it is my instinct to trust."[133]

With regard to opening up further land in NK reserve to mining, Hailsham asserted that the Report of Sir Albert Kitson, the best authority on mining, had been made public. Although the CO had not reached a final decision, Hailsham considered it difficult to persuade the S of S to depart from Kitson's recommendation. He believed that the CO would accept and implement the recommendations of the Report. In response to Lugard's suggestion that a portion of the gold royalties be devoted to improving the welfare of the Africans, Hailsham argued that since the gold belonged to the colony as a whole, the state would use the royalties for the benefit of all without any racial consideration.[134]

TLGOAA continued to hold meetings, pass resolutions and write directly to the S of S over the NLT(A)O of 1932, challenging the imperial government against taking any definite decision before the KLC reported. The group also demanded a halt to the issuance of prospecting permits in Kakamega; suggested entrusting the mining venture to a government enterprise; called for the repeal of the NLT(A)O of 1932, and the inauguration of honorable discussions with the Luyia landholders.[135] This storm of protest did not alter the imperial view, however.

THE CO DEFENDS THE NLT(A)O OF 1932

The CO's response to the immense criticism of the amending ordinance is a pointer to the significance with which both the imperial government and the colonial state held the gold discovery in Kenya within the context of the great depression years of the 1930s. As already shown, the S of S, Sir Philip Cunliffe-Lister, issued an authoritative statement in the House of Commons in response to scathing press criticism of the Kakamega land question on December 20, 1932. In his speech, Sir Philip Cunliffe-Lister reiterated the Crown's right to minerals in Kenya, a right that the creation of African reserves had in no way abrogated. He argued that the NLTO of 1930 endorsed this right. He also emphasized the coincidence of the gold discovery in Kakamega with the investigation of the KLC, the only legitimate body that was to resolve the land question in Kenya. The Commission would investigate the implication of gold mining for Luyia landholders.[136]

Sir Philip Cunliffe-Lister viewed the discovery of gold as an occurrence of greatest importance to the colony as a whole. As such, it had to be exploited, the circumstances notwithstanding. Moreover, the S of S was full of praise for the procedure the colonial state had adopted in dealing with the advent of mining. For example, the colonial state had enlisted the services of the eminent geologist of worldwide experience, Sir Albert Kitson, and that the Central Lands Trust Board, the CNC, the KLC, and the missionary representing African interests in the LegCo, had approved the amending ordinance. In fact, Sir Philip Cunliffe-Lister emphasized the pragmatism and temporariness of the NLT(A)O as a tool for meeting the urgency of developing the Kakamega goldfield. He then "assured" Parliament that the Luyia had been informed of the position by the governor and a statement drawn up by the CNC, translated into the local vernacular had been widely circulated.[137] Apparently, the colonial state had not stabbed the Luyia in the back. However, as already stated, this was an untrue statement.

Another justification for the change in land laws in the Kakamega case was the alleged smallness and the temporary nature of the envisaged exclusions. The S of S expressed enthusiasm for the little disturbance the dispossessed Luyia would suffer. Those displaced by mining activities, he argued, would receive ample compensation, be allowed to remain on the same land or make new homes among their neighbors. Besides, reports from colonial officials in Kenya demonstrated that relations between the Luyia and the European miners were satisfactory. In his view, therefore, the CO had approached a very difficult situation in the most pragmatic and sympathetic manner. Since the question of the adequacy of African reserves in Kenya was the *raison d'etre* of the KLC, Sir Philip Cunliffe-Lister, argued that no breach of faith had been committed against the Luyia landholders in Kakamega.[138]

The colonial state and the imperial government legitimized their actions in the name of the colonial land and mining laws operational in Kenya colony. The supremacy of exploiting the valuable gold resources was an uncontested issue. The colonial legal instruments, including the mining and land laws, justified the alienation of land in Buluyia for the extraction of gold. While the Luyia enjoyed unquestioned surface rights to the land, the colonial state had unhampered rights to the gold underneath. In addition, the alienation of land for mining was a temporary one, resting largely on the gold deposits underground. The colonial officials emphasized the adequacy of cash compensation while downplaying the looming danger of physical disturbance to Luyia households.

Despite the rhetoric, however, one can argue that the colonial land and mining laws were diametrically opposed to and critically incongruent with the Luyia land rights. The Kakamega case lucidly demonstrated the dichotomy between the Luyia concept of land proprietorship and the British perception of the Crown's uncontested rights to minerals. While it was legal for the colonial state to violate Luyia land rights in pursuit of the development of gold, the Luyia nursed deep grievances for what they rightly considered a breach of faith. This demonstrated cultural conflict at its height. Indeed, Jomo Kenyatta poured scorn on the ingenuity of British justice when he wrote, "On the one side was the Government solemn pledge, on the other was gold. Faced with this choice, the Government tore up its pledge and passed a Bill . . . ."[139] In fact, the KLC endorsed the official view and offered no lasting security for the Luyia reserve.

THE KLC ON THE NLT(O) OF 1930

The KLC and all the Kenya PCs reviewed the working of the NLTO of 1930 at a meeting in Nairobi on March 10, 1933. The general tenor of the debate demonstrated the Commission's support for the NLT(A)O of 1932. The proposals dealing with the amendment of the NLTO of 1930 drawn up by the CNC elicited much controversy. For example, the entire provincial administration decried the great difficulty that allegedly attended to the management and control of African reserves. The PCs desired a policy that entrusted African reserves to the provincial administration. According to the PCs, such a procedure limited the role of the Central Lands Trust Board and ensured the integrity of reserve boundaries. Yet, they agreed that the LegCo retain the power to legislate and approve any additions of land to African reserves.[140]

The suggestion of retaining consultation of Africans and the LNCs in all land matters quickly became the most contested issue. With the exception of the CNC, and the PCs of Nyanza and Maasai, who favored the abolition of the Local Land Boards and the retention of LNCs' consent, the rest of the PCs unanimously sought to preserve the former while bypassing the latter in land matters. At the center of this controversy was the question of the composition of the existing Central Lands Trust Board and Local Land Boards. European settler politicians and colonial officials who could not historically be divorced from both settler and colonial state interests in Kenya dominated the former. The three members of the Board who were in theory most representative of African interests, namely the CNC, the PCs, and the European representative of African interests in the LegCo, often found themselves in the minority. As a result, the PCs recommended a "small board, with the Chief Justice as chair."[141]

At the conclusion of this debate, the provincial administration agreed to the retention of the Local Land Boards staffed by members of the LNC nominated by the PCs after consultation with the LNCs. The PC in consultation with the Local Land Board was empowered to issue minor leases, set apart land for social services, and issue permits for quarrying of sand and lime. In all these cases, however, Africans could appeal to the Central Lands Trust Board or the CNC.[142] In any case, the meeting unanimously agreed to securing African reserves with an Order-in-Council, rather than by a local Ordinance, which could be readily amended.[143]

While the colonial administration and the KLC marshaled massive evidence to justify the amending of the NLTO of 1930, some voices emphasized its inadequacies. Watkins, members of the KMC, particularly, Archdeacon Owen and Rev. R. G. M. Calderwood, and William McGregor Ross, found many faults in the provisions of the Ordinance as it stood. Watkins was critical of christening African reserves Crown lands. He articulated that these lands were never Crown lands, but African lands and would always be in perpetuity. Owing to his pro-African stance, Watkins recommended that all lands in African occupation by the time of the establishment of the East Africa Protectorate in 1895 be declared "native Lands"; that lands already gazetted as African reserves with such additions as the Commission recommended be referred to as African lands; and that all other lands be called Crown lands. The latter reverted to African use and occupation as "native Reserves" with proper safeguards.[144] Convinced that the colonial state was not performing an act of beneficence by reserving lands for the Africans, Watkins declared:

> Do not promise them the land forever. Acknowledge it is already theirs. Give it an adequate guardian, and let their rights sink or swim on an equality with the rights of other landholders. If it is unwise to transfer land from one race to another, the Government has the power of veto. If it is wise, as it may be, why make it impossible?[145]

Watkins's long-standing dissatisfaction with the composition of both the Local Land Board and the Central Lands Trust Board was evident in his recommendation to abolish the former while limiting the Executive representatives on the latter to two. The ideal Central Lands Trust Board was to consist of the CNC as chair, the PC of the area in question as vice president, five unofficial members nominated by the CNC, not more than one to be a member of the Executive Council and co-opted Africans.[146] Such a Board would have enhanced the position of the CNC, who occupied a peripheral role in Kenya's political equation.

Evidence exists to demonstrate the colonial state's persistent manipulation of the CNC. In fact, the CNC's voice rose and fell rhythmically in accordance to the wishes and whims of the colonial administration. For example, CNC, A. de V. Wade, submitted a separate memorandum to the Commission after the PCs watered down his initial proposal at a meeting with members of the KLC. In this memorandum, Wade advocated the need to strengthen the Local Land Boards by granting them executive power in order to allow Africans an appreciable say in the internal management of their land. Membership to these boards was to be restricted to nominated African members of the LNCs. He particularly opposed the inclusion of elected European members on these boards and demanded the right for Africans to appeal to the S of S, thus paving the way for direct access to the imperial government.[147]

Likewise, the KMC expressed distrust for the Central Lands Trust Board as constituted. The Board represented an official body incapable of taking action against the colonial state. Archdeacon Owen specifically attacked the colonial state's legislation against direct African representation in important decision-making bodies. Africans had competently and admirably represented their views to the Joint Select Committee in London in 1931, and were more able to represent their opinion on the Central Lands Trust Board in Kenya. This was enough evidence for appointing at least one African to the Central Lands Trust Board and more on the Local Land Boards.[148]

While Owen was not critical of the provisions of the NLTO of 1930, McGregor Ross focused on its shortcomings. Ross believed the ultimate exorcism of unrest among Africans with regard to land lay outside the provisions of the NLTO. He contended that the only solution entailed the definition of the reserves by an Order-in-Council and not entrusting them to the colonial state and its officials. In addition, Ross considered the Ordinance a contemptible instrument. For example, the number of lines devoted to the security of African lands was trivial compared to the terms and conditions under which they were susceptible to alienation for non-Africans. Although he was paternalistic toward Africans, Ross brilliantly argued this out when he wrote:

> It is only by reading into the text of the Ordinance the unwarranted suppositions that land hunger among Europeans has ceased to exist and will never recur, that all members of public bodies will be virtuous, that no Africans will be stupid and timid, that no officials will be venal and pliable, and that the Secretary of State

will be fully and honestly informed in every turn of the precise conditions prevailing in Kenya, that the Ordinance bears even a shadowy semblance of a protection of African rights.[149]

Clearly, such protest did not change the Commissioner's favorable disposition to the NLT(A)O of 1932.

THE RECOMMENDATIONS OF THE KLC

At the conclusion of the investigation in Britain and Kenya, the Commissioners commenced the preparation of their report, which they submitted to the governor on July 7, 1933 and to the CO later in the year. The Report became public on May 14, 1934.[150] The Commission's recommendations were largely disappointing to the Kenya Africans and the Luyia specifically.

First, the Commission recommended the addition of 1,474 square miles to all African reserves in satisfaction of claims of right, 896 square miles on ground of economic need and 259 square miles as "temporary Reserves."[151] While the Commission admitted that the population of NK was both dense and increasing, they attributed the land problem in Buluyia to poor animal husbandry. The commissioners asserted that the rich fertile soils in the district would adequately support the growing population if the Luyia adopted better agricultural practices. As the commissioners tersely put it, without agricultural innovativeness, further additions of land to the reserve served no fruitful purpose.[152]

Closely related to this was the issue of overstocking. The commission argued that African livestock, especially cattle, constituted "a debased currency . . . valid for one purpose only (sic)- the purchase of wives." The Commission viewed the insignificance of meat in Luyia diet, the absence of a buoyant livestock market in Buluyia, and the complex trust system that governed cattle ownership among the Luyia, as indicators of the unprofitable nature of African animal husbandry. The commission criticized the colonial state's policies that allegedly aggravated the pervasiveness of overstocking in African reserves. The Commission contended that although veterinary services had significantly checked the spread of killer diseases among African herds, the colonial state's decision to terminate such services to Africans threatened European livestock. It recommended immediate action to cull African livestock. As the commissioners succinctly put it, "It [overstocking] is definitely not a problem which can be solved by an increase of land. If the uncontrolled increase of [live]stock [would] be permitted to continue, then the whole of Africa would be insufficient to satisfy the wants of the future."[153]

Nevertheless, the question of Luyia land rights remained fragile. Due to the storm of protest, the NLT(A)O of 1932 had aroused both in Kenya and England, the Commission was hamstrung to examine the impact of the gold mining industry on the Luyia community. In a special section dedicated to Kakamega, the Commission recommended the addition of 1,500 acres from either the Kakamega Forest or Elgon Forest Reserve to NK for individual and "tribal" requirements, before effecting its general recommendation.[154]

With regard to the development of African reserves, the Commission recommended the training of Africans in better methods of land management. Worse still, the Commission never conceived of the possibility of a European-African partnership in the development of the nascent rural capitalist industrial project in NK based on gold mining. Although the gold was located in the Luyia reserve, the Commissioners best articulated their pro-European disposition in the matter when they declared, "mineral wealth, . . . etc *can be properly exploited only by Europeans*."[155] Indeed, colonial mining laws made it extremely difficult for Africans to obtain mining licenses.

Second, the KLC christened the former African reserves and the newly added lands "Native Lands", and lands added to the reserves for economic needs, became "Reserves." The Commission also designated Special African Leasehold Areas for the so-called "detribalized Africans" who desired some form of "private ownership."[156] The Commissioners contended that the NLTO of 1930 had created watertight compartmentalized and unchanging African reserves whose boundaries were not coterminous with any "tribe." In addition, the Ordinance failed to distinguish between the so-called "protection" and "management" aspects of African reserves. As a result, they concluded that the Ordinance represented a serious hindrance to progress in the reserves. The Commission further observed that Kenya needed a new legal framework to ensure the security and permanence of African reserves. This meant the adoption of an Order-in-Council to secure "Native Lands", that had hitherto been Crown Lands.[157]

Third, the Commission recommended that a Native Lands Trust Board, composed of "trusted and detached men resident in England" be responsible for Native Lands in Kenya.[158] A London-based Board was free from local Kenyan politics, and was thus capable of making independent decisions.[159] Nevertheless, the striking absence of Africans on the envisaged London based Board epitomized the superficiality of the changes. In fact, the S of S for the Colonies, Sir Philip Cunliffe-Lister, rejected the recommendation and approved a Kenyan-based Board dominated by Europeans.[160]

Fourth, the Commission upheld the principle of the sanctity of the Kenya Highlands. All lands occupied by Europeans and areas that Europeans were interested in became the "White Highlands" where Europeans were to enjoy a privileged status. Like the Native Lands, the White Highlands," were to be secured by an Order-in-Council.[161] Ultimately, the Commission expunged all African rights to land in the White Highlands.[162] As Rosberg and Nottingham fittingly observed, the Carter Commission invented the European "tribe", with its "tribal lands" comprising some 16,700 square miles in the center of Kenya.[163] The commission rejected Luyia claims to Kaimosi and Kipkarren, and subsequently incorporated the two areas into the Highlands.[164] The only positive recommendation of the Commission fell outside its terms of reference. The Commission urged the CO to release £50,000 owed to Kenya Carrier Corps who either perished or went missing during WWI for the development of African areas. The Commissioners contended that since Africans were "legally protected persons," forced to fight in a foreign war from which they had gained nothing, the Regimental Debts Act of 1893 was applicable to African soldiers too.[165] After a protracted struggle between the CO on the one hand, and the Treasury and the War Office on the other, the imperial government agreed to an *ex gratia* payment of £50,000 to the Kenya government to cover the expenses of implementing the Commission's recommendations.[166]

CONCLUSION

The KLC failed to resolve the land issue in Kenya. Although the Luyia articulated their grievances and demonstrated intense fear for the security of their lands, the Commission used the economic argument to approve the NLT(A)O of 1932, and endorsed the alienation of land in the Luyia reserve for European industrial capitalists in the newly established gold mining industry. Like the imperial government, the colonial state, and the CNC, the supposed guardian of African interests in Kenya, and the KLC maintained that the gold industry would benefit the colony and the Luyia specifically. Arguing that the NLT(A)O of 1932 was a temporary measure intended to help the Crown exercise its right to the mineral resources of the colony, the Commission alleged that dispossessed Luyia households would receive generous cash compensation, enjoy unprecedented economic prosperity and suffer little physical disturbance as a result of gold mining. Indeed, apart from the Luyia who consistently opposed gold mining in their locations, even champions of African interests concurred on the inevitability of exploiting gold. While the Luyia enjoyed surface land rights in the reserve, the Crown had unhampered rights to the gold underneath. Luyia surface rights did not nullify the Crown's subterranean rights. Moreover, while there was consensus that the measure was unpopular to Luyia landowners, economic pragmatism dictated its adoption.

In addition to negating Luyia land rights, the KLC was perpetuating the cliché that poor agricultural practices were entirely responsible for economic retardation in African reserves.[167] Indeed, forced soil conservation measures and culling of African livestock became twin evils that Kenya Africans had to contend with when the Commission's recommendations took effect later in the decade.

Obviously, the KLC Report was disappointing to the Kenya Africans in general and the Luyia in particular. In addition to being too legalistic and voluminous, the report exposed the bankruptcy of British jurisprudence. It legally sanctioned territorial segregation on racial lines, giving the Europeans a pride of place in an African country. Arguably, the Commission's work was a futile, self-defeating exercise that made a mockery of the acclaimed British justice. Its composition and terms of reference restricted its suitability as the instrument of radical change in the controversial land question in Kenya. By freezing the frontier and restricting 1,029,422 Africans to 7,114 square miles in Nyanza, while granting 16,700 square miles to 17,000 Europeans, the Commission condemned Africans in Nyanza Province to a permanent state of congestion.[168]

Moreover, the acceptance of the much contested report by both the imperial government and the colonial state, demonstrated lack of commitment to a radical transformation of colonial land policy in Kenya. Africans emerged from the British judicial gymnastics as the losers, permanently relegated to a second-class status in their own homeland.[169] As Adam Ashforth succinctly puts it, "the work of Commissions of inquiry can be characterized as 'reckoning schemes of legitimation', whose significance lies in the elaboration of the 'idea of the state.'[170] This reassertion of the supremacy of the colonial state facilitated the rural capitalist industrial transformation of Western Kenya.

NOTES

1. Priscilla M. Shilaro,"Colonial Land Policies: The Kenya Land Commission and the Kakamega Gold Rush, 1932–4," in *Historical Studies and Social Change in Western Kenya. Essays in Memory of Professor Gideon S. Were,* ed. William R. Ochieng' (Nairobi: EAEP, 2002), 110–12.

2. Ibid.

3. Ibid. 112; Col. & Prot. of Kenya, *Report of the KLC,* III, 1.

4. NKDAR 1931, KNA: DC/NN/1/12; NKLNC, Minutes of a Meeting Held at Matungu on March 23 & 24, 1931, KNA (Kakamega): HW/13/1; NPAR 1931, KNA: PC/NZA/ 1/26; *Habari* (News), *Juni* (June) 1931.

5. Great Britain (GB), *Joint Select Committee on Closer Union in East Africa*. Vol. I. (London: His Majesty's Stationery Office (HMSO), 1931), 44; For a detailed discussion on Closer Union in East Africa see Robert Gregory, *Sidney Webb and East Africa: Labour's Experiment With the Doctrine of Native Paramountcy* (Berkeley and Los Angeles: University of California Press, 1962), 64–76; George Bennett, *Kenya, A Political History: The Colonial Period* (London: OUP, 1963), chapters 6 and 7; Jidlaph. G. Kamoche, *Imperial Trusteeship and Political Evolution in Kenya, 1923–1963: A Study of Official Views and the Road to Decolonization* (Washington, DC.: University Press of America, 1981), chapter IV.

6. Shilaro,"Colonial Land Policies," 111; *Report of the KLC,* 1–2; Mwangi Wa-Githumo, *Land and Nationalism: The Impact of Land Expropriation and Land Grievances Upon the Rise and Development of Nationalist Movements in Kenya, 1885–1939* (Washington, DC. : University Press of America, 1981), 350; Rita Mary Breen, "The Politics of Land: The Kenya Land Commission (1932–33) and Its Effects on Land Policy in Kenya," (Ph. D. dissertation, Michigan State University, 1976), Abstract, 45; Carl G. Rosberg and John Nottingham, *The Myth of "Mau Mau", Nationalism in Kenya* (Stanford, New York: Praeger, 1966), 136; Marjorie Ruth Dilley, *British Policy in Kenya Colony* (New York: Thomas Nelson and Sons, 1937), 128; M. P. K. Sorrenson, *Land Reform in the Kikuyu Country: A Study in Government Policy* (Nairobi, London: OUP, 1967), 22. Johnstone Kenyatta (later Jomo Kenyatta) and P. G. Mockerie of the Kikuyu Central Association, were refused a hearing as they were considered "disgruntled radicals." See Sorrenson, *Land Reform,* 22; Bennett, *Kenya,* 75.

7. Shilaro, "Colonial Land Policies," 112; *Report of the KLC,* 1–2; GB, *Kenya Land Commission: Evidence and Memoranda.* Volume III (London: HMSO, 1934), 2177 (hereafter *KLC,* III); The CO Official Land Enquiry Announcement, April 11, 1932, PRO: CO 533/416; Parliamentary File: House of Lords Debate, PRO: CO 533/428; Breen, "The Politics," 204; Wa-Githumo, *Land and Nationalism,* 350–51; Rosberg and Nottingham, *The Myth,* 144; Dilley, *British Policy,* 19; Tabitha Kanogo, "Kenya and the Depression, 1929–1939," in *A Modern History of Kenya.* ed. W. R. Ochieng' (Nairobi: Evans Brothers, 1989), 128; R. M. A. van Zwannenberg with Anne King, *An Economic History of Kenya and Uganda 1800–1970 (New Jersey: Humanities Press, 1975), 43;* Jidraph Kamoche, *Imperial Trusteeship,* 161–62; Christopher Leo, *Land and Class in Kenya* (Toronto, Buffalo: University of Toronto Press, 1984), 42.

8. Shilaro, "Colonial Land Policies," 112.

9. Pitts to Hooper, Letter for Presentation to Executive Committee, June 8, 1933, Lugard Papers, RHO, MSS. Afr. 77/1; Memorandum, P. D. Master, *KLC,* III, December 15, 1932, Ibid., 3254; *Parliamentary Debates,* 84 H. L. Deb. 5S., 1932, Col. 306; Wa-Githumo, *Land and Nationalism,* 361; Breen, "The Politics," 57–58; Rosberg and Nottingham, *The Myth,* 144–45.

10. Pitts to Hooper, Letter for Presentation to Executive Committee, June 8, 1933, Lugard Papers, RHO, MSS. Afr. 77/1; Memorandum, P. D. Master, *KLC,* III, December 15, 1932, Ibid., 3254; *Parliamentary Debates,* 84 H. L. Deb. 5S., 1932, Col. 306; Wa-Githumo, *Land and Nationalism,* 361; Breen, "The Politics," 57–58; Rosberg and Nottingham, *The Myth,* 144–45.

11. *KLC*, III, 2273, 2901; *Parliamentary Debates*, 84 H. L. Deb. 5S., 1932, Cols. 309, 311–16; Rev. W. A. Pitt Pitts, Secretary, Kenya Missionary Council (KMC) to Rev. H. D. Hooper, Private and Confidential, May 7, 1932, Lugard Papers, Rhodes House Oxford (RHO), MSS. Afr. 77/1; Wa-Githumo, *Land and Nationalism*, 350, 361; Breen, "The Politics," 50, 51, 57, 58; L. P. Mair, *Native Policies in Africa* (New York: Negro Universities Press), 1936. Reprinted 1969), 87; Rosberg and Nottingham, *The Myth*, 145; Dilley, *British Policy*, 19.

12. Shilaro, "Colonial Land Policies," 112; Sir Philip Cunliffe-Lister, S of S, *Parliamentary Debate Official Report, House of Commons, 1933–34* (London: HMSO, 1934), 292. H. C. Deb. 5S., 1933–34, Col. 561; Cabinet Memorandum, Secret, March 1934, enclosure in Boyd to J. D. Fergusson, April 5, 1935, PRO: CO 533/442/1; *Parliamentary Debates*, 84 H. L. Deb. 5S., 1932, Col. 317.

13. Shilaro, "Colonial Land Policies, " 112–13; *Parliamentary Debates*, 84 H. L. Deb. 5S., May 4, 1932, Cols. 307–8, 1305; Allen to Bottomley and Wilson, Minute, April 28,1932; Bottomley to Wilson, Minute, April 28, 1932 and Wilson, Minute, April 19, 1932, PRO: CO 533/424; *Parliamentary Debates*, H. C. Deb. July1, 1932, Col. 1915; *Parliamentary Debates*, 304 H. C. Deb. 5S., July 25, 935, Col. 2062. Opposition to Wilson's choice intensified when it became public that he had not compensated Africans on his farm when the land was alienated. See Breen, "The Politics," 60.

14. Shilaro, "Colonial Land Policies," 113; *Report of the KLC*, 2–3, *KLC*, III; Cabinet Memorandum, Secret, March 1934, PRO: CO 533/442/1; *Parliamentary Debates*, 292 H. C. Deb. 5S., 1933–34, Col. 560; NPAR, 1932, KNA: PC/NZA/1/27; Breen, "The Politics," 67–68; Kamoche, *Imperial Trusteeship*, 161–63.

15. *KLC*, III, 2221; NKDAR 1931, KNA: DC/NN/1/12; NPAR 1931, KNA: PC/NZA/1/26. See Rosberg and Nottingham, *The Myth*, 155 for such demands elsewhere in the colony.

16. Memorandum by the Natives of Kisumu Location, n. d., in *KLC*, III, 2142–43; Memorandum, H. W. Innis, 10 September 1932, Ibid., 2189; Evidence, The Natives of the Luo Tribe, September 9, 1932, Ibid., 2170, 2172; Memorandum, Provincial Commissioner Nyanza, September 6, 1932, Ibid., 2163–5; Memorandum, Members of the Kanyakwar Tribe, September 7,1932, Ibid., 2151; Evidence, Natives of the North Kavirondo District, September 12, 1932, Ibid., 2222; Memorandum, C. T. Cogle, September 13,1932, Ibid., 2223; Memorandum, Lt. Col. H. F. Stoneham, October 6, 1932, Ibid., 2307.

17. Shilaro, "Colonial Land Policies," 113; Evidence, The Natives of the North Kavirondo District, September 12, 1932, *KLC*, III, 2221; Shilaro, "Kabras Culture," 173–74,177.

18. Wa-Githumo, *Land and Nationalism*, 262–63, 275.

19. Evidence, Rev. Monsignor Brandsma, September 10, 1932, *KLC*, III, 2179.

20. *KLC*, III, 2136.

21. Evidence, Natives of the North Kavirondo District, September 12, 932, *KLC*, III, 2222; Memorandum, Elders of Masana, Ibid., 2143; Evidence, Natives of the Luo Tribe, n.d., Ibid., 2172; Memorandum, H. W. Innis, September 10, 1932, Ibid., 2189; Memorandum, Stoneham, October 6, 1932, Ibid., 2307.

22. Evidence, Natives of the North Kavirondo District, September 12, 1932, *KLC, III*, 2222; Evidence, Brandsma, September 10, 1932, Ibid. 2177–9; Report of the *KLC*, 297.

23. Memorandum, The Elders of Masana, n.d., *KLC, III,* 214–42; Memorandum, The Elders of the Kisumu Location, n.d., Ibid. 2142; NKDAR 1930, KNA: DC/NN/1/11; NKDAR 1931, KNA: DC/NN/1/12; NPAR 1931, KNA: PC/NZA/1/26; NPAR 1932, KNA: PC/NZA/1/27.

24. Memorandum, Stoneham, October 6, 1932, *KLC,* III, 2306.

25. Memorandum, P. D. Master, December 15, 1932, *KLC,* III, 3257; Evidence, E. B. Hosking, Acting Commissioner of Mines, March 1, 1932, Ibid., 2327; Provincial Commissioner Nyanza, Evidence, March 8, 1933, Ibid., 2340; Memorandum, The Very Rev. Archdeacon Owen, September 2, 1932, Ibid., 2198; Memorandum, Kavirondo Taxpayers Welfare Association (KTWA), n.d., Ibid., 2140; *Parliamentary Debates,* House of Lords, February 8, 1933; Copy of Letter from Rev. W. A. Pitts, Nairobi, n. d., For Presentation to Executive Committee, June 8, 1933, Lugard Papers, RHO, MSS. Afr. 177/1.

26. Evidence in London, Charles William Hobley, June 17, 1932, *KLC,* III, 350, 3355, Ibid. Memorandum, n.d., Ibid. 353.

27. Kavirondo is a pejorative term that was used in the colonial era to refer to the region of Nyanza Province inhabited by the Nilotic Luo and Bantu-Luyia communities. It is used here for historical consistency. Owen was born in Birmingham, England in 1879, and was raised in Belfast, Ireland. He attended the CMS training college at Islington in 1903 and became a deacon in 1904. Shortly, he was posted to Uganda. In 1918, he was transferred to Western Kenya, where he remained until his death in 1945. Throughout his tenure in East Africa, Owen was the most feared/hated critic of the colonial edifice in Kenya that his critics in official circles christened him, the "Archdemon of Kavirondo". See John Lonsdale, "Political Associations in Western Kenya," in *Protest & Power in Black Africa*, eds. Ali Mazrui and Rotberg, (New York: Oxford University Press), 608; Opolot Okia, "In the Interest of Community: Archdeacon Walter Owen and the Issue of Communal Labour in Colonial Kenya, 1921–1930," in J*ournal of Imperial and Commonwealth History* 32, no. 1 (January 2004): 25–26.

28. Shilaro, "Colonial Land Policies," 114; Memorandum, Owen, September 2, 1932, *KLC,* III, 2198.

29. Negley Farson, *Last Chance in Africa* (New York: Harcourt Brace and Company, 1950), 111.

30. Memorandum, Stoneham, October 6, 1932, *KLC,* III, 2306; Evidence, A. S. Hartley, Agricultural Officer (AO) CK, September 10, 1932, Ibid. 2207.

31. Memorandum, M. H. Grieve, AO NK, November 12, 1932, *KLC,* III, 2258.

32. Memorandum, The Native Chamber of Commerce, September 6, 1932, *KLC,* III, 2145–56; Memorandum, KTWA, n.d, Ibid. 2140.

33. Memorandum, R. O. Ney, September 6, 1932, *KLC,* III, 2161; Memorandum, N. J. Desai, n.d., Ibid. 2173.

34. Memorandum, Owen, September 2, 1932, *KLC,* III, 2201, 2202.

35. Memorandum, KTWA, n.d., *KLC,* III, 2140; Evidence, Natives of the North Kavirondo District, September 12, 1932, Ibid. 2222.

36. Throughout this study, the term miners will refer to European prospectors and companies engaged in extracting gold. All employees in the industry will be referred to as mineworkers, laborers or labor force

37. Shilaro, "Colonial Land Policies, 114–15; Memorandum, Copy of Petition Addressed to the House of Commons by Kavirondo Natives, n.d., *KLC,* III, 2138.

38. Ibid; PQ, H. C, Native Lands Trust Bill, PRO: CO 533/375/2 A, PQ, House of Lords, PRO: CO 533/375/2 C. Ibid.

39. Fearn, *An African Economy,* 144–45.

40. Memorandum, KTWA, n.d., *KLC,* III, 2139–41; Wa-Githumo, *Land and Nationalism,* 338.

41. Shilaro, "Colonial Land Policies," 114–15: Memorandum, KTWA, n.d., *KLC,* III, 2139–41.

42. *Report of the KLC,* 423; Memorandum, The Native Chamber of Commerce, September 6, 1932; Evidence, Natives of the Luo Tribe, September 9, 1932, *KLC,* III, 2144–45; 2169–71; NPAR 1932, KNA: PC/NZA/1/27.

43. Evidence, Natives of the Luo Tribe, September 9, 1932, *KLC,* III, 2166.

44. Evidence, Natives of the North Kavirondo District, September 12, 1932, *KLC,* III, 2223; Evidence, Natives of the Luo Tribe, September 9, 1932, Ibid., 2167; NPAR 1932, KNA: PC/NZA /1/27.

45. Rosberg and Nottingham, *The Myth,* 162; Kanogo, "Kenya," 130. NKCA's political protest is treated fully in Chapter 4.

46. Evidence, Brandsma, September 10, 1932, *KLC,* III, 2181.

47. Ibid. 2180–85.

48. Evidence, Owen, October 10, 1932, *KLC,* III, 2194, 2200.

49. Ibid, 2201; An Asian ex-chief clerk also rejected individual titles for Africans on account of the intricate legalities involved and the cost. See Memorandum, N. J. Desai, n.d., *KLC, III,* 2174. From 1919, laborers were required to carry the *Kipande* on which the owner's previous and current labor history was entered, including the nature of employment, date of engagement, length of contract and wages paid. The *Kipande* became a mechanism for controlling labor and regulating wages. See Kanogo, *Squatters,* 38. Tiyambe Zeleza calls the *kipande* "a badge of slavery." See Tiyambe Zeleza, "The Colonial Labour System in Kenya," in *An Economic History of Kenya,* ed. W. R. Ochieng' and R. M. Maxon (Nairobi: East African Educational Publishers, 1992), 181.

50. Shilaro, "Colonial Land Policies," 115; Evidence, Owen, September 10, 1934, *KLC,* III, 2201.

51. Shilaro, "Colonial Land Policies," 115; Memorandum, "Native Land Tenure in North Kavirondo," E. B. Hosking, n. d., *KLC,* III, 2243–44; NPAR 1932, KNA: PC/NZA/1/27. See also "Notes on the Colonial Office Statement," Lugard Papers, RHO, MSS. Afr. 77/1.

52. Memorandum, "Native Land Tenure," *KLC,* III, 2244.

53. Ibid.

54. Ibid. 2245.

55. Evidence, C. B. Thompson, September 13, 1932, *KLC*, III, 2232.

56. Memorandum, "Land Tenure Report: North Kavirondo," n.d., C. B. Thompson, *KLC*, III, 2241.

57. Ibid. 2242.

58. Excerpt, Owen, April 14, 1932 in Private & Confidential Correspondence From Kenya for Presentation to Executive Committee (Missionary Council), June 8, 1932, Lugard Papers, RHO, MSS. Afr. 77/1.

59. Memorandum, Native Chamber of Commerce, September 6, 1932, *KLC*, III, 2146; Wa-Githumo, *Land and Nationalism*, 377; Rosberg and Nottingham, *The Myth*, 161; Leo, *Land and Class*, 56.

60. Evidence, Natives of the Kavirondo District, September 12, 1932, *KLC*, III, 2222–23.

61. Ibid.

62. Memorandum, Copy of Petition Addressed to the House of Commons by Kavirondo Natives, n.d., *KLC*, III, Ibid.

63. Memorandum, H. R. Montgomery, PC Nyanza, September 6, 1932, *KLC*, III, 2162–64.

64. Memorandum, H. R. Montgomery, "Notes Regarding Evidence Given in Central Kavirondo," September 15, 1932, *KLC*, III, 2189.

65. Memorandum, Montgomery, September 15, 1932, *KLC*, III, 2289–90, Report of the *KLC*, 297.

66. Ibid.; Shilaro, "Colonial Land Policies," 116; See also NPAR 1932, KNA: PC/NZA/1/27.

67. Evidence, A. S. Hartley, AO CK, September 10, 1932, *KLC*, III, 2204.

68. Ibid. 2219–20.

69. Memorandum, Montgomery, September 15, 1932, *KLC*, III, 2291.

70. Shilaro, "Colonial Land Policies," 116; NKDAR 1931, KNA: DC/NN/1/12; NPAR 1931, KNA: PC/NZA/1/26; Fearn, *An African Economy*, 126; C. G. B. Dubois, *Geological Survey of Kenya: Minerals of Kenya* (No Place of Publication: No Publisher, 1966), 26; Wa-Githumo, *Land and Nationalism*, 86.

71. NPAR 1931, KNA: PC/NZA/1/26.

72. Fearn, *An African Economy*, 132; Wa-Githumo, *Land and Nationalism*, 327; Breen, "The Politics," 24; Kanogo, "Kenya," 127.

73. Wa-Githumo, *Land and Nationalism*, 327–28.

74. Gregory, *Sidney Webb*, 50–52; Maxon, *Struggle for Kenya*, 279; Maxon, *East Africa*, 200.

75. Passfield was Member of Parliament (MP) for Seaham 1922–29; President of Board of Trade 1924; Secretary of State for Dominion Affairs 1929–30 and for Colonies 1929–31. He was also a member of the Socialist Fabian Bureau. He was elevated to peerage as Ist Baron Passfield in 1929. Chris Cook & John Stevenson, *The Longman Handbook of Modern British History 1714–1995*. 3d ed. (New York: Longman, 1996), 396, 400. For an authoritative study of Lord Passfield and his attempt at implementing the policy of African paramountcy See Gregory, *Sidney Webb*.

76. Passfield to Grigg, Confidential Despatch, 1 August 1929, and Passfield to Officer Administering Government (OAG), Confidential Telegram, 1 August 1929, PRO: CO 533/375/3.

77. *LegCo Debates, 1929*. Vol. II., 820. Views on settler reaction were polarized. For example, the pro-settler newspaper, the *East African Standard* characterized the settler reaction "a critical political position," called the S of S's position "a hypothetical basis of fairness and justice" and castigated the passage of the Bill without the complete assent of all parties in the House. See *East African Standard: Uganda Argus*, 21 December 1929. Similarly, both the *Mombasa Times* and the *East Coast Herald* labeled the reaction "a protest and a peril" and condemned Passfield's "altruistic and idealistic efforts to give the savages of East Africa an automatic equality of status with Europeans." See *Mombasa Times* and *East Coast Herald*, 22 December 1929. The anti-settler *The Times of East Africa*, on the other hand, deprecated the exaggeration of the settler reaction as a declaration of war between the CO and the elected members of the Kenya LegCo. See *Times of East Africa*, 21 December 1929 and *Times*, 20 January 1930.

78. *LegCo Debates, 1929*.Vol. II, 826; *Times*, 20 January, 1930; ASAPS to Lord Passfield, January 9, 1930, PRO: CO 533/395; *Daily Herald*, 16 January 1930; *Times*, 20 January 1930, PRO: CO 533/396.

79. Parkinson to Bottomley, Minute, n.d., Bottomley to Wilson, Minute, February 26, 1930, Wilson to S of S through Drummond Shiels, Minute, February 26, 1930, PRO: CO 533/395.

80. Passfield to Grigg, P & P Telegram, March 6, 1930; See also Parkinson to Bottomley, Minute, March 3, 1930; Bottomley to Wilson, Minute, March 12, 1930, and Wilson to Passfield through Shiels, Minute, March 12, 1930, PRO: CO 533/395; Grigg to Passfield, P & P Telegram, March 10, 1929, PRO CO533/395; Passfield to Grigg, P & P Telegram, March 15, 1930, PRO: CO 533/395.

81. Grigg to Passfield, P & P Telegram, March 28, 1930. Passfield's despatch of March 6 was subsequently published as Col. & Prot. of Kenya, *Native Lands Trust Bill*. Confidential. Passfield to Grigg. March 6, 1930 (Nairobi: Government Printer, 1930), PRO: CO 533/395.

82. *LegCo Debates,* 1930. Vol. I (Nairobi: Government Printer, 1931), 164. For the views of the other elected members, See *LegCo Debates*, 1930, 159–68.

83. *LegCo Debates* 1930, 168–70.

84. *LegCo Debates*, April 11, 1930 (Nairobi: Government Printer, 1930), 184–95. See also *East African Standard*, 12 April 1930.

85. Grigg to Passfield, Despatch, May 23, 1930; Col. & Prot. of Kenya, *The Native Lands Trust Ordinance, 1930* (Nairobi: Government Printer, 1930) encl. in Ibid, PRO: CO 533/395. See also Parliamentary File: House of Lords Debates, PRO: CO 533/428; NPAR) 1930, KNA: PC/NZA/1/25.

86. Shilaro, "Colonial Land Policies," 117.

87. NKDAR 1930, KNA: DC/NN1/11; NPAR 1930, KNA :PC/NZA/1/25; Memorandum, A copy of Petition Addressed to the House of Commons by Kavirondo Natives, n.d., *KLC*, III, 2137; Fearn, *An African Economy*, 125; Dilley, *British Policy*, 258–61.

88. Gregory, *Sidney Webb,* 108; Breen, "The Politics," 42.

89. GB, *Memorandum on Native Policy in East Africa Cmd. 3573* (London: HMSO, 1930), 3–10, PRO: CO 533/404; Gregory, *Sidney Webb,* 108–12.

90. Gregory, *Sidney Webb,* 115–22; Breen, "The Politics," 42.

91. NKDAR 1929, KNA: DC/NN/1/10; NKDAR 1930, KNA: DC/NN/1/11; NPAR 1931, KNA: PC/NZA/1/26; Joseph Daniel Otiende, Oral Interview (OI), August 20, 1998. For a lucid expose of the effects of the locust devastation in Western Kenya and the control measures undertaken, see Odhiambo Ndege, "Struggles for the Market," 94–113; Maxon, *Going Their Separate Ways,* 57–59, 102–03.

92. Alfred F. Havighurst, *Twentieth-Century Britain.* 2d ed. (New York: Harper & Row Publishers, 1962), 220–24.

93. Gregory, *Sidney Webb,* 136.

94. Passfield to the King of England, September 5, 1930; Passfield to OAG, Confidential Telegram, October 2, 1930 (Destroyed by Statute); Passfield to OAG, Despatch, October 14, 1930, PRO: CO 533/403; *The Daily Herald* called him "a strong man for a tough job," quoted in Bennett, *Kenya,* 73. Bennett observes that Byrne was a man to take orders and to enforce policy. Unlike his predecessor, he was neither an orator nor a politician but a man who only wanted to do a job he was appointed to do. See *East African Standard,* 14 February 1931.

95. *Times,* 26 July 1932 quoted in McGregor Ross to Lord Lugard, October 28, 1932, Lugard Papers, RHO, MSS. Afr. 77/1. See also Byrne quoted in Lugard, Motion on Native Rights in Kenya. *Parliamentary Deb. House of Lords. Official Report* (Unrevised) Vol. 86. 16 (London: HMSO, 1933), 547. See also *H. C. Deb. 5S, 1931–1932; East Africa,* January 12, 1933, 418; Haruni Litavakha, OI, March 9, 1998; Beti Joseph Isenjia, OI, February 24, 1998; Peter Likhaya Mbwabi, OI, February 26, 1998; Otiende, OI, August 20, 1998.

96. Shilaro, "Colonial Land Policies," 119.

97. Sir Philip Cunliffe-Lister, CO Memorandum, PRO: CO 533/428; Rosberg and Nottingham, *The Myth,* 161; Dilley, *British Policy,* 67. The Nyanza PC, Montgomery, had already laid out suggestions of cash payments to the Luyia in anticipation of the amendment. See Memorandum, H. R. Montgomery, June 24, 1932, *KLC,* III, 2309–12.

98. Col. & Prot. of Kenya, *Legislative Council Debates (LegCo Deb),* December 19, 1932 (Nairobi: Government Printer, 1932), 477; Sir Philip Cunliffe-Lister, CO Memorandum, PRO: CO 533/428. The Mining Ordinance, 1931 Section 13 (in), provided that land within an African reserve could be prospected with the consent of the Native Lands Trust Board. See also Fearn, *An African Economy,* 128; Breen, "The Politics," 24.

99. *LegCo Deb., 1932,* 509–10; *East African Standard,* 24 December 1932.

100. Shilaro, "Colonial Land Policies," 118; *LegCo Deb., 1932,* 511; *East African Standard,* 24 December 1932; See also "Notes on the Colonial Office Statement," Lugard Papers, RHO, MSS. Afr. 77/1.

101. *LegCo Deb.,* 1932, 512.

102. Ibid.513.

103. Ibid.511.

104. Byrne to Sir Philip Cunliffe-Lister, Telegram, January 11, 1933, PRO: CO 533/ 428; *LegCo Deb., 1932*, 538; Col. & Prot. of Kenya, *An Ordinance to Amend the Native Land Trust Ordinance, 1930. Ordinance LI of 1932 (Native Land Trust (Amendment) Ordinance, NLT(A)O)* (Nairobi: Government Printer, 1932), 1, encl. in Colonial Secretary (CS) to the Under-Secretary of State for the Colonies, Despatch, January 17, 1933, PRO: CO 533/428; Sir Philip Cunliffe-Lister, CO Memorandum, PRO: CO 533/428; NPAR 1932, KNA: PC/NZA/1/27; Wa-Githumo, *Land and Nationalism*, 336.

105. *NLT(A)O, 1932*, 1–2; "Problems in Kenya," Lugard Papers, RHO, MSS. Afr. 77/1; NPAR 1932, KNA: PC/NZA/1/27; Wa-Githumo, *Land and Nationalism*, 338; Mair, *Native Policies*, 87; Kanogo, "Kenya," 130.

106. Byrne to Sir Philip Cunliffe-Lister, Despatch, January 14, 1933, PRO: CO 533/428; T. D. H. Bruce, Solicitor General, Legal Report: Native Land Trust Bill, 1932, December 24, 1932, encl. in Ibid. Byrne submitted that North Kavirondo reserve was 2,394 square miles with a population of 346,000 and, therefore, an average population density of 144. This figure was misleading for it obscured regional differences in population distribution in the reserve.

107. L. B. Freeston to Bottomley, Minute, January 24, 1933, PRO: CO 533/428.

108. Bottomley, Minute, January 25, 1933, Ibid.

109. S of S to Byrne, Private and Personal Telegram, February 3, 1933, PRO: CO 533/428; Howard G. Elphinston for Colonial Secretary, Kenya Proclamations: Rules and Regulations, 1933, PRO: CO 533/428.

110. Byrne to S of S, Telegram, November 22, 1932, PRO: CO 533/427.

111. Archdeacon Owen assisted the KTWA with the compilation and the subsequent publication of these grievances in the *Manchester Guardian*, 12 January 1933.

112. Byrne was alleged to have already caused the CNC Wade to circulate a pamphlet among the Africans of Nyanza Province as early as November 22, 1932. See Memorandum for Translation and Circulation to the Natives in the Nyanza Province, 17 October 1932; Memorandum: Chief Native Commissioner's Pamphlet and Note Thereon, PRO: CO 533/428; Also quoted in the *Manchester Guardian*, 26 January 1933. Emphasis is mine.

113. Memorandum for Translation and Circulation to the Natives in the Nyanza Province, October 17, 1932, PRO: CO 533/428. Reproduced in *East Africa*, 12 January 1933, 420.

114. Ibid.

115. Ibid.

116. Memorandum, Copy of Petition Addressed to the House of Commons by Kavirondo Natives, n.d., *KLC*, III, 2137; Petition by North Kavirondo Natives Presented to the House of Commons by Sir Robert Hamilton on May 9, 1933, PRO: CO 533/436/7; Lugard Papers, RHO, MSS. Afr. 77/1. The seventeen signatories were from North, Central and South Kavirondo.

117. Memorandum, Copy of Petition Addressed to the House of Commons by Kavirondo Natives, n.d., in *KLC*, III, 2137.

118. *Manchester Guardian*, 5 September 1933; Acting Governor's Deputy to S of S for the Colonies, October 7, 1933, PRO: CO 533/436/7.

119. Honorary Secretary, The London Group On African Affairs (TLGOAA) to Sir Philip Cunliffe-Lister, November 23, 1932; Harold Allen to Honorary Secretary, TLGOAA, December 1932, Lugard Papers, RHO, MSS. Afr. 77/1.

120. Owen to Rev. H. D. Hooper, Secretary, Missionary Council, Africa Section, December 16, 1932, Lugard Papers, RHO, MSS. Afr. 77/1; Notes for the Archbishop of Canterbury, January 27, 1933, Ibid.

121. W. Arthur Pitt Pitts, CMS to Rev. Hooper, December 23, 1932, Lugard Papers, RHO, MSS. Afr. 77/1.

122. Shilaro, "Colonial Land Policies," 120–21; Gregory, *Sidney Webb*, 36.

123. *Times*, 4 January 1933, Lugard Papers, RHO, MSS. Afr. 77/1.

124. Lord Lugard to the Most Rev. the Lord Archbishop of Canterbury, December 30, 1932, Lugard Papers, RHO, MSS. Afr. 77/1.

125. Ibid.; Shilaro, "Colonial Land Policies, 120–21; Lugard was a member of both the TLGOAA and ASAPS.

126. Lord Lugard to Most Rev. The Lord Archbishop of Canterbury, December 30, 1932; Archbishop of Canterbury to Lord Lugard, January 1, 1933; *Times*, 2 January 1933, Lugard Papers, RHO, MSS. Afr. 77/1.

127. Drummond Shiels to Lord Lugard, December 27, 1932, Lugard Papers, RHO, MSS. Afr. 77/1.

128. *Parliamentary Debates*, H. L. Vol. 86. no. 16, Official Report (Unrevised) (London: HMSO, 1932), Col. 550, PRO: CO 533/428; Drummond Shiels, Medical Secretary, British Social Hygiene Council to Sir Philip Cunliffe-Lister, encl. in Drummond Shiels to Lugard, December 27, 1932; James W. C. Dougall, Kampala to J. H. Oldham, December 23, 1932; Humphrey Leggett to Lord Lugard, January 3, 1933, Lugard Papers, RHO, MSS. Afr. 77/1. This view was articulated vigorously by Lords Passfield and Lugard, and the Marquess of Reading in support of Lugard's motion. See *Parliamentary Deb*. H. L. Vol.86. No.16, Cols. 580–82. This criticism was also expressed in evidence before the KLC. See Memorandum, Master, December 15, 1932, *KLC*, III, 3257; KMC, Evidence, February 22, 1933, Ibid. 2983.

129. Sir Humphrey Leggett was the managing director of the British East Africa Corporation, a firm with considerable commercial interest in Kenya and Uganda. He was also head of the East Africa Section of the London Chamber of Commerce. Leggett's views were significant in influencing economic policies between the CO and the Kenya administration. Maxon, *Struggle for Kenya*, 28.

130. Leggett to Lord Lugard, January 3, 1933; Louis Leakey to Lord Lugard, January 22, 1933 and *Times*, 12 January 1933; Dougall to Oldham, January 13, 1933, Lugard Papers, RHO, MSS. Afr.77/1.

131. *Times*, 21 January 1933; V. M. Fisher, Principal Labour Inspector, Kenya, to Drummond Shiels, January 27, 1933, Lugard Papers, RHO, MSS. Afr. 77/1. Sir Edward Grigg, former governor of Kenya, strongly articulated this view in articles to *Times* of 6 January and 12 January 1933.

132. Lord Lugard to Sir Philip Cunliffe-Lister, February 4, 1933, PRO: CO 533/428; *Parliamentary Debates*, H. L., February 8, 1933, 548, Cols. 579–82; See also Lord Lugard to Sir Herbert Samuel, February 6, 1933; Ibid., February 8, 1933; Ibid., February 18, 1933, Lugard Papers, RHO, MSS. Afr. 77/1.

133. *Parliamentary Debates,* H. L., February 8, 1933, Cols. 592–603; Lugard's views were opposed by Lords Onslow, Chairman of the Joint Select Committee on Closer Union in East Africa, 1931 and Moyne who had recently returned from East Africa at the head of a Committee enquiring into financial issues in Kenya in 1932. Moyne denied any broken pledge, hailed the benefits of gold mining in injecting "progress" in the allegedly backward Luyia. He stated thus: "the backwardness of the native is difficult to realise for those who have not seen them." Ibid., 590.

134. *Parliamentary Debates,* H. L., February 8, 1933, Cols. 603–06, PRO: CO 533/428. After this terse statement, Hailsham asked that Lugard not press the issue further. Although Lugard was not satisfied with the replies, he begged to withdraw the motion. Two days later, however, he blamed his ineffectiveness in the House on his weak debating abilities. In a letter to Oldham, Lugard wrote: ". . . I am no good as a talker, and have no debating ability, and could not formulate on the spur of the moment what should have been said." See Lugard to Oldham, February 10, 1933, Lugard Papers, RHO, MSS. Afr. 77/1. In fact Lugard's attempt to have Sir Herbert Samuel press the issue further in the Commons was unsuccessful. Both Sir Herbert Samuel and Sir Robert Hamilton sought to await the Report of the KLC. See Lugard to Sir Herbert Samuel, February 18, 1933; Sir Herbert Samuel to Lugard, February 23, 1933; Hamilton to Lugard, encl. in Ibid.; and Sir Herbert Samuel to Lugard, February 27, 1933, Ibid.

135. Lirie Noble, Honorary Secretary, TLGOAA to Sir Philip Cunliffe-Lister, February 17, 1933; April 26, 1933, Lugard Papers, RHO, MSS. Afr. 77/1.

136. CO, Gold Mining and Native Land in Kenya: Authorised Statement of the Press, January 18, 1933, PRO: CO 533/428; "Notes on the Colonial Office Statement," Lugard Papers, RHO, MSS. Afr. 77/1; McGregor Ross, "Gold in Kenya and Native Reserves," *Nature* 131, no. 3298 (January 14, 1933), 37, Ibid; *Scotsman,* 19 January 1933.

137. CO, Gold Mining and Native Land in Kenya," Ibid. 8; *Scotsman,* 19 January 1933. These views were also expressed in the S of S's transmission to Governor Byrne earlier in the year. See Memorandum: Mining Leases in Kenya encl. in Sir Philip Cunliffe-Lister to Byrne, Telegram, January 6, 1933, PRO: CO 533/428. It is important to note that while the S of S vehemently argued that all those best qualified to speak on the Kakamega land issue had been consulted (emphasis is mine), not a single Luyia was consulted on the issue. This illustrates the British colonial mind set and protectionist trusteeship by which Africans were perceived as intellectual minors to be gradually guided along the path of Western civilization. In this mode of thinking, Africans had no views, and were not, therefore, "best qualified" to speak for themselves.

138. CO, Gold Mining and Native Land in Kenya," Ibid.; Also quoted in Lord Lugard's motion in the House of Lords, February 8, 1933, PRO: CO 533/428.

139. J. Kenyatta, *Kenya: The Land of Conflict,* quoted in Wa-Githumo, *Land and Nationalism,* 337.

140. Evidence on the Administrative Aspects of the Land Problems, Memorandum of a Meeting between The Kenya Land Commission and the Provincial Commissioners, March 10, 1933, *KLC,* III, 2903–04.

141. Ibid. 2902–04.

142. Ibid. 2904.

143. Ibid. 2902. This move was a technical compromise by the colonial administration. Watkins, then deputy CNC had suggested as early as May 1926 that African reserves be secured by Order-in-Council. Since the area known as the "White Highlands" which constituted an exclusive white enclave in Kenya had not been delimited, a local ordinance, the NLTO of 1930 was preferred when securing African reserves. Such an ordinance was liable to amendment to allow further alienation of African reserves to white settlers. The KLC was to define the "Highlands" where Europeans were to enjoy a privileged position. These European "tribal" lands were also to be secured by Order-in-Council, hence the PCs compromise.

144. Evidence, Lt. Col. Watkins, *KLC,* III, February 3, 1933, 2907–14; Watkins, Memorandum Attachment No. I. Being a Memorandum on Measures Necessary to Amend the Native Lands Trust Ordinance, n. d., *KLC,* III, 2913. Elizabeth Watkins notes that her father prepared the memorandum as early as 1930. See Watkins, *Oscar*, 207.

145. Evidence, Lt. Col. Watkins, *KLC,* III, February 3, 1933, 2914.

146. Evidence, Watkins, *KLC,* III, Ibid.

147. Memorandum, CNC, March 14, 1933, *KLC,* III, 2924–25.

148. Evidence, KMC, February 22, 1933, *KLC,* III, 2976–77.

149. Memorandum, William McGregor Ross, n.d., *KLC,* III, 3373. Ross was well acquainted with the colonial situation in Kenya. He was also an authority on the limitations and contradictions that were the hallmark of colonial land policies in Kenya. Ross arrived in Kenya in April 1900 as an Assistant Engineer on the Kenya-Uganda Railway. He remained in the Uganda Railway service until March 31, 1905, when he assumed responsibility as Director of Public Works of the East Africa Protectorate. Between 1916 and 1922, Ross sat in the LegCo and retired on pension after eighteen years of service on April 5, 1923. See Evidence, Ross, June 20, 1932, *KLC,* III, 3337; McGregor Ross, *Kenya from Within: A Short Political History* (London: Frank Cass, 1968, New Impression. First Published in 1927); Gregory, *Sidney Webb*. 35.

150. Shilaro, "Colonial Land Policies," 121; *Manchester Guardian*, May 15, 1934; *Times*, 15 May 1934; Bennett, *Kenya*, 82; Kamoche, *Imperial Trusteeship*, 166; Breen, "The Politics," 81; Dilley, B*ritish Policy*, 274.

151. The Commission categorized land in Kenya as follows: Class A: Permanent additions to the reserves based on historical rights or use; Class B (1): Permanent additions based on economic needs; Class B (2) temporary additions based on economic needs; Class C: Land for which special provision was made for individual natives (not on tribal basis) usually on lease; Class D: Land on which no special privilege of race were to obtain. *Report of the KLC*, 7; See also CO, Summary of Kenya Land Commission Report, PRO: CO 533/434/1; *Parliamentary Debates,* 292 H. C. Deb. 5S. 1933–34, Col. 561, Wa-Githumo, *Land and Nationalism,* 351.

152. Additions under Class A and B were sixty-six and nil square miles respectively. See CO, Summary of Kenya Land Commission Report, PRO: CO 533/434/1.

153. Shilaro, "Colonial Land Policies," 121; *Report of the KLC*, 494; Cabinet Memorandum, Secret, 1934, PRO: CO 533/442/1. Ironically, the Commission con-

sidered NK relatively free from overstocking and commended the Luyia and the Kikuyu of Central Province for their progressive agricultural practices.

154. Shilaro, "Colonial Land Policies," 121; *Report of the KLC*, 299; Cabinet Memorandum, Secret, 1934, PRO: CO 533/442/1. The S of S, Sir Philip Cunliffe-Lister considered these additions "equitable and generous." See Sir Philip Cunliffe-Lister to Byrne, Telegram, May 4, 1933 quoted in J. E. W. Flood, Minute on Cabinet Memorandum examined together with Mr. Freeston at the CO, March 27, 1933, PRO: CO 533/442/1; Fearn, *An African Economy*, 132; Dilley, *British Policy*, 273.

155. CO, Summary of Kenya Land Commission, PRO: CO 533/434/1. Emphasis is mine.

156. *Report of the KLC,* 418, 420; Breen, "The Politics," 88–9; Wa-Githumo, *Land and Nationalism*, 353.

157. Shilaro, "Colonial Land Policies," 122; *Report of the KLC*, 420; CO, Summary of the Kenya Land Commission Report, PRO: CO 533/434/1; *Parliamentary Debates,* 292 H. C. Deb. 5S., 1933–34, Col. 564; Breen, "The Politics," 89; Mair, Native Policies, 89.

158. Shilaro, "Colonial Land Policies," 122; *Report of the KLC,* 429–32; CO, Summary of Kenya Land Commission Report, PRO: CO 533/434/1; *Parliamentary Debates,* 292 H. C. Deb. 5S., 1933–34, Col. 564; Breen, "The Politics," 96; Dilley, *British Policy*, 273; Wa-Githumo, *Land and Nationalism*, 354.

159. *Report of the KLC*, 433.

160. S of S for the Colonies, Cabinet Memorandum, Secret, March 1934, PRO: CO 533/442/1; *Parliamentary Debates,* 292 H. C. Deb. 5S., 1933–34, Cols. 564–47. The Board comprised the CNC, president, two European members, and two European appointees representing Africans. See Wa-Githumo, *Land and Nationalism*, 399; Kamoche, *Imperial Trusteeship*, 171–72. Carter later changed his mind and supported a Kenya Board. See *Times*, 19 July 1934.

161. Shilaro, "Colonial Land Policies," 122; *Report of the KLC*, 483–93; Cabinet Memorandum, Secret, March 1934, PRO: CO 533/442/1; *Parliamentary Debates,* 292 H. C. Deb. 5S., 1933–34, Col. 563; Wa-Githumo, *Land and Nationalism*, Ibid., 389, 391–12; Kamoche, *Imperial Trusteeship*,166. For debates on the definition of "White Highlands" and "privileged position" See Byrne to S of S, December 3, 1932, Byrne to S of S, December 20, 1932, PRO: CO 533/424/6; Moore, Governor's Deputy to S of S, January 1933, PRO: CO 533/428; *Parliamentary Debates,* 296 H. C. Deb. 5S., 1934–35, 961; *Parliamentary Debates*, 297 H. C. Deb. 5S., 1934–35. Cols. 2077–80.

162. *Report of the KLC*, 534; Wa-Githumo, *Land and Nationalism*, 351; Rosberg and Nottingham, *The Myth,* 157; Breen, "The Politics," 99. The question of evicting Africans from the "White Highlands" was hotly debated in the British Parliament. See *Parliamentary Debates,* 298 H. C. Deb. 5S.1934–35. Cols. 534–35.

163. Rosberg and Nottingham, *The Myth,* 99; *Report of the KLC*, 533; Wa-Githumo, *Land and Nationalism*, 351–53; See also van Zwannenberg with King, *An Economic History*, 43.

164. *Report of the KLC*, 492; Kamoche, *Imperial Trusteeship*, 167.

165. *Report of the KLC*, 412–4; Rosberg and Nottingham, *The Myth*, 99; Wa-Githumo, *Land and Nationalism*, 351–53; van Zwannenberg with King, *An Economic History*, 43; Geoffrey Hodges, *The Carrier Corps: Military Labor in the East African Campaign, 1914-1918* (New York: Greenwood Press, 1986), 84.

166. Donald Fergusson to E. B. Boyd, April 19, 1934; Boyd to Fergusson, April 20, 1934; Conclusions of a Cabinet Meeting Held on May 2, 1934; S of S to Governor, Telegram, May 2, 1934; Memorandum by the S of S for the Colonies on Unclaimed Balances of Pay to Native Porters of the Military Labour Corps in the East African Campaign, May 4, 1934; Kenya Land Commission: Cabinet. Memorandum by the Chancellor of the Exchequer and The Secretary for War, May 4, 1934; Extract from Conclusions of Cabinet (20) 34 Held on May 9, 1934; S of S to Governor, P&P Telegram, May 9, 1934; Great Britain (GB) *Kenya Land Commission Report: Summary of Conclusions Reached by His Majesty's Government, May 1934 Cmd. 4580* (London: HMSO, 1934), PRO: CO 533/442/1; *Parliamentary Debates*, 293 H. C. Deb. 5S., 1933–34, Col. 861.

167. Wa-Githumo, *Land and Nationalism*, 361; Kamoche, *Imperial Trusteeship*, 174–45; Shilaro, "Colonial Land Policies," 121–22.

168. Shilaro, "Colonial Land Policies,"122.

169. Ibid. 123.

170. Adam Ashforth, "Reckoning Schemes of Legitimation: On Commissions of Inquiry as Power/Knowledge Forms," in *Journal of Historical Sociology* 3, no.1 (1 March 1990): 1–17.

Chapter Three

Rural Industrialization, 1931–52

INTRODUCTION

This chapter traces the stages in the development of the gold mining industry in NK district. The focus is the nature of initial mining activities by the "small man," gold recovery methods, the emergence and development of large mining concerns, and the decline and eventual demise of the gold mining industry in Kakamega. The Kakamega goldfield attracted both individual prospectors with little capital, and large companies representing a wide range of interests, with substantial investments.

The discovery of the Kakamega goldfield may be attributed indirectly to A. D. Combe of the Uganda Geological Survey, who when concluding his report in 1930 on a geological reconnaissance of parts of northern Nyanza Province, recommended that prospecting be carried out in Maragoli and Nyang'ori in NK district.[1] L. A. Johnson of Eldoret made the actual discovery of alluvial gold in October 1931, on his way home from an unsuccessful search for gold in the Musoma area of Tanganyika.[2]

The Kakamega gold rush came at an opportune time. Environmental disasters such as drought and locust infestations, as well as the collapse of commodity prices had battered Kenya's economy and the European settler morale. The Kenya settler community considered the dawn of gold mining in Kakamega to be heaven sent. This significant find occurred against a backdrop of economic distress stemming from the global depression that begun in 1929, persisting into the better part of the 1930s. The value of gold to a depressed European population was extremely significant. Many disillusioned European settlers perceived Kakamega as an "Eldorado".[3]

The year 1931 was an unhappy one for NK district economically, just as it was for Kenya colony as a whole. The global depression had severely reduced the market for primary agricultural products, which constituted the backbone of the district's peasant producers just as in the fragile European settler economy in Kenya. Luyia peasants faced low prices for their crops, and with unprofitable prices being offered for their farm produce, many settler farmers became heavily indebted, thereby being forced to reduce their farming operations as well as cutting back on their African labor. The Luyia, who were accustomed to seasonal labor on European farms, faced reduced employment opportunities. Following quickly on the heels of the economic depression was a major locust infestation of the NK district toward the end of January 1931. By the end of the year, the entire district was writhing under the devastating impact of these unwelcome visitors. Although most parts of Buluyia escaped a general famine situation, serious belt tightening became inevitable in some locations.[4]

Arguably, the locusts made more serious an already difficult economic crisis. Mr. Lynne Watt, the Agricultural Officer (AO) in charge of the province summed up the situation as follows:

> This year has been one of difficulty and privation for the natives. Climatically it was excellent, but the locust infestation overshadowed all agricultural effort, and the damage done to crops both in native and settled areas has been appalling. . . . Great areas were planted out to the customary crops of cereals, pulses and some roots and when . . . some [were] almost fully developed; the locust hopper infestation made itself felt and great areas were completely destroyed. The *mtama* [sorghum], *wimbi* [millet] and maize [corn] crops were in turn severely attacked by the flying locusts which created enormous havoc.[5]

While the severe economic crisis forced many European settlers to cut back their farming operations, others stopped farming altogether and turned their energies toward the search for valuable minerals as an alternative survival strategy. Evidently, the Kakamega gold discovery occurred within this context of looking for other economic options.

Kenya European settlers who sought to see whether seeking gold was more remunerative than growing coffee were the first to invade the Kakamega goldfield.[6] Elspeth Huxley vividly captured the sense of despair that had engulfed most settlers who flocked to Kakamega "at a time when most of the farmers could do little but watch locusts eat their crops; and if they did manage to harvest anything, could sell their produce only at a loss. The gold discoveries offered hope of survival and release from despair."[7] Hugh Fearn correctly adds that "the 1929 world slump knocked the bottom out of the European settler farming in Kenya", so that one month after the gold discovery, about 200 European miners had converged on Kakamega.[8] The Kakamega goldfield provided temporary relief to European farmers from dire poverty and gave them a fresh interest in life. Its industrial progress was consequently significant to both the colonial settler economy and the colonial state.

The development of the gold mining industry proceeded in stages. Fearn distinguishes three phases in the development of the gold mining industry in Nyanza Province. The period between 1931 and 1934 constituted the era of prospecting, and the subsequent period of 1934 to 1944 involved the activities of large mining companies. The third and terminal phase of the industry was between 1944 and 1952. This was the era of diminished output and rapid decline in the industry.[9] This study adheres to Fearn's categorization in discussing the major trajectories in the evolution and development of the gold mining industry in NK district.

PROSPECTING, 1931–34

Although it was known as early as 1904 that gold existed in the neighborhood of Lake Victoria, by 1925, only four companies were undertaking mining activities, in the Lolgerian area to the south of Nyanza Province. In the period between 1926 and 1930, gold production in Kenya focused on a few small reef mines in the far south-west, close to the Tanganyika (now Tanzania) border.[10] The 1931 discovery, however, triggered a "rush" by European prospectors into the NK reserve.

When the rush started toward the end of 1931, mining activities concentrated on the areas surrounding the original discovery. Accordingly, the early prospectors focused on panning alluvial gold in the major rivers and streams around Kakamega. Rivers Yala (Lukose), Isikhu and Edzaba (Edzawa) and their tributaries, and streams such as Lutonyi, Shitoli, Shikokho, Eguiri, and Feradzi, were rapidly pegged, and the panning and sluicing of alluvial gold began in earnest.[11]

The areas most affected included Isukha, Idakho, Butsotso, Maragoli, Tiriki, and Marama locations of NK district. With time, however, mining activities quickly spread out to many parts of Buluyia. Since prospectors, initially, won alluvial gold using a saucepan and a sieve; capital requirements to commence mining were low. This served to attract more prospectors to Kakamega.[12]

In most cases, prospectors washed gold bearing materials without crushing. Usually, miners shoveled the material to fall through a coarse screen in order to remove pebbles into a "sluice box" which collected the gold. Sometimes, the pebbles were secured and sent to a crushing plant (which many companies owned by 1935). In nearly all cases, gold smelting took place in a crucible in a small forge, heated by wood, and cast into bars. Gold was not refined in Kenya, but sent, through Barclays Bank, the Standard Bank of South Africa, or the National Bank of India, to bullion refiners in England.[13] With time, as viable gold reefs materialized, reef mining surpassed alluvial mining. Unlike alluvial mining, reef mining required sophisticated extraction methods, including diamond drilling, as large mining companies with sizeable capital resources moved in.[14]

A would-be prospector obtained a one-year Prospecting Licence from the Commissioner of Mines in Nairobi or the Warden of Mines at Kisumu on payment of twenty shillings (Shs. 20). The license entitled the holder to prospect for minerals, other than in a prohibited area, without permission, in accordance with the NLTO of 1930. By this ordinance, Africans had rights to land they occupied while the colonial state retained mineral rights. To circumvent this restriction, a new ordinance, the Mining Ordinance of 1931, provided for prospecting in African reserves, with the consent of the Native Lands Trust Board, which in practice meant obtaining a permit from the Provincial Commissioner (PC).[15]

The licensed prospector proceeded to peg claims in accordance with the ordinances and registered them with the Warden of Mines. While the Mining Ordinance of 1925 allowed a prospector to hold multiple claims, the Mining Ordinance of 1931 restricted alluvial claims to one claim per person. Nevertheless, a claim holder could work on a second claim if adverse climatic conditions negated prospecting on one claim. Claims were valid for one year, renewable for further annual periods. Claims secured under a prospecting license had no rent obligations.[16]

Prospectors pegged claims closely, usually in the shape of a parallelogram, and in the case of a reef claim, the claim never exceeded 1,500 feet in length along the supposed strike of the reef and 600 feet in width. The maximum size of an alluvial claim was 300 feet by 100 feet. Most of the claims registered between 1931 and 1934, were alluvial claims.[17]

By mid-February 1932, more than 700 prospecting licenses had already been issued, 1,700 claims registered, and possibly some 400 prospectors were already in the neighborhood of Kakamega. About 80 percent of the miners concentrated on winning gold from streambeds over an area of 70 square miles, while the remaining 20 percent combined alluvial and reef mining.[18] A succession of different prospectors pegged and re-pegged claims repeatedly throughout the goldfield. Consequently, capturing the history of each block of claims is almost impossible in view of the scanty records available. For example, records show that between 1932 and 1933, the Kakamega goldfield had more than 1,000 active prospectors, a majority of whom left no permanent record of their workings.[19]

From 1932 onward, a rapid expansion of the mining industry ensued with intensive prospecting of the gold-bearing Nyanzaian and Kavirondian rocks. Increased interest in the gold mining industry was a result of the publication of a favorable interim report on the Kakamega goldfield by Sir Albert Kitson toward the end of October 1932. Kitson was a famous geologist and former head of the Geological Survey of the Gold Coast (now Ghana) colony in West Africa. "From all the evidence I have seen," he stated, "I am strongly of the opinion that prospects of the field quickly becoming a useful goldfield and later developing into an important field are distinctly encouraging. . . ."[20] This report led to a further influx of European prospectors into Kakamega. Miners besieged the recently established Mining Office in Kakamega as commercial and professional men, who saw opportunities for themselves in Kakamega Township, swarmed the District Commissioner's (DC) office.[21] At the end of the year, Nyanza Province had 4,573 reefs and 2,785 alluvial claims, most of them staked in Kakamega.[22]

The Spectator vividly captured the scenario as it was in December 1932:

> Since the publication of Sir Albert Kitson's report, the population of the Kakamega goldfields has doubled. . . . Lorries are arriving laden with household goods as evidence of an intention to take up more permanent residence. Three types of prospectors have arrived. The first is the man who is willing to work hard with his hands in the hope of making a reasonable living; the second is the businessman anxious to acquire plots in the township for the purpose of trading; and the third is the type anxious to obtain options on properties.[23]

Sir Albert Kitson, who had returned from Kenya, wrote this in *The Spectator*: "The road to Kakamega now resembles a miniature 'trail of '98' without the snow. Old mining men, from ex-Klondyke pioneers to Australian backwoodsmen, are hurrying to the spot."[24] Harold Pemberton, the special correspondent to *The Daily Express,* corroborated this view:

> Men and women are seeking feverishly for fortunes. . . . Into this veritable Garden of Eden the vultures are fast descending from all parts of the world . . . people in all walks of life are giving up practical things for golden dreams. *Millionaires are seeking further riches, and clerks have forsaken their office stools, hoping to become millionaires.* There are company promoters galore hoping for the gold feast, hard bitten prospectors who have trekked from far-flung goldfields; an American clergyman, full of faith; former schoolmasters; and all manners and kinds of men, and even women, . . . There are gold camps-de-luxe. I saw the camp of an American dollar millionaire. . . . He is mining in grand style, with full equipment. His camp is electrically lighted. Tinned caviar and champagne

are brought to him from the base. He even has an ice-making plant. In the same neighbourhood are to be found two young clerks from Johannesburg who left their jobs in search of a fortune. . . . I came across a tough old prospector from Abyssinia [Ethiopia]. His sight had almost gone looking for gold in the gold rushes of the past fifty years. . . . He was examining the dirt through a large magnifying glass. He implored me not to tell his native boys of his infirmity, otherwise they would not work for him. . . . A former schoolmaster was about to give up in despair. Then his eyes lighted up as he got sight of a golden glitter. It was one of the biggest gold nuggets found in the field—worth over £400.[25]

It appears that the Kakamega goldfield did not attract only the seasoned miner. As early as 1932, it was becoming increasingly clear that most of the new arrivals were completely ignorant of the most elementary principles of prospecting. Moreover, sensational rumors filtering from the goldfield regarding rich finds were partly responsible for the rush.[26] Kitson's personal report in the press augmented such rumors. He wrote: ". . . Luck has favoured women settlers who have turned prospectors. Two friends, wives of the discoverers of the field, recently struck so rich a patch of alluvial in one stream that in three weeks no day yielded less than 30 ounces of gold, while two days produced 250 and 350 ounces."[27] The special correspondent to *The Daily Express* in Nairobi corroborated Kitson's views in his account of a lucky female ex-farmer turned prospector:

> It was a woman who first struck lucky in this new Garden of Eden. She was the wife of a Kenya settler. Locusts had devoured their farm produce. Faced with ruin, they came to Kakamega. Both staked claims. . . . She plucked from the surface a gold harvest worth £2,000. She sold out at a big price to an American Syndicate. . . .[28]

These sensational outpourings served to increase the rush. By December 1933, there were 495 adult miners in the goldfield. Mining tended to concentrate on the Bukura Ridge eastward, with some concerns spreading westward toward the Kakamega Forest. Chief Milimu's location of West Isukha became the center of the greatest mining activity. Alluvial mining by a few of the smaller, hand worked, undertakings in East Kakamega flourished, as well as some on the Isikhu and Yala Rivers.[29] Toward the end of 1933, reef mining was becoming more important, especially in the area east and west of Kakamega town and in Idakho location. As the goldfield expanded, the colonial government faced a significant dilemma. In April 1933, the colonial government, acting out of fear that gold mining could interfere with the Kaimosi settled farms, and the more densely peopled regions of NK reserve including Maragoli and Bunyore, moved swiftly and closed these areas to prospecting.[30] Table 3.1 shows the annual figures of claims registered at December 31,

Table 3.1. Number of Claims Registered Between 1931 and 1934

Year	Number of Claims Registered
1931	1,074
1932	7,358
1933	18,529
1934	23,158

Sources: *MGDARs 1931–34*. Although the figures are colony wide, most of the claims were registered in Kakamega.

each year, as given in the annual reports of the Kenya Mines Department for the period 1931–34.

From 1934, there was a decline in alluvial gold mining as reef mining took the center stage. Reef mining brought to the fore the contradictions between the existing mining laws and the laws governing the rights of the Luyia to land within their reserve. Eager to retain its right to the mineral resources of Kenya colony and, thereby, ensure the exploitation of such minerals, the colonial state had introduced a new Mining Ordinance in 1930. By enacting this law, the colonial state wanted to attract larger companies, with substantial capital investments. The Mining Ordinance of 1930 empowered the colonial state to afford greater protection to prospectors than existed under the prospecting license, in which different prospectors pegged claims close to each other.[31] The LegCo introduced, debated and passed the Bill in 1930. Although the Bill passed into law on February 11, 1931, its provisions became operational on March 16, 1932.[32]

The Mining Ordinance of 1931 provided for the issue of Exclusive Prospecting Licences (EPLs), to a company during the initial stage of prospecting before it applied for a mining lease. To receive an EPL, applicants had to show sufficient means at their disposal to ensure the proper prospecting of a concession.[33] With the adoption of this legislation, companies outside Kenya became interested in the Nyanza goldfields, resulting in the injection of significant overseas capital into the nascent gold mining industry. Subsequently, several companies were floated between 1932 and 1936. Most of the large companies that began operations in Kakamega between 1932 and 1935 were registered in London, and some were subsidiaries of larger mining companies operating elsewhere in the world. There were also some locally registered companies, which frequently raised capital in London, and Johannesburg, South Africa.[34]

The adoption of the Mining Ordinance of 1931 led to numerous applications for EPLs by both local and foreign companies. The first EPL was granted to L. A. Johnson (who had made the strike at Kakamega in 1931) of Eldoret Mining Syndicate in October 1932. This company was registered in

Kenya on February 23, 1933 with a registered office at Eldoret and a London Committee. It had a capital of £400,000 in 1,600,000 shares of five shillings each. The concession involved 35 square miles of land in the Kaimosi-Maragoli area of the Kakamega goldfield. (See Map 4). Mr. Johnson carried out a considerable amount of prospecting as well as geological work, before Kentan Gold Areas was formed in 1934 to take over the Eldoret Mining Syndicate and other holdings. Five more EPLs were issued in Kakamega in 1932.[35]

In March 1932, Tanganyika Concessions Limited applied for an EPL over an area of approximately 5,900 square miles in Nyanza Province. The company offered to expend in prospecting a minimum sum of £20,700 during the first two years. Pending approval of the application, the colonial administration closed the area to mining interests on March 19, 1932.

Tanganyika Concession's application aroused much hostility from Kenya miners who were already in the field. Both organizations and individuals operating in North and South Kavirondo (SK) goldfields objected to the granting of this concession. They abhorred the stifling of competitive licensing and questioned the adequacy of the initial capital the company proposed to expend on the concession.[36]

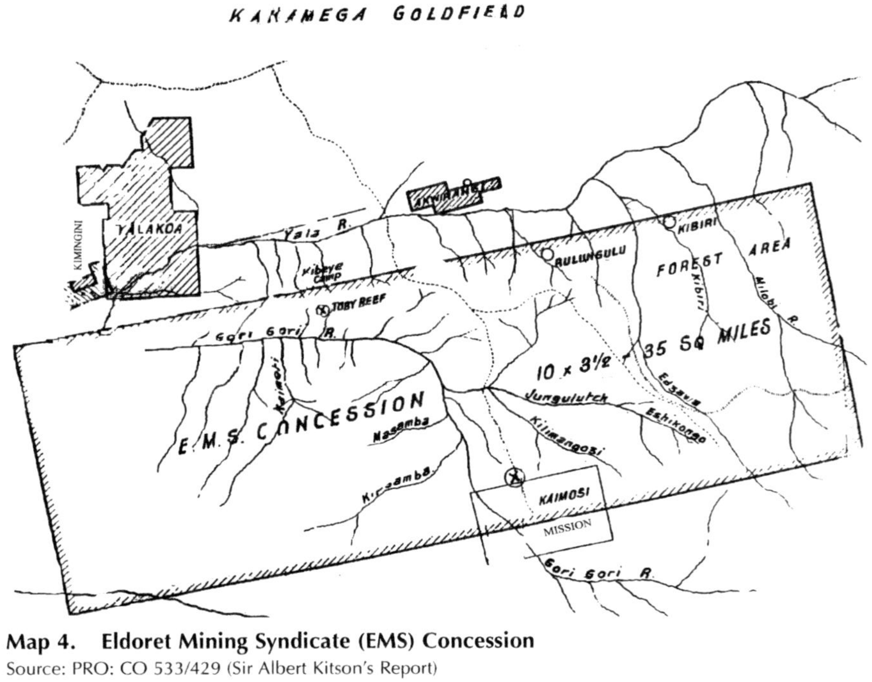

Map 4. Eldoret Mining Syndicate (EMS) Concession
Source: PRO: CO 533/429 (Sir Albert Kitson's Report)

It was against this background that the colonial state commissioned Sir Albert Kitson to make a geological examination of nearly 10,000 square miles of Nyanza Province. On November 2, 1932, Kitson submitted his second report dealing with the Tanganyika Concessions, sole prospecting licenses application to the colonial government. Kitson rejected the granting of an EPL over such a large area to one company. Instead, he recommended the division of the province (excluding Kakamega) into five mining areas, suitable for either large mining company operations or small mining enterprises.[37] (See Map 5).

Kitson was critical of the terms of the offer of Tanganyika Concessions. The company's envisaged minimum expenditure of £20,700, with a proposed staff of eleven Europeans and 330 African laborers, was inadequate to explore such a large area. Kitson contended that if the government granted the area to 200 local prospectors, this could have injected more capital into the goldfield, with a potential African labor force of 3,000. Kitson, therefore, recommended that conditions governing the grant of EPLs include a significant minimum expenditure, a large number of African laborers, and a definite detailed program of development.[38] Ultimately, in July 1933, Tanganyika Concessions received an EPL over No.1 Area, encompassing 1,550 square miles of land in north-west Kavirondo. By September, the company had commenced prospecting in the area.[39] (See Map 6).

In the meantime, Tanganyika Concessions focused on the development of its properties elsewhere in the goldfield. In 1932, working jointly with the Zambesia Exploring Company and the Rhodesia-Katanga Company, Tanganyika Concessions acquired a one-year controlling interest in the holdings of Eldoret Mining Syndicate. This included promising properties, such as the 35 square mile Eldoret Mining Syndicate's EPL, 250 reef claims, and 11 miles of exclusive prospecting license on River Yala. Subsequently, the company commenced work on the Kibiri, Rulungulu and Shikokho properties within their 35 square mile area. More developments occurred on the company's Kimingini reef-bearing property in Idakho, where the company begun sinking two vertical shafts, while running a diamond drill at its Shikokho property in Isukha. By October of 1932, the company had forty-three European employees; it had installed electricity at its headquarters, and planned to erect a telegraph line over its concession.[40]

In spite of the initial success, by February 1934, Tanganyika Concession sought to float a new company, Kentan to carry on its gold mining operations. Nonetheless, Kentan expected a momentous hurdle in raising funds from London due to mismanagement and the lack of technical skills other leading companies had demonstrated. Likewise, stakeholders in Kentan anticipated great difficulty in reaching reasonable terms with local companies operating in the goldfield.

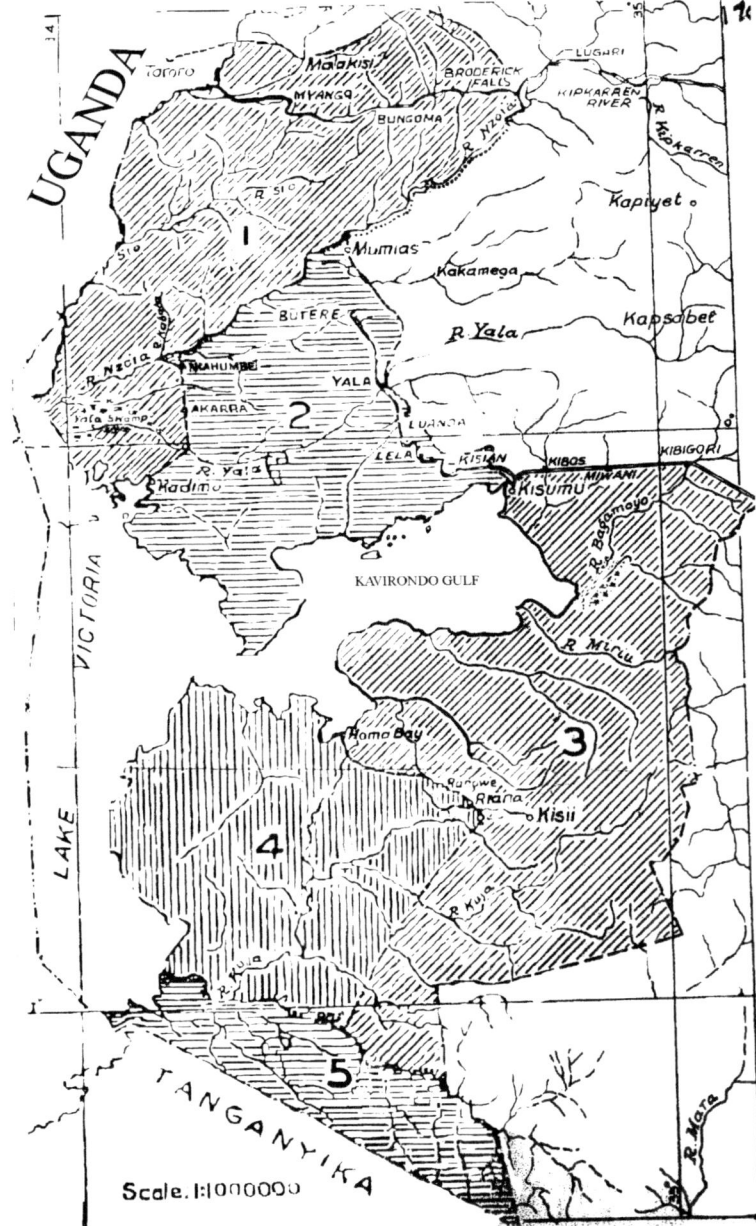

Map 5. Nyanza Province Mining Area Zones
Source PRO: CO 533/427 (Sir Albert Kitson's Report)

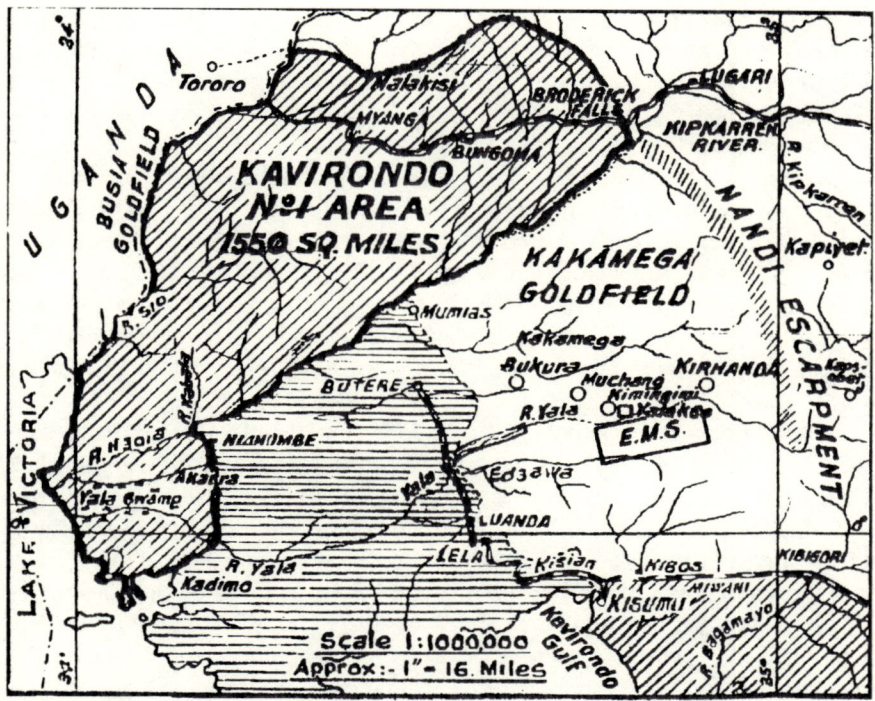

Map 6. Tanganyika Concession (1,550 Sq. Miles) (Sir Kitson's report)

By July 1933, the colonial government had granted EPLs to Mr. W. P. Aldersons, the Tanami Gold Mining Syndicate and Tanganyika Concessions on River Yala. In August, the colonial administration had six more EPL applications, covering about 16 square miles of land under review.[41] By October 1933, the Commissioner of Mines restricted the granting of EPLs to exceptional applicants. This change of policy drastically reduced the number of applications. As a result, during October, the colonial government approved only three EPLs for Tanganyika Concessions Ltd., S. M. Syndicate, and Kenya Development Ltd., covering about 23 square miles of land.[42] In 1934, the government granted more EPLs to Feza Ltd. (3 square miles); Eldoret-Kakamega Mining Venture (2 square miles); Plateau Transport (2.6 square miles); and M. de. Bord. Union Miniere on rivers Isikhu and Lugusidza and its tributaries.[43]

Most of the larger undertakings in Kakamega first obtained EPLs or took over options from individual prospectors and syndicates, and having established worthwhile claims three of them applied for mining leases. The

Mining (Amendment) (No. 2) Ordinance of 1932, facilitated lease grants to European prospectors. This ordinance provided for, *inter alia*, mining by means of registered locations and leases, rather than by claims and EPLs. In addition, the ordinance sought to prevent unauthorized representation and irregular pegging.[44]

Kimingini Gold Mining Company was the first to apply under the terms of this ordinance. The company applied for a mining lease of 113.5 acres in Idakho location on December 10, 1934, and received approval for a twenty-one year lease on August 1, 1935.[45] In November 1936 and June 1937, respectively, Rosterman Gold Mining Ltd. received a surface lease of 137.25 acres, and a subterranean lease of 515 acres south of West Isukha.[46] The government granted more leases as the gold mining industry progressed.

Despite this type of activity, the rush to Kakamega had leveled off by January 1934 to the extent that there were 439 European miners in the goldfield. This figure fell rapidly to 384 in February, before rising slightly to 393 in March. The colonial administration attributed the rapid decline of the European mining population in Kakamega to the disappearance of the "small men". In the words of the DC, NK, ". . . most had exhausted their resources but were holding out for fancy prices for their 'finds.' Gradually the large Companies . . . acquired options on more reasonable terms and many independent prospectors became employees of such Companies."[47]

The press supported the DC's observation:

> Two English lads, who sought their fortune, have had their dreams shattered. Ignorant of prospecting, with only small savings and ill-equipped, they slaved under the equatorial sun. Kenya folks helped them to buy equipment to ward off mosquitoes, but one of them was struck down with fever, and the other, after dreaming and plodding, was forced to give up.[48]

Most of the miners tended to concentrate near larger undertakings. The large companies operating in the Kakamega goldfield in 1934 included Kentan Gold Area Ltd., which was associated with Kimingini Gold Mining Company, the Eldoret Mining Syndicate and East African Concessions and Tanganyika Concessions. Others were Anglo-Continental Mines Company Ltd., Tanami (later Rosterman) Gold Mining Company Ltd., Kenya Consolidated Goldfields Ltd., Risks Ltd. (later Kavirondo Gold Mining Company), Kenya Development Ltd., and S. M. Syndicate. Smaller companies, such as Kenya Reefs, Edzawa Ridge, Blue Reefs, and Kenya Ore Reduction, were also in the field. The result was considerable development of reef prospecting and mining in comparison to alluvial mining, which was then confined to River Yala.[49]

DEVELOPMENT BY LARGE COMPANIES, 1934-44

The gold mining industry in NK district was definitely established by 1934. As large companies with substantial capital investments moved into the goldfield, they acquired properties of promise and commenced operations on a more ambitious scale. This led to the first stages of deep level mining with the industry attaining its adolescent stage at a tender age; its early growth into maturity seemed pleasing in prospect.[50]

During this time, surface trenching was largely overtaken by diamond drilling and shafts, and many large mines in the goldfield were approaching the production stage. The major center of mining activity remained chief Milimu's location, then christened Piccadilly Circus (after Piccadilly Circus business district in the heart of London). Considerable mining activity included the Bukura Ridge in Butsotso, Tanami (Rosterman) in Isukha, near Kakamega Township and, the area near Maragoli trading center. In fact, when No.2 Area was opened to mining on May 17, 1934, it had little effect on the mining population in NK district, which fluctuated between 360 and 420 persons during the year.[51]

Although numerous small and large companies established themselves in the Kakamega goldfield, only a handful became successful enterprises. In the analysis that follows, the study focuses on several case studies that yielded significant research findings. The analysis has three major objectives: first, to highlight the wide range of capital interests that converged on the Kakamega goldfield; second, to illustrate the major trajectories in the mining activities of these companies; and third, to illuminate the challenges these companies faced in their gold exploitation endeavors in Buluyia.

As noted before, the Kakamega goldfield attracted a wide range of companies, both local and foreign. Most of these concerns applied for EPLs, which entailed liberal land grants for their exclusive use and optioned properties from earlier prospectors. From 1933 onwards, most companies embarked on consolidating their position by absorbing or merging with smaller concerns. The resultant large companies dominated the goldfield as they squeezed out individual miners and financially disadvantaged firms. For instance, in August 1933, the Kenya Consolidated, the Kenya Development Ltd., the S. M. Syndicate and Messrs Risks and Ventures Ltd. had acquired the holdings of some five mining companies along the Bukura Ridge on option.

The Anglo-French Kenya Development Ltd., chaired by Sir Robert Horne, former Chancellor of the British Exchequer, and Sir Edward Grigg, former governor of Kenya as vice Chair, was by far the most active company in this respect. In 1933, the company submitted a two-year development program to the colonial state in which it envisaged spending £50,000 on its Sabatia claims in Maragoli. The company was importing machinery and engineers, had employed about forty men locally and was in the process of erecting a central camp on its property.[52]

Mr. Haskel of Johannesburg, South Africa, on the other hand, promoted Kenya Consolidated Goldfields Ltd. This company was registered in Nairobi, Kenya in February 1933, with a London office and transfer office in Johannesburg. The company's original capital of £60,000 in 240,000 shares of five shillings each was increased to £275,000 in June 1934 by the creation of 860,000 shares of five shillings each. In 1936, the company increased its capital to £450,000, £100,000 of which was earmarked for underground development of its valuable holdings in the 934 square mile concession area in Kakamega and in Lolgerian.[53] In addition, Kenya Consolidated Goldfields controlled 4,600 acres of claims in the Kakamega goldfield.[54] Later, most of its properties were taken over by Gold Corporation Ltd., about which scanty information exists.

Gold Corporation Ltd. represented foreign capital investment in the Kakamega goldfield. The company was registered in London in May 1933, with agents in Johannesburg, and Australia to acquire gold mining properties in Kenya. In 1934, Gold Corporation Ltd. had acquired several blocks of claims aggregating approximately 5,400 acres in NK district, besides its other interests in South Kavirondo (SK). While no information exists on the company's activities in Kakamega after 1934, its operations in SK lasted until March 1952, when the company voluntarily wound up.[55]

In the meantime, Tanganyika Concessions, one of the early foreign concerns in Kakamega, was pushing forward with its development agenda. By May 1934, the company had spent £15,000 on claim purchase and was waiting for the formation of two companies to manage its existing claims, including the 35 square mile concession. The company subsequently floated adequately capitalized subsidiary companies.[56] One subsidiary company, Kentan Gold Areas Limited, with a capital of £1,000,000, acquired all the Tanganyika Concessions gold interests in East Africa, involving a 51 percent controlling interest in the Eldoret Mining Syndicate and certain options in Tanganyika and Uganda. In order to deal with the Kakamega properties, Tanganyika Concessions formed two more companies, namely, the East African Concessions Ltd., and the Kimingini Gold Mining Company Ltd. The latter company embarked upon an intensive development program in Kakamega.[57]

Kimingini Gold Mining Company Ltd. was registered in London on July 25, 1934, and Kentan, East African Concessions and the Eldoret Mining Syndicate took all its shares. The Company was formed to purchase the properties of Eldoret Mining Syndicate that included Kimingini and Musgraves properties in Idakho location, north of River Yala and east of the main

Kisumu-Kakamega road.[58] The company was incorporated in July 1934, with an authorized capital of £600,000 divided into 1,200,000 shares of ten shillings each. Company subscriptions were distributed as follows: East African Concessions, Ltd. 404,000 shares; Kentan Gold Areas Limited 391,880 shares; Tanganyika Group 12,120 shares; and Eldoret Mining Syndicate 392,000 shares. By November 1934, the company had commenced underground development. The company's two major shafts had reached the 200 feet level, and in June 1935, the acting Governor accompanied by the CNC formally opened its milling plant.[59]

By 1936, Kimingini was mining at the 550 feet level, and the company had commenced treating ores at its Musgraves mine. By 1937, underground development had reached 660 feet, with an average production of 900 oz of bullion per month.[60] Although the company kept the average monthly production of 885 oz for eleven months of the year, all mining and milling operations ceased in November 1938. Consequently, on December 15, 1938, the company went into liquidation.[61] From June 1935 to November 1938 when the mill closed down, some 102,000 tons of ore had been milled from both Kimingini and Musgraves. These yielded 29,650 oz of fine gold, valued at £209,000.[62] Several factors contributed to the demise of Kimingini Gold Mining Company.

The Warden of Mines attributed the closure of the mines to the exhaustion of gold ores, coupled with large volumes of water encountered underground. The surveys carried out by the senior government geologist, Stansfield Hitchen, in July 1938, confirmed the impossibility of continued operations at Kimingini. Hitchen concluded thus, ". . . One can only foresee either the salvage of the mine or its salvation through a 'miracle'—The former prospect is a dismal one and the latter need not be contemplated."[63] While officials in the mining fraternity blamed nature for the company's closure, the provincial administration blamed it on internal factors. The PC Nyanza considered the closure "a great disappointment." In his view, the closure resulted from overcapitalization and extravagance.[64]

Following the closure, all Kimingini staff moved to Geita in Tanganyika territory. In 1939, Kerebe Mines Ltd. took over Kimingini mining lease and produced 1,629 oz on this property during the year.[65] In the first half of 1941, Kerebe Mines procured 2,500 oz of fine gold. The company changed hands again, with the new owners producing small amounts of gold until November 1951, when company operations ceased.[66]

Another important company in the gold field was Anglo-Continental Mines Limited (formerly Anglo-Continental Gold Syndicate (1899) Limited). The company was registered in London on December 23, 1909, as a finance company. Anglo-Continental represented foreign capital investment in

Kakamega, with an initial capital of £150,000 in 300,000 shares of ten shillings each.[67]

Anglo-Continental entered the Kakamega goldfield in November of 1933. The company's early operations in Kakamega focused on the Koa-Milimu (also known as Risks) claims in Idakho location. Initially, the company expanded its activities through options, until November 1933, when it obtained an EPL over 20 square miles of land between River Yala and the Kakamega-Kapsabet road. During this time, the newly formed East African Mining Areas Ltd was supporting Anglo-Continental, which it eventually took over. The East African Mining Areas Ltd was founded by Sir Edmund Davis, representing Rhodesian (now Zimbabwe) and South African interests, Professor Lawn of the Bearcroft Group, and Mr. Carl Davis of a powerful American combination, the Consolidated Miners Selection Trust. Although the company had a nominal capital of £100,000, it represented a real capital of £10 million.[68]

Another company, the Kavirondo Gold Mines, Ltd., was floated in London to take over and develop the holdings of Risks Limited and the Koa-Milimu Mines. Risks Ltd. was formed in 1933 when several claim holders transferred their properties to the company. Risks Ltd. operated near Shikonde in Kakamega, and remained active until December 1935. In April 1933, the company erected a 6 head stamp battery of 350 pounds falling weight, a rock breaker and a three feet saw bench. In addition, Risks Ltd. installed portable compressors for operating jackhammers. Due to the scarcity of drill steel in Kenya colony, the company hardly utilized the compressors.[69] Risks continued with intensive work in Isukha, concentrated along the Butere road throughout the year. By January 1934, however, the company was abandoning most of its properties. At the time, Risks had installed and was operating a 70-ton a day mill.[70]

It was against this background that the new company, the Kavirondo Gold Mines Ltd., was floated in 1936 to take over the assets of Risks. These included 1,376 acres in the Kakamega goldfield, comprising 169 mining claims, an EPL and a timber license over 500 acres in the Kakamega Forest Reserve. With an authorized capital of £375,000 in 750,000 shares of ten shillings each, Kavirondo Gold Mines proceeded to engage in active mining operations at Sigalagala in Isukha location. By the end of 1938, the company had become the second producer of gold in Kenya colony behind Rosterman Gold Mines. From 1935 to 1946 when the company's active mining operations ceased in Kakamega, some 30,000 oz of gold, valued at £240,000 had been produced from workings on numerous veins, and a large number of shafts and adits. These excavations reached a depth of 500 feet. The company went into liquidation in 1950.[71]

Several factors accounted for the closure of the Kavirondo Gold Mines Ltd. As early as 1938, the company was experiencing considerable financial and technical problems. Nonetheless, the company continued to produce steadily, attaining a monthly profit of £2,000 in 1939, prompting the company to install new machinery. Nevertheless, company fortunes dwindled, forcing it to retrench its staff and to rely on low-grade ore.[72] Thus, financial and technical difficulties, staff retrenchment, and low ore deposits, precipitated the closure of Kavirondo Gold Mines Ltd. in 1946.

Rosterman Gold Mines Ltd., on the other hand, was by far the most important mining concern in the Kakamega goldfield. Rosterman was registered in London on February 3, 1935 under the management of Bewick, Moreing and Co. Ltd., with an authorized capital of £400,000 in 1,600,000 shares of five shillings each. The company was formed to acquire gold mining claims over 350 acres on the bank of River Isikhu in Kakamega previously owned by Tanami Gold Mining Syndicate Limited. Tanami was an Australian company, which was registered in London in October 1932 with a capital of £100,000. In addition to its large share interest in Rosterman, Tanami had other share interests in Pakaneusi Prospecting and Development Company in Kenya as well as companies operating in Tanganyika and Transvaal in South Africa.[73]

The Ross Mining Syndicate initially worked the claims, located in Isukha about two miles south-west of Kakamega before Tanami acquired them by option. Subsequently, Rosterman applied for an EPL over 4,000 acres and commenced milling in the latter half of 1935. Rosterman remained active until 1952 when the mine closed down. Rosterman milled 655,000 tons of ore, and recovered 259,142 oz of fine gold, constituting over half of the recorded total gold production from the area. Rosterman remained the largest producer of bullion gold in Kenya colony throughout its existence. For instance, in 1936, the company installed a production plant, quickly topping the list of gold producers in the colony. In April 1936, the company increased its capital to £515,000 by creating 460,000 new shares at five shillings each.[74] The company's fortunes improved when it received a 137.25-acre surface lease and a 515-acre subterranean lease in 1936 and 1937, respectively. In 1937, Rosterman maintained an average production of 1,200 oz of gold per month. Production figures rose to 1,400 oz in 1938 and leapt to 1,800 oz in 1939. With its newly installed plant, the company concentrated on underground diamond drilling so that by 1936, development had reached 500 feet. The mine reached the 660 feet level in 1937 and by 1944, the company attained the 1,800 feet. By 1946, however, when Rosterman operations were at 1,940 feet below the surface, gold production begun to decline. Despite falling productivity, Rosterman continued to pursue an intensive development and drilling program.⁷⁵ In 1940, Rosterman recorded the highest gold output consisting of 23,915 oz of gold, worth about £200,000. This translated into a 5 percent interim dividend for the company's shareholders.⁷⁶ Early in 1952, Rosterman Gold Mines, still by far the largest producer in the colony, closed down.⁷⁷

While large mining companies dominated the decade 1934 to 1944 as individual prospectors declined, it appears that even before the outbreak of World War II (WWII) in 1939, smaller firms tended to be more viable than larger ones. Notwithstanding, by 1944, several companies in both categories had either ceased production or were struggling on, at much reduced production capacities. The next section focuses on the major trends, the impact of war(s) and the vast array of factors that negated the growth of the gold mining industry in NK district between 1934 and 1944.

TRENDS IN MINING, 1934–44

As in 1934, the decline of the activities of individual prospectors in Kakamega persisted throughout 1935. Thus, the year 1935 witnessed a considerable reduction in the number of European prospectors and miners and the merging of those that remained into a few large companies. In the main Kakamega area, Rosterman, Kimingini and Risks were the principal mining companies, although several small companies remained active. Several companies including Kenya Consolidated Goldfields, East African Mining Areas and S. M. Syndicate, discontinued gold mining operations during the year.

Rosterman Gold Mines, Gold Areas of East Africa, and the East African Goldfields, were floated early in 1935. Yet, the grave international situation, caused by the outbreak of the Italian-Ethiopian War, played havoc with mining finance in Kenya due to the proximity of the war theater to the Kenya goldfields. Therefore, before the end of 1935, it was almost impossible to float development companies. As capital temporarily dried up, a number of companies cut back their development programs, particularly gold exploration-related ventures. The hopeless economic situation intensified as individual prospectors' ceased operations. For example, the Kakamega goldfield witnessed significant exodus of individual prospectors, largely alluvial workers to No. 2 Area in CK. With this exodus, the once vibrant alluvial mining declined appreciably, and became restricted on Rivers Yala and Isikhu, the initial foci of the Kakamega goldfield.⁷⁸

When the N0. 2 Area was opened to prospecting on May 17, 1934, the seventy-four pioneer prospectors came from Kakamega, in syndicates of five or four. On arrival, they erected a Protection Notice, and withdrew, leaving one man to prospect. These pioneers included well-known prospectors such as

Messrs L. A. Johnson, Starnes, Musgrave, the Anglo-Continental, Tanganyika Concessions Ltd., Sunshine Syndicate and Stepney.[79]

The expansion of the goldfield sustained the significance of gold in Kenya's colonial economy. Gold continued to be by far the most important mineral exploited in Kenya colony, ranking fourth in value among the colony's exports in 1936. Notwithstanding, investment in gold exploration diminished significantly, as prospectors abandoned large numbers of claims haphazardly pegged during the early stages of the goldfield. In spite of this, underground developments realized sound progress when two mines reached depth levels of between 500 and 600 feet, with a third one approaching 450 feet. Accordingly, reef mining became the most important source of gold, with production reaching roughly ten times that obtained from alluvial mining. In 1936, three mills produced 53 percent of the total gold (reef) output, seven produced 18 percent and the rest of the field accounted for 29 percent.[80]

The major producing mines working at depth in the Kakamega goldfield were Rosterman Gold Mines Ltd. and Kimingini Gold Mining Company Ltd. Edzawa Ridge Mining Company Ltd., owned by a European settler, produced just short of the above category, and the Kavirondo Gold Mines Ltd. were at the development stage, having taken over Risks Ltd. and Koa-Milimu.[81]

The growth of the gold mining industry was evident from the number of mills operating in Nyanza Province. While there were forty mills in 1936, the number had increased to fifty-three in 1937. By this time, two mines reached between 600 and 700 feet deep, respectively. These developments enabled Kakamega to maintain its position as the chief gold producer in the colony. Although the goldfield experienced revitalized activity by small producers in 1937, reef mining expanded to the point where alluvial gold constituted less than 6 percent of the year's total gold output.[82]

Such extensive development in underground mining made gold an important export from Kenya colony. For example, in 1937, gold was the fourth most valuable export commodity. In 1938, it was the second.[83] Kakamega goldfield remained the biggest producer of both reef and alluvial gold in Kenya colony. In 1938, the Kakamega goldfield registered a 14 percent increase in gold output over its 1937 record.[84] Apart from increased production, the gold industry witnessed remarkable financial gains. In 1938, the value of production surpassed the combined expenditure on mine development and operating costs. During the last half of the year, for the first time in the history of the industry in Kenya, income was considerably in excess of expenditure on all items, including prospecting, machinery, and capitalization, by approximately £84,000.[85]

Despite this remarkable growth in the gold mining industry in Kenya, by early 1937, there were intermittent "scares" of the possibility of the larger

Table 3.2. Percentage Increase in Gold Production, 1933–38

Period	Percentage increase
1934 over 1933	12 percent
1935 over 1934	84 percent
1936 over 1935	66 percent
1937 over 1936	43 percent
1938 over 1937	31 percent

Source: MGDAR 1938, 9.

companies folding up. Although this did not occur immediately, most of the smaller firms did close down while others amalgamated into large firms. A case in point was the merger between Bukura Mining Company, and Smallwood and O'Brien Property.[86] By the end of 1938, the DC, NK acknowledged the centrality of small workers in the Kakamega goldfield. He noted, "The future of course lies with the small worker who alone discovers gold and works economically not with the two surviving over-capitalized large Companies, whose financial operations have made it so difficult to raise money for [the] mining enterprise in Kenya."[87]

The district administrator also hailed the so-called medium companies as the most encouraging part of the gold mining industry. These included Bukura Mining Company, Edzawa Ridge Mining Company, and S. Everett and Sama Syndicate, belonging to former European settler farmers who engaged in reef mining. Similarly, Kenya Reefs Ltd. and Mrs. Stitt's H. S. F. Syndicate specialized in alluvial mining. In this genre of companies, Edzawa Ridge Company was the most promising. In 1936 and 1937, Edzawa mining company paid out 100 percent dividends to its shareholders. In 1937, the company erected a cyanide plant, and realized a 125 percent dividend in 1940.[88]

Apart from Edzawa, Bukura Mining Company paid 40 percent dividends to its shareholders in 1938, and invested large reserves in gilt-edged securities. This company made profits of over £1,000 per month during the last part of the year, and boasted sufficient payable ore reserves. Its founder and manager, Mr. Smallwood, who employed nine European workers, had lived for a long time on less than £2 per month.[89] Smallwood exemplified a settler who made a fortune in the gold mining industry.

At the start of WWII in 1939, Bukura Mining Company Ltd. was on the verge of joining the ranks of larger mines, with an average gold output of 560 oz per month. In 1939, the company's profits represented approximately 60 percent of its general capital investment, and the company paid out 45 percent dividends during the year. By 1940, its future seemed bright with an average production of 6,811 unrefined oz. Bukura Mining Company also increased its investment with the purchase of the holdings of Owombo Mining Company Ltd., in Tiriki location of NK district.[90]

As the war progressed, however, the company's fortunes began to falter. By 1942, Bukura Mining Company had either abandoned or transferred all its claims, except the Owombo block on which all mining had ceased altogether. Operations on the Taylor section of its property near Lisulu stopped in 1945, and by the middle of the year, the company ended all operations in Kakamega. The company's closure resulted from water pumping problems, diminishing payable ores, and financial difficulties.[91]

In addition to medium size companies, in 1938, the Kakamega goldfield had twenty-eight small producers or prospectors in the field, producing between 5 and 60 oz of gold per month. Such concerns belonged to experienced ex-European settler farmers who continued to extract gold in payable quantities, in claims abandoned by large mining companies such as Tanganyika Concession and its subsidiaries.[92]

When No. III and IV Areas of SK opened for prospecting in 1939 the Kakamega gold industry suffered a slight setback as prospectors and small miners left for SK. Here the dominant companies included the Kenya Consolidated Goldfields, W. P. Alderson, and L. A. Johnson of Eldoret Mining Syndicate.[93] This means, most of the large and small companies in the Kakamega goldfield, remained relatively stable.

Thus, the Kakamega gold mining industry was relatively sound at the start of WWII. Nevertheless, at the end of 1939, only two of the large mining companies continued their operations in Kakamega namely: Rosterman Gold Mines and Kavirondo Gold Mines. These companies produced an average of 1,400 and 590 oz of gold per month, respectively. Of the smaller mines, Bukura Mining Company produced 433 oz per month and a dozen others produced between 50 and 150 oz per month. With regard to alluvial mining, Colonel Stitt and his wife remained the principal producers. While the gold industry showed a steady healthy improvement, the DC NK contended that it "had yet to prove itself financially."[94]

All together, there were fifty-three operating mills in Kenya colony in 1939, most of which were in the Kakamega goldfield. In spite of this, the war brought difficulties to the gold mining industry, chiefly the shortages of machinery and skilled labor. These shortages led to declining production for the next twelve years. In 1952, the rate of decline accelerated sharply, forcing Rosterman Gold Mines, the largest producing mine in Kenya to cease production.[95]

WWII brought in its train difficulties regarding European mine employees. Most perceived the War as a chance to change their employment by either joining the army or going to Northern Rhodesia, to replace many mineworkers joining the army. There was, therefore, an unsettled feeling, as the Assistant Warden of Mines, W. H. Bailey, noted:

> Mr. Griffiths of Kenya Gold Mining Syndicate Ltd. called and in course of conversation stated that all the men at their mine wanted to join [the army] and they did not want a 'Certificate of Exemption' which they regarded as a stigma. I suggest[ed] that they might look on things differently if they had something in the nature of a 'certificate for essential War Service.' I also hinted that perhaps their men might be keener to remain if conditions of employment were improved.[96]

By December 1939, the increasing dearth of skilled European workers threatened to curtail gold mining activities. The following year, only three large mining companies, namely Rosterman Gold Mines Ltd., Bukura Gold Mining Company and the Kavirondo Gold Mines Ltd., remained active in the Kakamega goldfield. These companies produced monthly averages of 1,000, 800 and 500 oz, respectively, with Mrs. Stitt as the sole producer of alluvial gold. Indeed, in April 1940, the gold mining industry began to look to imported labor as an alternative. The Assistant Warden of Mines reported that a small number of European and South African unskilled workers were available for employment. This state of affairs persisted, forcing the Assistant Warden of Mines to urge employers to improve wages in order to attract skilled European underground workers from Southern Rhodesia and South Africa. The dearth of a skilled European work force had peaked by September 1940. For example, during September 1940, Edzawa Ridge Mine was under the management of two Europeans, with several Seychellois underground workers.[97] In 1941, the company temporarily ceased production due to shortage of and high cost of mining materials.[98]

Ironically, the war situation created immense pressure for increased gold production in Kenya colony. The Assistant Warden of Mines summed up the situation as follows: "It would appear that a great deal more could be done to bring the gold production up to be worth at [least] *50 Bombers* or 100 Pichter Aircraft annually. Gold means sorely needed armaments to Great Britain, and *it is* possible to considerably increase the output from this country."[99]

The tremendous build up of economic pressure on both the imperial exchequer and the satellite colonial economies made the production of gold an economic imperative. In Kenya colony, the colonial state stepped up new initiatives aimed at invigorating the gold mining industry. For example, between 1941 and 1942, the colonial state commissioned several colony wide geological surveys aimed at opening up new goldfields.[100] Yet, it was clear that the gold mining industry was under siege. The East African Chamber of Mines (EACM) was sounding a warning: "The time is approaching when, through lack of equipment, stores, deterioration of working conditions, certain of the mines will have to suspend operations during the period of the war."[101]

It is not surprising, therefore, that the year 1942 witnessed significant declining fortunes in the gold mining industry in Kenya colony. In the case of NK district, the Kavirondo Gold Mining Company ceased production in September, and Saramanji, a small property, also went out of production. Rosterman, the major gold producing firm, reduced output after retrenching its staff. This downward trend continued unabated as more mines closed down in 1943, principally on account of the difficulty of obtaining working materials from Europe or North America. The mines most affected included Owombo, Zawadi, Kimingini and Bukura. It was anticipated, however, that such mines would re-open when the supply position became favorable. Even at reduced production capacity, Rosterman Gold Mining Ltd. remained the principal gold mining concern in Kenya.[102]

The contraction in gold mining activities was apparent by 1943. Mining officials acknowledged the impact of WWII on policy and the trend of mining activities in their annual report for the year. Wartime taxation, higher costs of production, insurance and freight, shortage of supplies and machinery, and a declining labor force, played a central role in the declining volume of gold mining. Large numbers of individual miners ceased mining operations, leaving the goldfield to a few large concerns, thereby accentuating a discernible decline in gold production. Indeed, the few companies operating in Kakamega proceeded at a much-reduced tempo, with the inevitable outcome that the number of mills drastically fell from fifty-three in 1939 to eleven in 1944. Likewise, gold output declined by 21 percent in 1942; 20 percent in 1943, and 6 percent in 1944.[103] The adverse war conditions colluded against the gold mining industry from reaping benefits from the prevailing wartime increase in the price of gold.

It is worth noting that the reduction in gold output was largely a result of the difficulties the industry was experiencing, and not so much from want of gold in the ground. The gold mining industry, therefore, retained the potential to be a major player in Kenya's economy. Indeed, during the war period, gold remained one of the most important exports of Kenya colony. Table 3.3 is a comparison of the value of gold exports and the four major agricultural exports of the colony between 1936 and 1945. It is clear from the table that gold held a favorable position among the colony's major agricultural exports, especially for the period between 1937 and 1944.

For example, between 1936 and 1937 gold ranked fourth among the colony's major exports. It moved to third place in 1938; second in 1939 and attained first rank in 1941 when it accounted for 17 percent of the total value

Table 3. 3. The Value of Gold Exports Visa-Viz the Value of the Four Major Agricultural Exports of Kenya Colony, 1936–1945

Year Export	Coffee (£)	Sisal (£)	Tea (£)	Maize (£)	Hides (£)	Gold (£)	Total value of exports (£)	Gold as a % of total value of exports
1936	969,804	690,459	335,690	254,087	—	269,947	3,278,662	8
1937	732,684	673,719	466,872	228,018	—	379,626	3,200,629	12
1938	767,789	436,259	501,099	289,212	—	499,601	3,045,507	16
1839	800,827	442,434	501,732	—	—	607,753	3,242,475	19
1941	575,259	399,934	—	168,039	152,705	606,042	3,543,329	17
1942	699,308	539,367	—	—	162,446	476,870	3,726,766	13
1943	562,651	555,563	—	—	185,547	378,989	3,220,828	12
1944	528,844	707,746	—	—	138,033	354,976	3,883,397	9
1945	639,433	769,494	—	465,501	—	333,259	4,720,215	7

Sources: Col. & Prot. of Kenya, *Department of Agriculture Annual Report (DofAAR) 1936* (Nairobi: Government Printer, 1937), 7; *DofAAR 1937* (Nairobi: Government Printer, 1939), 6; *DofAAR 1938* (Nairobi: Government Printer, 1939), 7, 9; *DofAAR 1945* (Nairobi: Government Printer, 1946), 6, 9–10; *DofAAR 1946* (Nairobi: Government Printer, 1948), 5; *DofAAR 1948* (Nairobi: Government Printer, 1950), 5. For gold export figures, see Table 3.6.

of the colony's exports. Between 1942 and 1944, gold maintained third place among the colony's exports before falling back to the pre-war position.

A colonial government geologist, Huddleston, estimated that between October 1931 and 1950, the Kakamega goldfield had yielded 437,000 oz of fine gold valued at £3,500,000.[104]

In addition, the colonial state received more income from mining-related fees as shown in Table 3.4. Conservative as the figures may be, it is clear that gold mining contributed significantly to the colonial economy, contrary to the conclusions of Fearn and Roberts who minimized the gains and maximized the costs of developing the industry to the colonial state and European capitalists.

"A FAILED ELDORADO," 1945–52

The gold mining industry in Kenya in general and Kakamega in particular, had shrunk to an appreciable extent by 1944. This trend persisted for the remainder of the period under study. Indeed, the DC, NK, F. D. Hislop, acknowledged this fact in 1945. While Rosterman remained the largest producer, the company faced immense labor problems. Generally, personnel-related problems in the industry resulted from several factors. Overall, the small European personnel lacked the necessary mining skills. Yet, they were expected to teach, direct and discipline locally recruited

Table 3.4. Government Revenue from Gold Mining Fees, 1933–1952

Year	Amount
1933	11, 512
1934	17,957
1935	20,197
1936	19,181
1937	20,381
1938	2,958
1939	3,541
1940	27,652
1941	33,577
1942	No figures available
1943	23,790
1944	24,540
1945	19,238
1946	16,164
1947	No figures available
1948	4,356
1949	3,803
1950	5,655
1951	No figures available
1952	2,606
Grand total	257,108

Sources: *MGDARs* 1933–52. These included mining fees, royalties, and assay fees.

African and Asian labor. Since mining companies offered low salaries, they were unable to attract good technicians. This acute shortage of European expertise was stretched further during the war period, as many were called for war service. For example, while there were 265 skilled European personnel in the Nyanza goldfields in 1939, their population had shrunk to ninety-seven by 1945. The inexperience of the African labor force engaged in mining, further compounded the management and supervision problems of mining operations.[105]

Moreover, although Edzawa Ridge and Bukura continued to function, the latter did so at a much-reduced scale. The situation was even worse with regard to alluvial mining. Mrs. Stitt's H. S. F. Syndicate, whose specialty was alluvial mining, closed down, leaving the field to illegal mining operations by Africans.[106] The industry suffered a further setback when both Edzawa Ridge and Bukura Mines folded up in 1946. Thus, Rosterman remained the only concern of importance in the Kakamega goldfield hereafter.[107] This state of affairs served to compound the financial troubles of not only Kenya colony, but also, the imperial government which was already experiencing immense economic strain.

Britain's involvement in WWII had significant implications for its post war economy. WWII completely shattered the British economy, forcing Britain to look to its colonies for the much-needed economic blood transfusion. The need to revamp the post-war depressed British economy, forced S of S for the Colonies, George Hall, to dispatch a priority telegram to Kenya in February 1946, pointing out the need for extending gold production to the African population. This initiative from the CO prompted the Commissioner of Mines to urge the colonial state to re-open a section of Maragoli, which the colonial administration had closed to mining in 1933, because of dense population. This led to a protracted war of words between the colonial state in Nairobi and the provincial administration in Western Kenya. This struggle for gold rights emanated from the diametrically opposed differences between the provincial administration's favorable disposition toward legalizing African gold mining activities and the secretariat's opposition to the same.[108]

While the colonial state was reluctant to incorporate the Luyia in the gold mining industry, the cessation of hostilities brought only dismal improvements in conditions affecting gold mining. For example, the colonial administration anticipated a dramatic boom in the gold industry at the end of the war. Regrettably, the expected post-war boom in gold production never occurred. A combination of factors including the difficult labor situation, the low ebb in prospecting and the spectacular developments in the opening of mineral deposits in neighboring Tanganyika territory, kept many pre-war miners from resuming their activities in Kenya colony. The result was that gold production declined by almost 23 percent in 1946 compared with the output for 1945.[109] Table 3.5 shows the general trends in the European mining population in Kakamega between 1932 and 1947, indicating the surge in European population in NK during the rush period and the decline thereafter.

The table shows that the Kakamega gold rush lasted between 1932 and 1933. From 1933 onward, the number of European miners plummeted and continued to decline up to the end of WWII. Although there was a slight increase in 1946, the following year witnessed a further reduction in the number of miners.

Several factors contributed to the decline of European mining population. In the initial years, the departure of individual miners was the most important reason, especially beginning in January 1934. For example, while the goldfield had 439 European miners in January 1934, the number fell drastically to 384 in February, 365 in May, and 349 in June. It slightly rose to 354 in July, 381 in August, before dropping again to 364 in September. This process continued unabated. In the 1936 annual report, the Nyanza PC confirmed that the gold industry had significantly shrunk following the "departure of alluvial workers, small miners and prospectors."[110] Most of these prospectors had come to the goldfields with the first rush, but had failed to make good.

With the onset of war, the number of miners diminished further as many entered active duty to defend their mother country. For instance, from the prewar figure of 173 adult male miners in 1938, the figure reached a paltry sixty-five miners in 1945.[111]

Gold mining in NK district continued very much along the same lines in 1947 as in 1946, with Rosterman maintaining its position as the largest and most important concern. Although the industry experienced a slight revival of mining activity along the Bukura Ridge in Butsotso, this did not prove to be permanent. In fact, gold mining remained at low ebb in subsequent years. Although one or two of the smaller mines re-opened intermittently in 1948, Rosterman maintained the lead.[112] The situation had deteriorated so much that in 1948, gold was now second to soda ash as the most valuable export mineral for Kenya colony. Nevertheless, the colonial state anticipated a small increase in gold production for the year although the industry continued to face

Table 3.5. European Mining Population in Kakamega, 1932–1947

Year	Men	Women	Children	Total
1932	1,078	110	109	1,297
1933	600	307	130	1,037
1934	No records	No records	No Records	No Records
1935	386	320	146	852
1936	242	144	77	463
1937	221	125	57	403
1938	173	117	83	373
1939	161	112	51	324
1940	121	100	52	273
1941	114	92	68	274
1942	98	83	66	247
1943	88	69	53	210
1944	75	70	62	207
1945	65	53	48	166
1946	105	78	63	246
1947	81	60	30	171

Sources: NPAR 1932, KNA: PC/NZA/1/27; NPAR 1933, KNA: PC/NZA/1/28; NPAR 1935, KNA: PC/NZA/1/30; NPAR 1936, KNA: PC/NZA/1/31; NPAR 1937, KNA: PC/NZA/1/32; NPAR 1938, KNA: PC/NZA/1/33; NPAR 1939, KNA: PC/NZA/1/34; NPAR 1940; KNA: PC/NZA/1/35; NPAR 1941, KNA: PC/NZA/1/36; NPAR 1942,KNA: PC/NZA/1/37; NKDAR 1943, KNA: DC/NN/1/25; NPAR 1943, KNA: PC/NZA/1/38; NKDAR 1944, KNA: DC/NN/1/26; NPAR 1944, KNA: PC/NZA/1/39; NKDAR 1945, KNA: DC/NN/1/27; NPAR 1945, KNA: PC/NZA/1/40; NKDAR 1946, DC/NN/1/28; NKDAR 1947, KNA: DC/NN/1/29.

insurmountable production costs, which only served to accelerate the declining fortunes of the gold mining industry.[113]

This trend of reduced mining activity persisted throughout 1949, in spite of the significant devaluation of the £ sterling which subsequently raised the price of gold from Shs.172.50 an ounce to Shs. 248.50. Although the devaluation caused a very welcomed increase in the earnings of gold producing mines, it neither stimulated prospecting to any appreciable extent, nor caused any noticeable increase in gold production throughout the year.[114]

The colonial state had hoped that the devaluation of the £ sterling would inject a new lease of life in the gold mining industry in Kenya in the form of yet another "rush" by European prospectors to the Kenya goldfields. Unfortunately, the expected gold rush never materialized. The colony's 1949, Annual Report noted that the increase in the price of gold had failed to restore the gold mining industry to its pre-war economic position.[115]

Indeed, in 1950, soda ash continued to enjoy a pride of place as the most important mineral produced in Kenya. For example, while the value of soda ash produced in 1950 was worth £806,000, the value of gold for the same year estimated at nearly £260,000. In spite of this, the actual value of gold produced, amounting to over 21,000 oz, half of which came from Rosterman Gold Mines in NN district, was consistent with the 1948 and 1949 production figures.

The dramatic decline in gold output resulted from increased costs of production, which continued to rise at an alarming rate. The outcome was that the industry lost any advantage that accrued from the increase in the price of gold due to the devaluation of the £ sterling.[116] Concisely, very little mining activity was carried on in NN district in 1950, and attempts at resuscitating the Kimingini Mines during the year were abandoned the following year when the company's plant was sold.[117]

By 1951, gold had definitely yielded its pride of place in Kenya's colonial economy to non-precious minerals, particularly soda ash and salt. Prospecting for new gold occurrences was minimal throughout the year, while the continuous and unrelenting rise in production costs as shown in Table 3.8 virtually obliterated any benefits obtained from the prevailing increased price of gold. This state of affairs was extremely alarming to the gold mining community in Kenya to the extent that the European Member for Commerce and Industry personally led a Kenya gold mining delegation to the CO in September to appeal for direct assistance from the imperial government. The delegation obtained a complete remission of all gold royalty for ten years, and sanction for a grant to aid gold mining development in Kenya. In addition, permission was also granted for colonial territories to sell up to 40 percent of their gold production on the free market.[118]

In spite of the CO's liberal considerations for the mining fraternity in Kenya, the gold mining industry suffered a severe blow when Rosterman Gold Mines stopped milling in 1952. Consequently, the output of gold for the year fell to approximately 14,800 oz of bullion valued at about £134,500.[119]

It is worth noting that the closure of Rosterman Gold Mines occurred in spite of the inauguration of the Gold-Mines Development Loans Ordinance of 1952. The ordinance granted interest-free loans for companies facing financial difficulties to facilitate underground gold mining in Kenya colony.[120] With the closure of Rosterman Gold Mines, the gold mining industry in Kakamega met its Waterloo. Commenting on the closure of Rosterman Gold Mines, the PC Nyanza wrote, "It is sad to record that the Rosterman Gold Mine at Kakamega which had operated longer than any other mine in the area, ceased production at the end of the year."[121] Table 3.6 shows gold production figures at Rosterman Gold Mine between 1936 and 1951. The centrality of Rosterman Gold Mining Company in the development of the gold industry in Kenya is evident when compared to colony-wide gold output figures as shown in Table 3.7.

Table 3.6. Output (milled, fine gold value) and Cost per Ton at Rosterman Gold Mine at Kakamega, 1936–1951

Year	Tons Milled	Fine Gold (Ozs)	Value (£)	Costs (£)	s.	d
1936	22,306	6,801	47,925			
1937	40,264	13,558	95,218	2	1	7
1938	34,489	14,848	106,464	2	2	4
1939	39,840	17,822	152,690	2	2	5
1940	46,429	23,915	199,212	2	2	10½
1941	54,000	22,099	185,633	2	0	6½
1942	46,600	18,960	159,250	2	4	3
1943	38,700	16,039	134,725	2	7	4
1944	47,800	18,832	158,184	2	5	7
1945	46,970	18,982	164,144	2	10	8
1946	41,700	14,890	128,469	2	17	0
1947	30,593	11,232	86,873	3	14	4
1948	30,156	12,753	14,116	3	17	9
1949	33,080	12,309	121,744	3	14	3
1950	42,195	13,171	163,487	3	1	8
1951	28,497	10,972	131,074	4	3	2

Source: Fearn, *An African Economy*, 259.

Table 3.7. Refined Gold Output for Kenya Colony, 1935–1952

Year	Fine Ozs Mainly from (NPAR)	Value (A) (£)
1926/1935	66,897	386,135
1936	41,753	269,947
1937	54,073	379,626
1938	69,435	499,601
1939	77,445	607,753 (B)
1940	77,243	648,783
1941	72,148	606,042
1942	56,771	476,879
1943	45,118	378,989
1944	42,259	354,976
1945	38,517	333,259
1946	29,892	257,942
1947	21,959	189,397
1948	23,429	202,076
1949	20,071	201,237
1950	22,945	284,806 (C)
1951	19,765	245,458
1952	10,209	131,881

Source: Adapted from Fearn, *An African Economy*, 127, 256; Fearn, "Gold-Mining," 48. See also *MGDAR* 1938, 9; *LMSDAR* 1946, 13; *MGDAR* 1952, 5.

Notes:
(A) The annual value of gold is the total gross amount received by the producers during the year.
(B) In October 1939, the price of gold on the London market stood at Shs. 168 per fine ounce as against the former price of Shs. 148.
(C) The devaluation of the £ sterling increased the price of gold per fine ounce from Shs. 172.50 to Shs. 248.50.

Several factors accounted for the steady decline and eventual demise of the gold mining industry in Western Kenya. Fearn identifies five major contributory factors: namely, the exhaustion of gold deposits; difficulties of mining finance; problems of labor management relations; peculiar problems of African labor supply and its instability; and increasing production costs.[122] According to this school of thought, the reduction of prospectors in Kakamega from 1935 onwards perhaps pointed to the fact that Kakamega would not develop into another Rand. In addition to inadequate local capital, the depressed economic conditions of the 1930s failed to draw foreign capital to Kenya. Without the discovery of a significant body of ores in the goldfield, Kakamega could not attract the necessary capital.[123] While the lack of capital theory is conceivable, oral evidence commonly referred to companies closing down "because of the prevalence of 'immature' gold deposits." Moreover, the failure of the Gold-Mines Loans to rejuvenate an important mine such as Rosterman, supports the "ore exhaustion" theory.

Table 3.8. Local Costs of Mining Requisites Supplied by John Riddoch, Esq., M. L.C

Item	Quantity	1942/45 Price Shs.	Cts.
Posho (Cornmeal for African food)	per bag	12	00
Mill Balls	per lb.	0	43
Mercury	per lb	13	00
Gelignite	per 50 lb Case	91	10 (1949)
Drill Steel 7"/8 Hex Hollow	per lb.	0	90
Cyanide	per 200 lb drum	187	00
Zinc Shavings	per 2 cwt. Case	240	00
Carbide	per 100 kilo drum	73	00
Belting 3" 4 ply	per foot	1	70
Crucibles	Each	38	00
Sledge Hammers (8 lbs.)	Each	7	25
Miners Hats 110	Each	8	50
Hose Rock Drill ½	per foot	1	65
Karais 18" 22 g	Each	3	25
Lamps Miner 61	Each	17	50
Litharge 28 lb—kegs	Each	25	00
Assorted Wire Nails	per lb.	0	53
Picks 5 lb.	Each	4	00
Wire Rope ½"	per foot	0	94
Mining Shovels	Each	8	55
Wheelbarrows	Each	45	00

Oil—Per Gallon		1939		1945	
		Shs.	Cts.	Shs.	Cts.
Motor Spirit		2	35	2	45
Gas Oil		0	80	0	90
Diesel Fuels		0	67	0	82
Automobile Oils		5	65	7	05
Industrial Oils:	Talpa	4	50	5	90
	Fiona	3	70	4	90
	Carnea	2	95	3	90

Source: Fearn, An African Economy, 260.

WWII also affected gold mining operations in several other ways. During the war, the industry experienced an acute shortage of tools and equipment, delayed delivery of spare parts, and high costs of production. For instance, during and after the war, costs of materials such as gelignite, drill steel, steel balls and carbide, rose sharply, while mining at great depth increased production costs.[124] Table 3.8 shows the local costs of mining requisites supplied during the war years by John Riddoch, a leading European entrepreneur in Kisumu and member of the EACM.

CONCLUSION

This chapter illustrates the contribution of the Kakamega gold mining industry to Kenya's colonial economy from the initial gold discovery in 1931 to the closure of the most important gold mining concern in Kakamega in 1952. Although records show a combined production figure of gold valued at £386,135 for the period 1926 to 1935, J. Clive, who served as DC NK between January 1937 and June 1938, reminisced that gold mining helped many settlers to tide over the slump years.[125] Between 1936 and 1937, gold ranked fourth among the colony's major exports. It moved to third place in 1938, second in 1939 and attained first rank in 1941, when it accounted for 17 percent of the total value of the colony's exports. Thereafter, it remained the third most valuable export until 1945, when it dropped to fourth place.

The Kakamega goldfield attracted both individual prospectors with limited capital resources and large companies with more substantial capital investments, representing diverse economic interests both local and foreign. Although alluvial mining dominated the industry in its infancy, by 1934, reef mining had become quite important. The development of deep mining by large companies necessitated the alienation of land from Luyia households for mining purposes. Nonetheless, significant problems such as increasing costs of production, lack of capital, exhaustion of ore deposits and the myriad problems that came in the wake of WWII, proved fatal to the nascent Kakamega gold mining industry. Although the gold mining era in NK district had ended by 1952, its most lasting legacies lay in the politics of land that emanated from the general European invasion of the reserve, the activities of prospectors, and the alienation of land for mining purposes. Chapter 4 details Luyia responses to the development of the gold mining industry and the intense land insecurity associated with these developments.

NOTES

1. *Manchester Guardian,* (n.d) March 1933; C. G. B. DuBois, *Geological Survey of Kenya: Minerals of Kenya* (Nairobi: Government Printer, 1966), 26; A. Huddleston, *Geology of the Kakamega District* (Nairobi: Government Printer, 1952), 22.

2. *East African Standard,* 30 January 1932, Rhodes House Oxford (RHO), MSS. Afr. S. 1281 (i); GB, *Annual Report on the Social and Economic Progress of the People of the Kenya Colony and Protectorate 1932* (London: His Majesty's Stationery Office (HMSO), 1934), 20. According to the *Manchester Guardian,* traces of gold were found in River Yala by Johnson and his companion, Major Starnes. See *Manchester Guardia*n, (n.d) March 1933.

3. Acting PC Nyanza to the Colonial Secretary (CS), November 30, 1931, KNA: PC/NZA/3/14/374. For details on the impact of the Depression on the economy of Kenya in general and Western Kenya in particular, See Tabitha Kanogo, "Kenya and the Depression, 1929-39," in *A Modern History of Kenya.* ed. W. R. Ochieng' (Nairobi: Evans Brothers, 1989), 112- 43; Odhiambo Ndege, "Struggles for the Market," 151-210.

4. NKDAR 1931, KNA: DC/NN/1/12; NKDAR 1932, KNA: DC/NN/1/13.

5. NKDAR 1931, KNA: DC/NN/1/12; NPAR 1931, KNA: PC/NZA/1/26. The devastation continued throughout 1932. See NKDAR 1932, KNA: DC/NN/1/13.

6. Dr. C. A. Marsh, "A Visit to the Kenya Gold field," Advance Proof of an Article Appearing in *Quaker Weekly Journal*, April 7, 1933.

7. Elspeth Huxley, *No Easy Way: A History of the Kenya Farmers Association and Unga Ltd.* (Nairobi: East African Standard Ltd., 1957), 11; See also Charles Chenevix Trench, *Men Who Ruled Kenya: The Kenya Administration, 1892-1963* (London: The Radcliffe Press, 1993), 99.

8. Fearn, *An African Economy*: 1961), 125. See also KNA: PC/NZA/3/14/374; A. D. Roberts, "The Gold Boom of the 1930s in Eastern Africa," *African Affairs* 85 (October 1985): 551.

9. Fearn, *An African Economy*, 126.

10. Ibid.123, 124; C. C. Wrigley, "Kenya: Patterns of Economic Life, 1902-1945," in *History of East Africa.* Vol. II. ed. Vincent Harlow and E. M. Chilver (Oxford: At the Clarendon Press, 1965), 248-9; DuBois, *Geological Survey*, 25; Roberts, "The Gold Boom," 551.

11. Trench, *Men Who Ruled Kenya*, 99; NKDAR 1931, KNA: DC/NN/1/12; GB, *Annual Report on the Social and Economic*, 20; Petro Mutondo, Oral Interview (OI), February 18, 1998; Andrea Mukabwa, OI, February 12, 1998; Gabriel Shivachi, OI, February 15, 1998; Erasto Ayiekha, OI, February 15, 1998; Gabriel Musalimwa, OI, February 17, 1998; Beti Isenjia, February 24, 1998; Josephat Amagadu, OI, March 3, 1998; Zablon Shikali, OI, March 3, 1998; Christopher Mavia, OI, March 3, 1998; Frederick Mutembei, OI, March 17, 1998; Peter Shitsukane, OI, March 19, 1998; Adriano Vihembo, OI, March 26, 1998; Enock Mwinamo, OI, March 26, 1998; Priscilla Misaki, OI, July 18, 1998; Samson Amahwa, OI, August 29, 1998; Priscilla Abwao, OI, September 11, 1998; Jairo Akibaya, OI, August 18, 1998.

12. Fearn, "The Gold-Mining Era," 45; Fearn, *An African Economy*, 126.

13. Fearn, *An African Economy*, 133; Acting Commissioner of Mines to PC Nyanza, January 5, 1932, KNA: PC/NZA/3/14/374; Mutondo, OI, February 18, 1998; Minishi Seleta, OI, February 15, 1998; Zachariah Andove, OI, February 28, 1998.

14. NKDAR 1934, KNA: DC/NN/1/15.

15. Fearn, *An African Economy*, 127-28.

16. Ibid.128.

17. Ibid.; Amahwa, OI, August 29, 1998.

18. Col. Swinton Howe, President, Miners Association, Memorandum for Consideration to H. E.: RE. Kakamega Goldfields, February 4, 1932, encl. in PC Nyanza to acting Commissioner of Mines, February 5, 1932, KNA: PC/NZA/3/14/374; L. F.

Streit, Assistant Inspector of Mines, Tanganyika, A Preliminary Report on Kakamega, encl. in PC Nyanza to the Chief Secretary, February 17, 1933, Ibid.

19. Huddleston, *Geology*, 25.
20. *Times*, 27 October 1932. See also GB, *Kenya Col. & Prot. AR 1932*, 18.
21. NKDAR 1932, KNA: DC/NN/1/13; *Daily Express*, 31 December 1932.
22. GB, *Kenya Col. & Prot. AR 1932*, 18.
23. *Spectator*, 12 December 1932.
24. Ibid. 19 January 1933.
25. *Daily Express*, 27 February 1933. Emphasis as in the original.
26. *Spectator*, 31 December 1932.
27. Ibid. 19 January 1933.
28. *Daily Express*, 27 February 1933.
29. NKDAR 1933, KNA: DC/NN/1/14; NPAR 1933, KNA: PC/NZA/1/28.
30. Col. & Prot. of Kenya, *Mining and Geological Department Annual Report (MGDAR) 1933* (Nairobi: Government Printer, 1934), 4; Fearn, *An African Economy*, 129; Priscilla M. Shilaro. "'A House Divided Unto Itself': The Colonial State and the Stillborn Maragoli Gold Mining Scheme, 1936–55." Paper presented at the annual meeting of the Historical Association of Kenya, Eldoret, July 1998, 1.
31. Fearn, *An African Economy*, 131.
32. Ibid.; *Annual Report on the Social and Economic*, 1932, 20; NKDAR 1932, KNA: DC/NN/1/13.
33. *News of the World*, 4 December 1932; Warden of Mines, Confidential Monthly Intelligence Report (CMIR), September 1933, KNA: PC/NZA/3/14/372; *MGDAR* 1933, 4; Fearn, *An African Economy*, 131. An EPL was valid for six years and required a paid registration fee of Shs.10 and a conveyance fee of Shs.150. *News of the World*, 4 December 1932.
34. Fearn, *An African Economy*, 131–6; A. A. Menkin, "Gold Mining in Kenya: A Review of 1935–36," *East African Annual* (1936–37): 123.
35. Fearn, *An African Economy*, 248–89; *MGDAR* 1933, 6; Huddleston, *Geology*, 43.
36. Col. & Prot. of Kenya, *Report by Sir Albert Kitson on an Application by Messrs. The Tanganyika Concessions, Ltd. for An Exclusive Prospecting Licence Under the Mining Ordinance, 1931, In Respect of An Area of 5,900 Square Miles Approximately in the North, Central and South Kavirondo Districts, Kenya Colony* (Nairobi: Government Printer, 1933), 1–7; Goldfields in the Kavirondo District, 1933, PRO: CO 533/428; DuBois, *Geological Survey*, 26; Governor Sir Joseph Byrne to the Secretary of State (S of S), Secret Despatch, November 19, 1932; Application for EPL by Barnard to Commissioner of Mines, n.d., encl. in Ibid, PRO: CO 533/427; J. Austen Bancroft, Report on Kenya Colony, PRO: CO 533/429/15; *Manchester Guardian*, March 21, 1933.
37. *Report by Sir Albert Kitson*, 12–16; *Times*, February 4, 1933; Fearn, *An African Economy*, 131–32. Although the S of S for the Colonies approved the publication of the report by secret telegram on December 23, 1932 (See PRO: CO 533/427), CO officials convinced him to postpone the publication to "avoid political embarrassment." Lord Lugard had a motion on the goldfield scheduled for debate in the House of Lords

on February 8, 1933; hence re-opening 5,900 square miles of land in NK reserve to prospecting was sure to cause a storm of protest at home. See chapter 4, and Freestone, Minute, January 29, 1933 and Cecil Bottomley, Minute, January 30, 1933, PRO: CO 533/428.

38. Report by Sir Albert Kitson, 8–12; *Times*, February 4, 1933.

39. Acting Governor, Henry Monck-Mason Moore to S of S, Despatch, July 8, 1933, PRO: CO 533/428; Second Progress Report of Sir Robert Williams & Co. on Work on the East African Goldfields: Eldoret Mining Syndicate's Properties, November 14, 1933, PRO: CO 533/429/15; Warden of Mines, CMIR September 1933, KNA: PC/NZA/3/14/372; GB, *Kenya Col. & Prot. AR 1933*, 15.

40. Warden of Mines, CMIRs September and October 1933, KNA: PC/NZA/ 3/14/372; Managing Director, Sir Robert Williams's Address to The Ordinary General Meeting of Tanganyika Concessions Held on July 27, 1933 in London, Reproduced in *Times*, 28 July 1933. The company's capital for all its activities in Africa stood at £10 m in 1933. Ibid.

41. Warden of Mines, CMIRs July and August 1933, KNA: PC/NZA/3/14/372.

42. Warden of Mines, CMIR October 1933, KNA: PC/NZA/3/14/372.

43. Warden of Mines, CMIR May 1934, KNA: PC/NZA/3/14/372.

44. *Kenya Col. & Prot., AR 1932*, 48.

45. NKDAR 1935, KNA: DC/NN/1/16; Nyanza Province Intelligence Report (NPIR) December 1935, KNA: PC/NZA/4/5/1; North Kavirondo Local Native Council (NKLNC), Minutes of a Meeting Held at Matungu on February 14, 1935, KNA (Kakamega): HW/13/2.

46. AO NK, April 7, 1936, KNA: PC/NZA/3/14/357; NKLNC, Minutes of a Meeting Held at Kakamega on April 15 & 16, 1936, KNA (Kakamega): HW/13/2.

47. NKDAR 1933, KNA: DC/NN/1/14; NPAR 1933, KNA: PC/NZA/1/28.

48. *Daily Express*, 27 February 1933.

49. Ibid; NKDAR 1934, KNA: DC/NN/1/15; NPIRs February and March 1934, KNA: PC/NZA/4/5/1; GB, *Kenya Col. & Prot. AR 1934* (London: HMSO, 1935), 15–16.

50. Menkin, "Gold Mining," 123.

51. NKDAR 1934, KNA: DC/NN/1/15; NPIR February and March 1934, KNA: PC/NZA/ 4/5/1; *Kenya Col. & Prot. AR 1934*, 15–16; Wagner, *The Bantu*, Vol. 1, 10. The area lay between the Kakamega goldfield and Lake Victoria, and extended over 1,000 square miles. GB, *Kenya Col. & Prot., AR 1934*, 15.

52. Warden of Mines, CMIR August and September 1933, KNA: PC/NZA/3/14/372; Sir Albert Kitson arrived in Kenya on January 17, 1934 as Director of Kenya Development Ltd. See Warden of Mines, CMIR January 1934, Ibid.

53. "Gold Mining Expansion in Kenya Colony," *Rhodesian Mining Journal* (January 1935): 69; Menkin, "Gold Mining,"126.

54. Warden of Mines, CMIRs January, February and June 1934, KNA: PC/NZA/3/14/372; Fearn, *An African Economy*, 251–2; Bancroft, Report on Kenya Colony, PRO: CO 533/429/15. Records on the company's activities in Kakamega after 1934 are lacking. See Fearn, *An African Economy*, 252.

55. Fearn, *An African Economy*, 251–52.

56. Warden of Mines, CMIRs May and June 1934, KNA: PC/NZA/3/14/372; *MGDAR 1934* (Nairobi: Government Printer, 1935), 3.

57. *MGDAR 1934*, 5; Huddleston, *Geology*, 43; "Gold Mining," 68; *Economist*, 16 November 1935.

58. *MGDAR 1934*, 6; Fearn, *An African Economy*, 253.

59. Fearn, *An African Economy*, 253; *Economist*, 16 November 1935; Warden of Mines, CMIR November 1934, KNA: PC/NZA/3/14/372; "Gold Mining," 68.

60. *MGDAR 1936* (Nairobi: Government Printer, 1937), 4; *MGDAR 1937* (Nairobi: Government Printer, 1938), 4–5.

61. Attorney for the Liquidator to DC NK, December 21, 1938, KNA: DC/KMG/2/18/10; *MGDAR 1938* (Nairobi: Government Printer, 1939), 4; Warden of Mines, CMIR December 1938, KNA: PC/NZA/3/14/372; NPIR November 1938, KNA: PC/NZA/4/5/2; NPAR 1938, KNA: PC/NZA/1/33; Inspector of Mines to DC NK, January 14, 1939, KNA: PC/NZA/3/14/372. Commenting on the prevalent rumors about the possible closure of Kimingini Gold Mines in May 1938, the Warden of Mines regretted, " 'This will leave a nasty taste in the mouth' of financiers who have invested money in Kenya mines." See Warden of Mines, CMIR May 1938, KNA: PC/NZA/3/14/372.

62. Huddleston, *Geology*, 40.

63. C. Stansfield Hitchen, Report on the Underground Geology of Kimingini Mine, Kakamega, July 25, 1938. Mines & Geological Department Library, Nairobi, Eldoret/20; See also Fearn, *An African Economy*, 253.

64. NPAR 1938, KNA: PC/NZA/1/33.

65. Warden of Mines to PC Nyanza, July 1, 1952, DC North Nyanza to P. M. Mahida, Midlands Provision Store, Kakamega, November 6, 1956, KNA DC/KMG/2/18/10; NPAR 1939, KNA: PC/NZA/1/34; Warden of Mines, CMIR March and June 1939, KNA: PC/NZA/3/14/372; KNA: PC/NZA/3/14/355.

66. Huddleston, *Geology*, 40.

67. Fearn, *An African Economy*, 248.

68. Ibid; *MGDAR 1934*, 3; Warden of Mines, CMIRs November 1933 and February, July, October, and November 1934, KNA: PC/NZA/3/14/372.

69. Warden of Mines, CMIR April 1933, KNA: PC/NZA/3/14/372. Fearn found no entry at all in the Mining Year Books for Risks Limited and the Koa-Milimu Gold Mining Company Limited, but they were locally registered companies in Kenya. Fearn, *An African Economy,* 250.

70. Fearn, *An African Economy,* 249–50; *MGDAR 1934*, 3; Menkin, "Gold Mining," 26.

71. *MGDAR 1936*, 4; NPAR 1936, KNA: PC/NZA/1/31; *MGDAR 1938*, 4; Huddleston, *Geology*, 40; Fearn, *An African Economy,* 249–50; A. W. Turton, Manager, K. G. M. Ltd., to DC NK, April 28, 1951, KNA: DC/KMG/2/18/11.

72. Fearn, *An African Economy*, 250; Inspector of Mines to DC NK, January 14, 1939, KNA: PC/NZA/3/14/372; Warden of Mines, CMIR December 1938, KNA: PC/NZA/3/14/372; NPAR 1938, KNA: PC/NZA/1/33; NPAR 1939, KNA: PC/NZA/1/34; NPAR 1940, KNA: PC/NZA/1/35.

73. Fearn, *An African Economy*, 254.

74. Huddleston, *Geology,* 25; *Economist,* 16 November 1935; *MGDAR 1936,* 5; NPAR 1936, KNA: PC/NZA/1/31; NPAR 1940, KNA: PC/NZA/1/35.
75. *MGDAR 1936,* 5; *MGDAR 1937,* 4; *MGDAR 1938,* 4; NKDAR 1938, KNA: DC/NN/1/20; Unpublished MGDAR 1944; Col.& Prot. of Kenya, *Land, Mines & Surveys Department Annual Report (LMSDAR) 1946* (Nairobi: Government Printer, 1948), 12.
76. NPAR 1940, KNA: PC/NZA/1/35; Huddleston, *Geology,* 25. According to Fearn, the company realized dividends as follows: 1940, 5 percent; 1942, 5 percent; 1943, 6 percent; 1944, 5 percent and 1945, 4 percent. No dividends were realized thereafter. Fearn, A*n African Economy,* 254.
77. *MGDAR 1952* (Nairobi: Government Printer, 1954), 2, 4; GB, C*olonial Reports, Kenya 1952* (London: Her Majesty's Stationery Office, 1953), 76.
78. GB, *Kenya Col. & Prot. AR 1935* (London: HMSO, 1936), 16; NKDAR 1935, KNA: DC/NN/1/16; NPAR 1935, KNA: PC/NZA/1/30; *MGDAR 1935,* 2–3; NPIR October 1935, KNA: PC/NZA/4/5/1; Fearn, *An African Economy,* 136; Menkin, "Gold Mining,"123. In fact, the main mining office was moved from Kakamega to Kisumu in 1935. NPIR November 1935, KNA: PC/NZA/4/5/1.
79. Warden of Mines, CMIR May 1934, KNA: KNA: PC/NZA/3/14/372; NPIR May 1934, KNA: PC/NZA/4/5/1. Only fifty prospectors remained in the area at the end of May 1934. See CMIR May 1934, KNA: PC/NZA/3/14/372.
80. GB, *Kenya Col. & Prot. AR, 1936,* 16–17; *MGDAR 1936,* 3.
81. NPAR 1936, KNA: PC/NZA/1/31.
82. NPAR 1937, KNA: PC/NZA/1/32; *MGDAR 1937,* 3–4; GB, *Kenya Col. & Prot. AR, 1937,* 20.
83. GB, *Kenya Col. & Prot. AR 1938* (London: HMSO, 1939), 22; *MGDAR 1938,* 4; NPAR 1938, KNA: PC/NZA/1/33.
84. *MGDAR 1938,* 4.
85. *MGDAR 1937,* 14; *Kenya Col. & Prot. AR 1938,* 22.
86. NKDAR 1937, KNA: DC/NN/1/19; NPIR October 1937, KNA: PC/NZA/1/5/2.
87. NKDAR 1938, KNA: DC/NN/1/20.
88. *MGDAR 1937,* 5; NKDAR 1938, KNA: DC/NN/1/20; NPAR 1938, KNA: PC/NZA/1/33.
89. NKDAR 1938, KNA: DC/NN/1/20; NPAR 1938, KNA: PC/NZA/1/33.
90. NPAR 1939, KNA: PC/NZA/1/34; NPAR 1940, KNA: PC/NZA/1/35.
91. Huddleston, *Geology,* 39; *LMSDAR 1946,* 11.
92. NKDAR 1938, KNA: DC/NN/1/20.
93. Warden of Mines, CMIR June 1939, KNA: PC/NZA/3/14/372; Fearn, *An African Economy,* 135.
94. NKDAR 1939, KNA: DC/NN/1/21.
95. DuBois, *Geological,* 26.
96. Assistant Warden of Mines, CMIR October 1939, KNA: PC/NZA/3/14/372. A tribunal set up to examine exemptions from the Kenya Defence Force visited Nyanza goldfields in 1939 and granted exemptions on a generous scale to mining employees. See *MGDAR 1939* (Nairobi: Government Printer, 1940), 2.

97. Assistant Warden of Mines, CMIRs December 1939; April, July, and September 1940, KNA: PC/NZA/3/14/372.
98. NPAR 1941, KNA: PC/NZA/1/36.
99. Warden of Mines, CMIR June 1940, KNA: PC/NZA/3/14/372.
100. Shilaro, "'A House Divided,' " 4.
101. *MGDAR 1939*, 2; Extract from Recommendations of The Chamber of Mines encl. in Commissioner of Mines to PC Nyanza, August 15, 1942, KNA: PC/NZA/2/2/68.
102. NKDAR 1942, KNA: DC/NN/1/24; NKDAR 1943, KNA: DC/NN/1/25; NPAR 1943, KNA: PC/NZA/1/38.
103. Unpublished MGDARs 1943 and 1944; NKDAR 1944, KNA: DC/NN/1/26; *LMSDAR 1945* (Nairobi: Government Printer, 1946), 16; GB, *Colonial Annual Reports (CARS) Kenya 1946* (London: HMSO, 1947), 47. At the start of the war, the Imperial Airways Service to England was curtailed and until the end of September when service resumed, no gold was shipped to England. Difficulty was also caused by the requisitioning of lorries (trucks) from the mines for military purposes. See *MGDAR 1939*, 2–3. The price of gold suddenly rose from Shs.148 to Shs.168 per fine ounce on 5 September and was fixed at that price in London on October 2, 1939. See Ibid. 3.
104. Huddleston, *Geological*, 1, 22.
105. GB, *CARS Kenya 1946*, 49–52; Fearn, *An African Economy*, 136–37.
106. NKDAR 1945, KNA: DC/NN/1/27; NPAR 1945, KNA: PC/NZA/1/40.
107. GB, *CARS Kenya 1946*, 47; NKDAR 1946, KNA: DC/NN/1/28. Edzawa Ridge closed down after exhausting its proved ore of reserves. See *LMSDAR 1946*, 11.
108. Shilaro, "'A House Divided,'" 7–16, 20–31.
109. *LMSDAR 1946*, 10–1.
110. NPAR 1936, KNA: PC/NZA/1/31. See also NPIRs February, May, June, July, September, and October 1934, KNA: PC/NZA/4/5/1.
111. NPAR 1939, KNA: PC/NZA/1/34; NPAR 1945, KNA: PC/NZA/1/40.
112. NKDAR 1947, KNA: DC/NN/1/29; North Nyanza District Annual Report (NNDAR), 1948, KNA: DC/NN/1/30. The name of the District was changed from NK to North Nyanza in 1948. This change was welcomed by the African population who regarded the term "Kavirondo" as a term of contempt. Ibid.
113. *CARS Kenya 1948* (London: HMSO, 1950), 43.
114. Ibid.
115. GB, *CARS Kenya 1949* (London: HMSO, 1950), 43.
116. Ibid. *1950* (London: HMSO, 1951), 35.
117. NNDAR 1949, KNA: DC/NN/1/31; NNDAR 1950, KNA: DC/NN/1/32; NNDAR 1951, KNA: DC/NN/1/33.
118. GB, *CARS, Kenya 1951* (London: Her Majesty's Stationery Office, 1952), 48.
119. NNDAR 1952, KNA: DC/NN/1/34; GB, *CARS, Kenya 1952* (London: Her Majesty's Stationery Office, 1953), 50; Andove, OI, February 28, 1998; Amagadu and Shikali, OI, March 3, 1998; Mutsotso, OI, March 4, 1998.

120. *MGDAR 1952* (Nairobi: Government Printer, 1954), 2, 4; GB, *CARS, Kenya 1952*, 50, 76. This Ordinance was a result of a LegCo resolution which was adopted on August 1950. It was approved by the S of S for the Colonies after the Kenya gold mining delegation's visit to London in 1951. See *MGDAR 1952*, 2, 4; NPAR 1952, KNA: DC/KSI/1/24.

121. NPAR 1952, KNA: DC/KSI/1/24.

122. Fearn, "Gold-Mining," 47. A fuller account of issues relating to African labor is found in Chapter 6.

123. Ibid. 47–48, 53.

124. Ibid. 51–52; NPAR 1939, KNA: PC/NZA/1/34.

125. John H. Clive, "A Cure for Insomnia, Reminiscences of Administration in Kenya, 1920–47," RHO, MSS. Afr. S. 678.

Chapter Four

The Politics of Land, 1931–52

INTRODUCTION

The discovery of gold in Western Kenya in 1931, the inauguration of the gold mining industry and the subsequent excision of land from the NK reserve to pave the way for gold mining, constituted a complex arena on which militant Luyia politics of land came to the forefront at individual and organized levels. The Luyia articulated their opposition to land alienation and the threat of rural industrial capitalist transformation in the 1930s and the 1940s in myriad ways. Three distinct phases are discernable: The first phase involved spontaneous individual opposition to gold mining. The second phase constituted organized political protest by the NKCA. The third phase involved opposition to soil conservation measures and colonial agrarian policies in general. This chapter examines why the Luyia opposed gold mining in their locations, the diverse forms that characterized the protest, and illustrates how the colonial state responded to Luyia anti-gold mining activities. The Luyia emerge as major historical players who resorted to diverse actions and reactions in making their grievances known to both the colonial state in Kenya and the imperial government in England. While such militant opposition proved ineffective in influencing colonial policies in Kenya, it nevertheless marked the birth of a Luyia ethnic consciousness, and a deeply rooted suspicion of government land policies that outlived British colonialism.

EARLY FORMS OF PROTEST

Spontaneous resistance to mining and prospecting in Kakamega was a common occurrence. This resulted from conflicts between Luyia landholders and

European prospectors who demonstrated a lack of respect for African land rights. European prospectors often descended on family holdings, pitched tents, and commenced prospecting and pegging in order to secure their interests without consulting local landholders.[1] Such actions alarmed Luyia households who naturally regarded the convergence of European prospectors into their reserve with both consternation and indignation. The difference between the British land laws and the African customary land tenure systems were a major source of conflict. Unlike the British who had title to land, Africans in Kenya colony had no title deeds to validate land ownership. The incongruence between British land laws, the colonial mining laws in Kenya, and the African customary land tenure systems, bred significant disagreement and distrust. Colonial mining laws reserved all mineral wealth to the Crown. That meant African claims of ownership were limited to the surface, at a maximum depth of three feet.[2] The Luyia landholders not schooled in the British laws did not understand such limits.

As early as December 1931, the colonial state was already aware of potential conflict between African landowners and European gold prospectors in NK reserve. These tenuous relations between miners and African landholders forced the acting CNC, Commissioner for Local Government, Lands and Settlement, and the acting Commissioner of Mines to meet with both groups separately. These meetings aimed at preempting any fears the Luyia were entertaining regarding the advent of the gold mining industry. In spite of this, the Luyia continued to treat the colonial state and the mining industry with great distrust.[3]

This distrust found expression in numerous obstructions, physical assaults, and cases where Luyia landholders confiscated mining paraphernalia from European prospectors. In January 1932, for example, a young woman beat a European prospector senseless for "knocking down her father" who was resisting the invasion of his land.[4] By June 1932, collisions between the Luyia and European miners had become widespread. Oral narratives are replete with heroic reminiscences of Ashiono Lumidi's assault on and expulsion of a European prospector, which earned him three months in prison. The prosecution and imprisonment of Luyia for obstruction bred extensive ill feeling toward all Europeans in Buluyia. The colonial state's enactment of stringent rules to coerce the Luyia to submit to the transforming industrial capitalist enterprise, only served to exacerbate the situation.[5]

The advent of mining and its attendant complications put pressure on colonial officials in NK district. This growing tension forced the colonial administration to appoint a Warden of Mines to oversee the mining areas. While the duties and responsibilities of this official were nebulous at first, his functions crystallized and came to encompass serving as the deputy to the DC when addressing Africans besides being deputy to the Commissioner of Mines in charge of technical issues pertaining to mining. In addition, the Warden of Mines arbitrated disputes between European miners. The presence of the Warden of Mines in the goldfield, coupled with a series of the DC's *barazas* (public meetings), offered an opportunity for ventilating grievances before they reached heightened proportions. Such intervention offered the possibility of early resolution of legitimate complaints.[6]

The colonial administration also looked to local Luyia functionaries in the conflict resolution process. Colonial chiefs including Milimu Amaidza of West Isukha, Sore of East Isukha, Paul Amiani of Tiriki, Mulupi Shitanda of Kabras, and Paul Agoi of Maragoli, were expected to aid the transition from the predominantly rural agrarian economy of Buluyia to a rural capitalist economy based on gold mining. These colonial functionaries, sometimes, found their assigned duties difficult to execute. Chief Milimu's West Isukha location was ferociously anti-mining and numerous struggles between miners and the Isukha occurred over time. Chief Milimu himself nursed anti-mining sentiment, but his role as a colonial functionary forced him to be discreet in his activities.[7] Colonial chiefs were instruments for facilitating European exploitation of the Luyia economy. As "eyes and ears" of the colonial state, their jobs were at stake in the event of any recalcitrant behavior by their subjects or themselves. It was normal for the DC to summon and have the so-called "hard headed" chiefs flogged before him.[8]

The European dominated colonial apparatuses of control curtailed overt Luyia protest against the gold mining industry. All police constables in the goldfield were Europeans, and explicit opposition to miners unleashed unimaginable police brutality.[9]

Apart from blunt police intervention, public meetings constituted the most important arena for assuaging Luyia opposition. The Warden of Mines and the DC orchestrated numerous *barazas* in Milimu's location and adjoining locations in the goldfield including North Maragoli, East Tiriki, Idakho, Kisa and Marama. At these meetings, the administration maintained that the European prospectors' interest was solely the extraction of gold and not taking over the land.[10] These pronouncements had little effect. Most Luyia continued to nurse a grudge against the colonial state and European miners. Indeed, when the KLC visited Kakamega in September of 1932, the issue of miners descending into NK reserve without the knowledge and consent of the local inhabitants formed the core of Luyia grievances.[11]

Luyia apprehensions resulted from the liberal interpretation European claim holders put upon a clause in the Mining Ordinance, granting them the right to reside on their claims. By this clause, European claim holders could legally erect substantial housing settlements, gardens, garages, workshops, nursing homes, stamp batteries and machine-houses. In a variety of ways, the precarious security of tenure of ancestral land became increasingly apparent to the Luyia.[12]

Trouble between the Europeans and the Isukha in particular also emanated from the changed status of Kakamega Forest. In 1931, a European forester arrived in Kakamega to take charge of the forest. Initially, opposition hardened against any form of government control whatsoever, forcing the acting CNC to consult the NK LNC before the Luyia accepted government proposals regarding the use of the forest.[13] In accordance with the recommendations of the KLC, the government earmarked 1,500 acres from the forest for Luyia households displaced by mining leases. The colonial state subsequently gazetted and excluded this block from mining operations. For political reasons, however, the Isukha rejected this offer. Official records show that government efforts designed to induce the Isukha to cultivate "the wonderfully rich forest soil" proved unsuccessful. The Isukha contended that cultivating their own land under any form of supervision was beneath their dignity. Although sixty cultivators ultimately accepted the land and planted crops in return for clear cultivation, they put little heart into the work with insignificant results. Although the Isukha complained of shortage of arable land, they blamed the government of robbing them of the forest, and refused to accept directions from the forester with regard to where to cultivate.[14] In a sense, the Isukha were cultivating a tradition of protest against state control of their agricultural activities.

Kakamega Forest comprised 61,700 acres, 13,000 of which comprised grassland and scrub. Between 1933 and 1934, three mining companies namely Messrs Risks Ltd., Tanganyika Concessions Ltd., Kenya Consolidated Ltd., and one trading company namely Mitchell Cotts received a total grant of 2,100 acres of concessions to cut timber for fuel. These timber concessions yielded Kshs.18, 398 and Kshs.32, 725 respectively to the colonial state in royalties.[15] The Isukha disliked this arrangement for some endeavored to sell their own timber to the mines for fuel and pit props.

The reorganization of holdings in Kakamega Forest and the timber concessions to mining companies exacerbated land matters in Buluyia. The Isukha in East Kakamega were particularly anxious about the ultimate destiny of the forest. Very quickly, a movement emerged in East Kakamega and North Maragoli advocating rejection of rents or compensation. Luka Mutsami, a former LNC member, Jonah Kahiya, an elder of the Appeal Tribunal, Johanna Lumwachi, a teacher at Irhanda FAM School, and Nathan Mbwabi, a teacher at Church of God, Bunyore and landowner in West Isukha, provided leadership for this movement. The first three residents of East Isukha dominated *barazas* in times of crisis to the extent that chiefs and elders remained silent in their presence, never venturing to criticize them. In North Maragoli, a location that was under strong FAM influence but with a small European mining population, some elders assumed a strong anti-mining attitude.[16] This state of affairs heightened with the expected visit of Sir Philip Cunliffe-Lister, S of S for the Colonies, early in 1934.

Although the visit of the S of S for the Colonies to Kakamega in January 1934 was postponed owing to illness, the governor, the Colonial Secretary (CS), and the acting Commissioner of Mines made an informal visit to Kakamega where they met a deputation led by chiefs Milimu and Sore. The Isukha demanded an assurance from the governor of the ultimate security of their lands to which the governor promised sympathetic consideration.[17] As a result of this calming reassurance, relations between European prospectors and Luyia land holders reportedly improved by the end of the year. Prosecutions for obstruction to mining on the one hand and for defrauding Africans on the other, diminished significantly. Notwithstanding, not all the Luyia in the gold mining area were reconciled to the European invasion of their locations. Politically minded Luyia continued to agitate against mining. The announcement of impending leases for mining companies only served to aggravate fear of losing land among the Luyia.[18]

Arguably, mining leases constituted the most contentious issue among the Luyia in the 1930s and 1940s. By August 1933, the issue was causing much anxiety, particularly in East Isukha where Messrs Risks Ltd. intended to apply for a lease. Members of the FAM in the location sent speakers throughout other locations in the mining area agitating against leases. To counter this, the Warden of Mines and the DC held joint and separate public meetings aimed at explaining the question of mining leases to the local population with little success. The Luyia disapproved the principle of leasing land to mining companies and objected to proposed water permits and labor camps.[19] In 1935, however, the colonial state set apart 113.5 acres of land in Isukha for the Kimingini mining lease in spite of opposition from the local inhabitants.[20]

Resistance to gold mining sometimes manifested itself in peg-pulling, fence removal and the destruction of beacons. Luyia landholders perceived the erection of wooden pegs, mining notices and fences by European miners as empirical evidence of European takeover of their land. Thus, the elimination of these symbols of European "ownership" of such land is only intelligible within this context. In a petition addressed to the S of S for the Colonies in 1934, the NKCA wrote, "We are afraid when we see the pegs set by the miners that our land will be taken."[21] The Isukha, Idakho, and Logoli in North Maragoli location, removed pegs and notices as soon as European prospectors erected them in spite of the heavy penalty of Shs. 6,000 or imprisonment for two years or both, for tampering with beacons.[22] To curb these unsettling acts of sabotage and to circumvent the difficulty of apprehending perpetrators, big mining firms resorted to stone beacons. Messrs Risks, for instance, erected concrete beacons with smaller concrete landmarks at short intervals on their property in West Isukha.[23]

The persistence of these acts of resistance prompted the provincial administration to intensify *barazas* in the locations concerned to warn against such activities. The administration pressured local headmen to check these developments. Notwithstanding, the Isukha objected to the erection of concrete beacons.[24] Besieged by heightening acts of defiance, the colonial administration intensified *barazas* in West and East Isukha, North Maragoli, Tiriki, Kisa, Marama and Butsotso at which government officials hammered home the need for cooperation from the local population. Despite this, East Isukha location continued pursuing an uncompromising anti-mining and anti-government attitude with the chief seldom intervening. Two core protagonists in this struggle were Jeanes teachers Lazaro Afwayi and Nathan Mbwabi. Afwayi had previously made offensive remarks at a governor's *baraza*, in addition to challenging the Warden of Mines on what attitude the colonial state would adopt if the Luyia declared war against miners.[25] In response to Afwayi and Mbwabi's anti-mining activities, the administration threatened to withdraw their salaries if they persisted with political activism.[26]

Nevertheless, the destruction of pegs and beacons persisted. In May 1935, for example, local peoples destroyed a survey beacon established during the 1932 triangulation of the goldfield.[27] In spite of DC E. L. B. Anderson's instructions to the chief of the location that the beacon "be left alone,"[28] local Nyore destroyed it again in July.[29] As a result, the district administration intensified *barazas* to educate Luyia households against interfering with beacons. Since the malicious destruction of pegs and beacons continued unabated, in 1936, the government refused to compensate Luyia landholders with missing or destroyed pegs and beacons.[30] Yet in 1939, the Isukha destroyed the Kibiriri trigonometrical station marking the boundaries of Rosterman Company's property.[31] Indeed, as late as July 1941, the mining community still complained of "wanton destruction of claim pegs." The administration advised miners to undertake regular inspection of their pegs.[32] In addition, a system of closer cooperation between claim holders, the police, and the "native" administration realized a somewhat marked improvement in the upkeep of pegs and beacons.[33] Where the administration failed to track the offenders, the DC continued to hold meetings with residents near the affected mines with a view to discouraging such actions.[34]

European-Luyia relations were, therefore, not always amicable. In July 1934, a European assaulted an African assessor accompanying two members of the LNC on a mission to examine a water permit. Although the mining authorities considered the assault "not serious," it proved "intensely irritating" to race relations in the goldfield.[35] With time, the spontaneous individual opposition eventually gave way to organized resistance.

ORGANIZED PROTEST: THE NORTH KAVIRONDO CENTRAL ASSOCIATION (NKCA)[36]

The NKCA came into existence in 1933,[37] shortly before a 1934-scheduled visit to Kakamega of S of S for the Colonies, Sir Philip Cunliffe-Lister. As an organized political protest movement, NKCA arose as a response to the fears that had emerged over Luyia land security following the discovery of gold in Kakamega in 1931. By February of 1934, the association had fifty members drawn from the more politically minded young men of Maragoli, Bunyore, Tiriki, Isukha and Idakho locations of Buluyia, mostly adherents of the FAM.[38] As individual resentment against gold mining subsided, it was natural that Luyia politicians desired organized protest as an efficacious vehicle for articulating the Luyia voice. Thus, the NKCA represented general Luyia resistance to the gold mining industry as opposed to sporadic individual discontent. With time, NKCA became the medium for articulating Luyia politics of land, among other grievances.

In its protest, NKCA adopted diverse methods including petitions, memoranda, and deputations in making Luyia grievances known to both the colonial state and the imperial government in England. In 1934, NKCA directed two petitions to the S of S for the Colonies complaining, among other things, about accidents involving people and livestock falling into open mining pits. Although the CO referred to the governor, the Warden of Mines cautioned that the association symbolized a formidable force to reckon with in the future. NKCA leaders were Andrea Jumba, a Logoli schoolteacher living in Tiriki as Chairman, Lumadede Kisala, Vice Chairman, and Erasto Ligalaba, Secretary. Other key members included James Shiberenge and Robert Jumba (Maragoli), Luka Mutsami, Moses Herbert Muhanga and Mmudi Lisudza, Nathan Mbwabi, and Jeremia Masidza (Isukha), John Adala (Nyore) and Andrea Wanakacha (Kabras).[39]

NKCA articulated a plethora of Luyia grievances within the colonial context. As a political protest movement, the NKCA represented a potent vehicle for achieving Luyia unity and a common Luyia voice against the colonial excesses. Its major concerns revolved around the question of a Luyia paramount chief, the much hated colonial *kipande* [identity card], the existence of Luyia squatters in the Uasin Gishu settled areas, and, most important for this study, the security of Luyia land rights.

Oral accounts made reference to NKCA as *Eshama sha malova*, literally "the Association of the [Luyia] soils" and interviewees commonly referred to its founding members as *watetezi* [advocates].[40] NKCA dreaded the looming prospects of evicting and resettling those displaced by the gold mining industry in Mount Elgon Forest Reserve in accordance with the recommendations of the KLC. In addition, the association aimed at building a common identity for the Luyia hitherto referred to as the "Kavirondo," in the colonial context. In Addition, NKCA demanded appropriate and just compensation for land alienated for gold mining purposes.[41] NKCA resented the European invasion of Luyia holdings without consulting owners, the general destruction of land resulting from mining activities such as trenching and drilling, and the high-handedness of European prospectors.[42] Later, the association's concerns came to encompass issues relating to water and air pollution, working conditions in the mines, mining accidents and compensation for victims of such accidents as above, and, most notably, the colonial soil conservation program.

In articulating these grievances, NKCA received assistance from FAM missionaries and later the *Guru* of writing petitions and memoranda and critic of British imperial policy, Archdeacon W. E. Owen of the Church Missionary Society (CMS), Maseno. More fundamentally, NKCA established a working relationship with the Kikuyu Central Association (KCA), a major organ of Kikuyu opposition to colonialism in Kenya.[43] Indeed, written and oral sources acknowledge a close link between NKCA and KCA. Although KCA articulated myriad Kikuyu grievances against the colonial state in Kenya, the central issue remained the alienated White Highlands. KCA was born in the womb of immense land insecurity brought about by the alienation of land for European settlers. The impending insecurity posed by the gold mining industry, therefore, provided a common cause for Luyia-Kikuyu cooperation. The initial bridge builder was Erasto Ligalaba of South Maragoli, former translator for the government Swahili newspaper *Habari*. As a former worker in Nairobi, Ligalaba collaborated with KCA members in highlighting the Luyia land question. In April 1934, for instance, the Warden of Mines reported:

> ... [T]he publication of the Bill to amend the Native Lands Trust Ordinance ... appears to have effected a working agreement between the North Kavirondo Central Association and the Kikuyu Central Association. Contact between the two associations is supposed to be maintained with the help of Erasto Likalaba [sic], a Maragoli former compositor of 'Habari', and the Indian 'Democrat' newspapers who was at the time an employee of Anglo-Continental Gold Mining Co.[44]

The close contact between the two associations involved resource pooling. KCA requested and received monetary contributions from NKCA for expenses of keeping in contact with London and paying fines. Jumba and Adala of NKCA visited Nairobi often to mail petitions to England, and paid different sums of money to KCA.[45]

Several sources validate the NKCA and KCA link. Eric Aseka attributes the strong NKCA and KCA link to Chief Joseph Mulama Shiundu of North Marama.[46] Bruce Berman shows that Jomo Kenyatta, Secretary of KCA, contributed articles to the Communist *Labour Monthly Journal* after 1933, the first of which dealt with the Kakamega gold rush. This was the first time that Kenyatta had written about the affairs of another people other that his own Kikuyu ethnic group. Moreover, in an article titled "Kenya," for the landmark anthology, *Negro*, published early in 1934, Kenyatta castigated the colonial government in Kenya for violating both the 1930 White Paper on Native Policy and the guarantees of the NLTO of 1930 in order to make room for European gold prospectors in Kakamega.[47] Moreover, in 1936, the provincial administration maintained that NKCA was modeled along the lines of KCA, and described it as "the most active and least useful of the Native Associations in the entire Province." The association claimed a membership of 300 so-called "vocal and semi-literate" members mainly drawn from the FAM and CMS, with a sprinkling of Muslim Wanga. By 1938, the figure had reached 800. Although the PC was convinced that NKCA was undoubtedly an offspring of KCA, he opined that its members did not have the ability and intelligence of the parent body.[48]

Both NKCA and KCA were most critical of the KLC Report of 1934 for approving racially designated lands; thereby extinguishing African land rights in the Kenya Highlands. In addition, the Commission approved the NLT (A)O of 1932 and recommended the extension of NK reserve to Mount Elgon to provide alternative land for Luyia households expropriated by the gold mining industry. In addition, the KLC dismissed Luyia claims to Kaimosi and Trans Nzoia settled areas, approved gold mining leases, as well as compensation rates and the channels for disbursing such compensation. In the case of NKCA, Moses Muhanga, a former clerk in the DC's office, emerged as the most energetic critic of the report as it related to mining. Muhanga won crowds over by publicly displaying the Volume III of the KLC Report, which dealt at length with the land question in Nyanza Province. In his sensational speeches, Muhanga accused the colonial state of planning to re-settle Luyia households from leased areas to Mount Elgon Forest Reserve. He also claimed that the government planned to make compensation to the NK LNC as opposed to individual dispossessed Luyia landholders, and blamed the Commission for fixing compensation rates independent of the real value of the property involved. These allegations considerably influenced the Luyia who voluntarily made monetary contributions toward NKCA.[49]

The promulgation and adoption of the Native Lands Trust Ordinance (NLTO) in 1938 energized the KCA-NKCA partnership. The NLTO of 1938 implemented the recommendations of the KLC. This move led to renewed opposition both in England and Kenya toward British land policies in the colony. Criticism and disaffection in England proved most opportune for Kenyatta's campaign. In January 1938, Kenyatta launched a series of lectures in England highlighting the Kenya land question. In a lecture to the Workers' Educational Association (WEA), Kenyatta condemned the loss of land to gold prospectors in Kakamega.[50] Months later, Kenyatta revisited this issue at another meeting in London where he exhaustively reviewed British land policy in Kenya colony. Kenyatta singled out the findings of the Ormsby-Gore and Hilton Young Commissions, the assurances of the 1930 White Paper and the NLTO of 1930, the shattering effects of the Kakamega gold discovery and the recommendations of the KLC on the security of African lands in Kenya.[51]

At the local level, the year 1938 was significant in the development of the politics of protest in Kenya. In accordance with the recommendations of the KLC and other impetus, the colonial state introduced compulsory soil conservation, soil reconditioning and destocking policies in African reserves. Closed out of the White Highlands and facing economic ruin, Kenya peasants rose in protest, notably in Taita and Ukambani. At the peak of this spontaneous rural protest, the Ukamba Members Association (UMA) led a strong 3,000 Akamba protest march, including women and children to Nairobi where they camped for over three weeks demanding an interview with Governor Brooke-Popham.[52] The rationale for checking environmental degradation in the colony seemed laudable. Nevertheless, the compulsory disposal of African livestock below market price to the European-run Liebig's packing plant resonated with the colonial state's policy designed to protect settler-produced meat from a competing parallel African meat industry.[53] The Akamba refused to take this vicious assault on their wealth while lying down.

In Buluyia, the NLTO of 1938 provided a new impetus to the NKCA-KCA link. Members of NKCA, an outspoken Asian representative to the Legislative Council (LegCo) with links to KCA Mr. Isher Dass, Joseph Kang'ethe, and Jesse Kariuki, President and Vice President of KCA respectively, jointly condemned the ordinance during a visit to ex-chief Mulama in October 1938. Mulama contributed two hundred and fifty shillings toward the KCA fund.[54]

The revitalized NKCA-KCA link produced little results. It was the North Kavirondo Taxpayers' Welfare Association (NKTWA), the Luyia branch of the original Kavirondo Taxpayers' Welfare Association (KTWA), which took up the matter with the district administration. At a meeting held at Butere on October 18, 1938 the DC's attempt to explain the NLTO of 1938 to the gathering failed dismally.[55] Ultimately, KCA and KTWA, not NKCA, submitted a joint memorandum directly to the S of S for the Colonies. The memorandum discounted the ordinance's pledge to secure African land rights in Kenya; it condemned the reservation of the Highlands to Europeans and criticized the passage and adoption of the NLTO of 1938 without consulting LNCs in the colony. The memorandum also took issue with the government's decision to limit African representatives on the Local Land Boards to four members, whose dissenting opinions were subject to the DC's veto power as chair of these boards. In addition, the memorandum criticized the government's decision to entrust compensation matters in the goldfield to the district and provincial administrations, and demanded that the Supreme Court hear compensation appeals. The petitioners also contested the arbitrary nature in which the colonial administration declared lands in African areas Forest Reserves, thereby closing these lands to African settlement or use. Although the petitioners expressed satisfaction with the new label of "African Lands" as opposed to "Crown Lands," they demanded that the colonial administration leave the management, development, and control of these lands exclusively to Africans rather than colonial officials whose decisions often conflicted with their administrative duties. A London based Native Lands Trust Board with educated African representatives such as Kenyatta was favored as opposed to one within the orbit of the Kenya government.

Lastly, the petition called for the registration of African lands as a protective measure and urged the CO to utilize money owed to African Carrier Corps for the expansion of both primary and secondary education, training Africans in land and animal husbandry, and human medicine. In essence, the Luyia rebuffed the KLC's recommendation to commit these funds on re-purchasing alienated lands to compensate dispossessed groups. As they tersely put it "Africans could not re-purchase their own lands with their own 'BLOOD' which flowed freely during the Great War (World War I)."[56]

Bode argues that the NKCA backed out because the land issue had faded, there was growing internal dissension among its leaders and general disillusionment with the benefits of their connection with KCA. They contended that Luyia monetary contributions to KCA had produced little tangible benefits.[57]

As expected, the protestations failed to influence both the colonial state and the imperial government from pushing through the legislation. The Native Lands Trust (Amendment) Ordinance (NLT(A)O) of 1938, encompassing the Kenya (Native Areas) Order-in-Council and the Kenya (Highlands) Order-in-Council securing Native Lands and the White Highlands respectively, was adopted and became operational on March 1,1939. This legislation effectively extinguished all African rights to land in the White Highlands.[58] The KCA and KTWA had jointly telegraphed the S of S for the Colonies expressing opposition to this legislation early in February. An excerpt of the telegram from the CO to the colonial governor in Kenya read:

> Kikuyu Central Association [KCA] and Kavirondo Taxpayers Welfare Association [KTWA] vehemently opposed to Order in Council Native Lands and reservation of Kenya Highlands for Europeans Stop Such policy partial and favoring the Whites must bring disastrous consequences both to Government and people alike Kenya Government notified.[59]

On March 29, 1938, such protest compelled the CO to consult the Kenya administration for clarification. The governor in turn sought the views of the Nyanza and Central Province PCs on the joint KCA-KTWA memorandum. When the governor responded to the CO in August, the Kenya administration agreed that the petition was certainly the brainchild of the Kikuyu of Kiambu District, and not the "Kavirondo native [Luyia]".[60] Close consultation between the governor's Deputy, Sir Walter Harragin, the Nyanza PC, S. H. Fazan, and the acting PC, Central Province, urged the CO to inform the petitioners that the NLT(A)O of 1938 was passed after a comprehensive investigation by a competent Commission which provided an opportunity for Africans throughout the colony to present their grievances. Under no circumstances would the colonial administration reopen discussion on the legislation.[61] Accordingly, on November 6, 1939, the S of S for the Colonies advised the secretaries of the KCA and KTWA, through their respective PCs of the finality of the NLT(A)O of 1938. In a curt statement, the Kenya administration emphasized the significance of Africans channeling their grievances through the "properly constituted colonial bureaucratic hierarchy."[62]

Despite the reversals, NKCA expanded rapidly in several locations in Buluyia where the association enjoyed veiled support from local chiefs and headmen. Except for Idakho location where Chief Ngorio's unfriendly disposition mired recruitment for the NKCA, the association broke the aloofness of Catholic adherents in West Isukha, and won the support of chief Milimu and his headmen. Although Milimu overtly assumed a pro-miners position, he surreptitiously made financial contributions to the NKCA. This response was common among other chiefs including Chief Sore of East Isukha, Chief William Odanga of North Maragoli, and Chief Mnubi of South Maragoli.[63]

THE COLONIAL STATE'S RESPONSE

While NKCA's concerns were legitimate to a certain extent, the colonial state was always dismissive. All levels of the colonial administrative hierarchy in Kenya colony often shunned the association. For instance, the DC, NK and the PC, of Nyanza Province considered the association's criticism of the KLC report as vicious. According to the PC, "NKCA's destructive criticism of the [KLC] Report was based on the members' inability to understand English or deliberate misrepresentation."[64]

Despite sustained negative publicity in official circles, the growing popularity and activities of the NKCA, particularly in mining locations, continued to be a concern for the administration. Chief Milimu's location of West Isukha, the matrix from which all mining activities radiated, and where several companies had already applied for mining leases, became the focus of the provincial administration. In 1934, the administration mounted numerous anti-NKCA spirited propaganda *barazas* in its strongholds of East Tiriki, East Isukha, and Maragoli to counter its activities.[65]

While NKCA's public meetings had previously proceeded with the chief's approval, the colonial administration put pressure on chiefs not to approve any meetings. Most notable, the colonial state devised plans to incapacitate the association's financial base. In an order to all chiefs in the district, the administration proscribed the collection of money for NKCA funds.[66] In November 1934, for instance, the administration convicted a headman in East Isukha for "robbery for taking by 'violence' subscriptions of twenty-cents from individuals who did not wish to pay." While the accused contested the charge, the administration resolutely believed the subscriptions were for NKCA.[67]

Sometimes the administration intervened directly. In September 1934, the governor and the acting PC, Nyanza, M. R. R. Vidal, visited Kakamega and held two meetings for chiefs and local people at the Government African School (GAS), Kakamega. In his address, the governor exhaustively discussed gold mining industry while condemning the NKCA's political agitation. These meetings, coupled with administrative intimidation, dismally failed to curtail opposition to mining. In fact, during the same month, a fracas occurred in East Tiriki when the assistant chief accompanied the Warden of Mines to inspect a site for the Tanganyika Concession Ltd's labor camp. The administration attributed the fracas to young men, most likely members of NKCA, and absolved the elders as most "agreeable." Thus, although the officials successfully inspected the labor campsite without the display of force, the local authority arrested the alleged ringleaders on the spot. Later, the administration concluded that the alleged "ringleaders" were "insignificant individuals," attributing the disturbance to the association's leadership.[68] Such a crackdown on the NKCA drove the movement into extreme secrecy.[69]

Concomitantly, the President of NKCA, Andrea Jumba of Tiriki, was gaining strong support in his location and in East Isukha where the colonial administration expected intense protest against the Kimingini lease application of 113.5 acres. NKCA, often doubling as a Trade Union, challenged the conviction of an African charged with assaulting a European miner, Mr. Hinds of the East African Concession Ltd. near Kaimosi. The administration was apprehensive of increasing African unrest in Kaimosi and Tiriki location with the anticipated expansion of the Tanganyika Concession's mining operations in these areas. Besides the Kimingini lease application, the colonial administration was expecting application for land in West Isukha for Risks and Rosterman Gold Mining Companies.[70] The colonial state had significant economic stakes in the successful granting of these mining leases.

As in 1934, the year 1935 witnessed close state scrutiny of NKCA's activities. In April 1935, the DC, NK reported that NKCA had engaged in misguided activity relating to compensation for death or disabilities caused by mining accidents.[71] The NKCA had taken on the mantle of a trade union; a role that served to heighten state suspicion of the association.

The colonial administration remained extremely uneasy with NKCA throughout 1935. Due to this hostile disposition, the colonial state put into place grand propaganda machinery aimed at discrediting the association. The government accused NKCA of relying on gross mistaken belief in petitions directed to the S of S for the Colonies. Apparently, the association had petitioned the S of S claiming the death of many women and children who fell in prospecting pits and the meager five shillings awarded as compensation to families of male victim of similar accidents. As the administration grappled with NKCA's activities, the fundamental concern remained taming the association. The colonial administration urgently needed to control the flow of damaging representations from the colony to England. Nevertheless, taming or educating the NKCA members in colonial compliance remained a daunting task. Indeed, in his 1935 Annual Report, the PC dejectedly noted his inability at guiding the NKCA into "a useful direction." The administrator accused the leaders of the association of making a living off the alleged deceptive activities.[72]

Notwithstanding, NKCA continued to attract many followers. When Mulama Shiundu openly joined the association on June 1, 1935 and the association unilaterally declared him Paramount Chief of Buluyia, the colonial administration suspended and subsequently sent Mulama into retirement.[73] In spite of this, NKCA successfully established branches in almost all locations of Buluyia by the end of the year.[74]

Before Chief Mulama was dismissed, the acting Governor and the CNC visited Kakamega to address the NKCA. The governor advised NKCA members to confine their activities to making constructive suggestions through proper channels instead of having the impertinence of meddling with the appointment of chiefs.[75] By January 1936, however, the colonial state had adopted imprisonment as a means of intimidating NKCA into subservience. For example, in January the district administration arrested six members of NKCA for collecting membership fees. Five of those namely Messrs Andrea Jumba, John Adala, Reuben Muhati, Lumadede Kisala and Mmudi Lisudza, were charged and convicted to serve two months in prison.[76] This prompted the NKCA to appeal the conviction in a petition directly addressed to the S of S for the Colonies.

In defense of state action, the governor maintained that the NKCA's fundraising program proceeded without the approval of the DC in line with the Secretariat Circular of 1931. Most importantly, the governor argued that these funds sought to fund the appointment of a paramount chief. The above circular intended to discourage "subversive agitation" of the NKCA type, and prohibited the collection of funds without permits.[77] Incidentally, the Luyia had quickly learned how to utilize the colonial state's apparatus of control. The convicts appealed to the Supreme Court, which subsequently reversed the convictions.[78]

Despite the intimidation, NKCA continued to be active. In June 1936, the NKCA's President, Andrea Jumba wrote what the administration considered a "rude letter" to the DC. As expected, the DC ignored this letter.[79] It is important to note that state indifference in no way signified the administration's denial of the danger NKCA posed. In the 1937 Annual Report, the Nyanza PC extensively quoted the DC's report on NKCA:

> For 20 months, and probably a vastly longer time . . . , every attempt has been made to induce this Association to work in line with the Government, but so far without avail. They have some intelligent members, Jeremiah Matsitsa [sic] and Eliakim stand out, but their President, Andrea Jumba a funny little man usually clad in a bottle-green dinner jacket and a sense of his own importance, and their Secretary, John Adala, formerly a master at the Government African School [Kakamega], but now to be found with a camera at barazas taking photographs of 'brutalities', are neither of them helpful, and the power behind the throne is Joseph Mulama, a man with a grievance. . . .[80]

Throughout the year, the provincial administration continued to portray the NKCA as a useless association with no legitimate grievances. The administration viewed NKCA as a radical anti-government organization, propped up and sustained by ex-chief Mulama. State censure did little to deter NKCA activism against colonial rule. In April 1938, NKCA dispatched another petition directly to the S of S for the Colonies in England. Once again, the colonial administration cautioned the association on the centrality of approaching authorities in England through the colonial governor in Kenya.[81] Following this rebuttal, a five-member NKCA deputation including LNC members John

Adala, Lazaro Afwayi, and Jeremiah Masitsa, visited the DC. The party's grievances included the land use question and compensation for those displaced by the establishment of Kakamega Township. As expected, the DC dismissed these claims.[82]

In 1939, NKCA renamed itself the Abaluhya Central Association (ACA). The district administration hailed this development as "its only constructive achievement." While devising the name Abaluhya for Bantu Kavirondo, NKCA, centered largely in the southern locations of Buluyia, maintained its critical stance against the colonial state. With time, the association's influence spread rapidly in the northern and western locations, as FAM schools increased in these areas. The DC, A. N. Bailward, voiced his unsuccessful attempt at converting the association to the service of the colonial state when he wrote:

> Like both my predecessors, I have tried to show this Association a means of serving a useful purpose, and like them I regret to record that it continues to confine itself to spreading untruthful and alarmist rumours about land policy and to the illegal collections of money. Its leaders are not able men but on the whole it spreads a considerable suspicion of Government throughout the Reserve. . . . The President's name is Lumadede and his principal assistants are Herbert Moses Muhanga and John Adala.[83]

Indeed, NKCA played a significant role in the anti-soil conservation campaign in Buluyia. To most colonial officials, NKCA comprised a bunch of "inconsequential ignorant trouble makers." When the colonial government banned KCA in 1940, the NKCA experienced a political jolt. Further, the war situation made NKCA's prospects of continued operations minimal and forced the association to go into voluntary dissolution. This brought NKCA's protest against the colonial state to a temporary halt. At government urging, NKCA shelved controversial issues in order to create a united Luyia front in the war effort.[84] This political lull paved the way for a period of a certain degree of collaboration between NKCA and the provincial administration.[85] Although Aseka maintains that NKCA subsequently became a body of harmless conformists that threw parties and praised the colonial administration, evidence exists to show its desire to reopen by January 1942.[86] F. D. Hislop, DC, NK, noted that NKCA was the only political association in the district and that most of its leaders were cooperating with the government during the war. Nonetheless, the administration found location branches most difficult to control. Although Hislop claimed neutrality in matters relating to the association, he quickly pointed out that it comprised "a mixture of bad and good" owing to "misapprehension of facts and policies." In his view, the association's cardinal error was its support of Mulama as paramount chief.[87] By 1946, the

NKCA, hitherto confined to the southern locations of Buluyia, with a particularly strong presence in Maragoli, was spreading its tentacles to the northern locations of Bukusu.[88] The common denominator remained anti-soil soil conservation measures.

THE ANTI-SOIL CONSERVATION CAMPAIGN

The greatest site of Luyia contest against gold mining in Kakamega and the attendant fear of loss of land the episode created, was perhaps the anti-soil erosion campaign. By the middle of the 1930s if not earlier, a shortage of land, especially in Bunyore, North and South Maragoli, declining soil fertility and dilemmas associated with soil erosion were clearly apparent. Against this background, the colonial state, with the support of the metropolitan government, prioritized food security and soil conservation.[89] These measures created a complex arena of conflict between state-imposed solutions and local Luyia interests.

Although sporadic soil conservation demonstrations in Buluyia began in 1934, the state made soil conservation its decided policy in 1935. In January 1935, AO M. D. Graham spent six weeks surveying soil erosion problems in South Maragoli where he initiated demonstrations on contour and rock terracing and planting terraces with sweet potatoes. Although the Logoli reportedly received the demonstrations with "keenness," few households and headmen turned up for the demonstrations. Shortly after, it became extremely difficult to get the Logoli to maintain conservation work without supervision.[90] For example, when AO T. Y. Watson visited the site in March, he found several Logoli at work who admitted to have turned up for work because they were aware of his visit.[91] This attested to the fact that Logoli households had not willingly embraced soil conservation measures. South Maragoli location in particular displayed outright hostility to, and a lack of interest in soil conservation efforts.[92]

Initial soil conservation work focused on the worst eroded regions of the district namely, part of North Tiriki and Maragoli, and smaller patches of the southwest locations. Usually AOs urged the people to understand the importance of maintaining soil fertility by preventing and combating soil erosion. However, all was not smooth sailing. It was against this background that the AO held a *baraza* in South Maragoli in January 1936 to discuss soil erosion issues. The Logoli, however, refused to adopt soil control measures until the LNC had commenced work on the badly eroded Red Ridge on the border of Maragoli and Tiriki. The Red Ridge was the most severely eroded part of the district, and the more antagonistic Logoli were in effect challenging the colonial state and its apparatus of control to restore its fertility. Only then would they listen to propaganda directed at their individual holdings. The AO considered the Logoli a "very stubborn crowd" over which their chief, Mnubi had no influence.[93]

The Logoli's demand that the LNC initiate soil control measures can be linked to a 1935 Resolution in which Council unanimously agreed that the DC:

> (M)ay issue orders through a Headman that all natives under his jurisdiction (or natives inhabiting such area in his jurisdiction as may be indicated) shall take such measures to conserve their soil from erosion either individually or communally as shall be recommended by himself or by an Agricultural Officer.[94]

In proposing the resolution, assistant chief Paul Agoi of South Maragoli stressed the increased incidence of hillside erosion in the district. Consequently, the LNC empowered local authorities to issue orders and to ensure compliance with the advice of agricultural officers. The target was to reclaim some 2,000 acres of once fertile land in Maragoli, which had become barren. Officials envisaged a great soil conservation campaign radiating from the hillsides of Maragoli and Tiriki to the rest of the district. Accordingly, the LNC proposed to contribute £303.47 for initial work.[95] The colonial state and the LNC took up the challenge and commenced conservation work in the Maragoli-Tiriki area on March 11, 1936. The project employed a paid labor force of 100 "boys," two *Nyaparas* (African supervisors), six temporary inspectors, and one instructor. The notorious Red Ridge was terraced and planted with trees and Napier grass, so that a year later, it was reportedly "green."[96] However, soil erosion persisted in the densely peopled areas of Maragoli and Bunyore due to intensive cultivation.[97] This led to intensified soil conservation measures throughout Nyanza Province.

To push agrarian change forward, the Native Authority Ordinance (NAO) of 1937 empowered chiefs and headmen to extract minor communal services from Africans under their jurisdiction. Under the NAO, all able-bodied adult males were required to undertake six days of unpaid labor per quarter on all soil conservation work. This measure proved rather unpopular among the Luyia as shown in the numerous prosecutions in both subordinate courts and Native Tribunals of Luyia who failed to meet this requirement.[98]

Throughout 1938, the colonial state continued to prioritize the prevention of soil erosion. Toward this end, the administration often coerced individual *shamba* (farm) owners to undertake soil conservation work. Alternatively, organized groups performed communal soil reclamation work on a selected day of the week in specific areas. Agricultural officials favored laying crop refuse in contour rows, digging contour ditches, planting live wash stops, placing stones along contours, and stopping boundary furrows, as the best anti-soil erosion measures.[99] Local peasants nonetheless avidly resented soil conservation measures, which entailed increased labor demands on peasant households. The persistent suspicion that land reconditioning could herald European settlement in Buluyia solidified the anti-soil erosion campaign.[100]

In addition, soil improvement concerns were partially a product of population pressure especially in the more densely populated southern locations of Maragoli and Bunyore. In 1938, population densities for Maragoli and Bunyore locations stood at 600–700 and 1,100 persons per square mile respectively. High population had led to land fragmentation, diminished pasturelands, and a general reduced length or complete elimination of fallow bush, which naturally facilitated soil fertility under pre-conquest customary farming practices.[101] Although a small migration of Logoli to Kisii highlands particularly in Nyaribari, Bassi and Kitutu locations, occurred in the decade, as a rule, the Logoli were reluctant to emigrate unless forced by acute land hunger.[102] For example, in 1941, due to population pressure and land dispute with their Gusii hosts, Logoli immigrants in Gusiiland moved to Kanyamkago in SK. As the case with Gusiiland, the arrival of large numbers of Logoli families in Kanyamkago alarmed local Luo inhabitants. By 1945, and certainly by 1950, a combination of local opposition and state intervention made Logoli immigration to SK untenable. Even then, most Logoli immigrants retained their ancestral lands in Buluyia, which did little to alleviate the land situation.

Faced with this dilemma, colonial officials fell back on land improvement measures. The three Nyanza Province LNCs drafted a standard soil erosion resolution for their respective districts. In the case of NK, the resultant NK LNC Resolution No. 5 of 1938 laid out conditions governing land use and the prevention of soil erosion.[103]

By this resolution, Luyia peasants had to carry out and maintain soil conservation measures including terracing, banking, strip-cropping, contour ploughing, stone walling, lining land with plant residues, planting and maintaining grass or plant strips, live-wash stops and other protective works. The resolution *empowered headmen to acquire able-bodied men to take such measures for dealing with soil erosion as may be necessary*. In other words, soil conservation work became mandatory communal service for all able-bodied Luyia males.[104] From then on, Luyia households encountered intense coercion in fulfilling such communal labor obligations.

Although experimental communal work in Buluyia commenced in 1937, by 1940, it had made little headway.[105] This was a result of intense suspicion among the Luyia of government policies regarding land. When the Geological and Mines Department commenced geological surveys in December 1939 in parts of Maragoli closed to gold mining since 1933, the survey educed opposition among the Logoli. In a letter to the PC, Nyanza, in early 1940, the government geologist, William Pulfrey averred, "I have today experienced a slight amount of interference with my work. A small party of youths . . . west of Mbale, persisted in telling me to 'get out', as this is not Kakamega! . . . They threatened to call up a party to fight us . . . The youths were armed with *rungus* [sticks] and *pangas* [machetes]. . . .[106]

Clearly, Logoli opposition to survey work was borne of the underlying fear of loss of land brought about by the gold rush. The Logoli believed strongly that the survey would lead to land alienation as seen in an excerpt from a demi official letter by D. K. Williams, a long-standing miner in Nyanza: "I gather the natives in Maragoli are being led to believe that the Government is preparing to take away their land. Evidence: Pulfrey is using Survey instruments there."[107]

In the minds of Luyia peasants, Kakamega represented the immediate classic point of reference for colonial land alienation policies. European mining interests dispossessed many Luyia. The use of land survey instruments in Buluyia often aroused the enduring fear and suspicion that the colonial state was laying the groundwork for taking their land.

At times, the Luyia anti-soil conservation stance manifested itself in deliberate destruction of soil conservation work, notably in Maragoli and Bunyore. In these locations, about 70 percent of conservation work carried out in 1938 and 1939 in individual *shambas* was destroyed as soon as European staff was withdrawn at the outbreak of World War II. The AO NK attributed this to the "foolishness [of the Logoli and Nyore] in listening to the Central Association [NKCA] advice."[108]

By March 1940, opposition to soil conservation measures had spread to Bukusu in northern Buluyia where the NKCA enjoyed a particularly strong following. The establishment of an anti-soil erosion pilot scheme and the attempt to begin terracing in Bukusu, consolidated opposition among local NKCA members, who believed the move signified the demarcation of land for European occupation. State officials' attempt to change such perceptions proved futile.[109] Indeed, some chiefs supposed local heightened popular resentment a basis for their tacit collaboration with the NKCA.[110]

As opposition to soil conservation measures persisted, the cultivation of maize increased on an unprecedented scale. In 1940, maize was reportedly the most "paying and effortless" cash crop. These developments spurred increased ploughing and planting of large areas in Malakisi, Bukusu, and Kabras locations, forcing the colonial administration to take additional steps to conserve the soil. The AO proposed new rules requiring the conservation of the existing large holdings as a precondition for cultivating more land.[111]

Subsequently, the NK DC, the AO and the Soil Engineer, held a *baraza* (public meeting) in Kimilili, Bukusu to inform Bukusu peasants about the proposal. Inopportunely, Bukusuland was rife with the rumor alleging that the colonial administration sought to limit maize production to three acres per household. Thus, although the meeting recorded phenomenal attendance, a majority of Bukusu peasant farmers remained obstinate, rejecting drastic and arbitrary agrarian measures. The DC utilized this forum to wage a derisive attack on Chief Mnubi of South Maragoli for purportedly allowing an alleged prominent agitator and member of the NKCA to disseminate propaganda at his official *baraza*. The DC argued that the said agitator spread rumors to the effect that the colonial government intended to limit all forms of cultivation, and planned to move a portion of the population to the Northern Frontier Province, to make room for European settlement. The DC emphasized that such rumors led to sporadic anti-soil conservation measures in Buluyia. The DC's poignant speech notwithstanding, one month after the meeting, Bukusu peasants in Malakisi location deracinated established coleus hedge.[112]

The NKCA persisted in its anti-soil conservation protest. At a NKCA meeting held at Mbale, Maragoli, attended by the DC, mission leaders, and local chiefs on May 15, 1940, a member of the association censured the proposed policy aimed at limiting land under cultivation and plans to commence culling the Luyia livestock in accordance with the NLTO of 1938. Predictably, the administration dismissed both charges as "undigested ideas" designed to project government agrarian policies unfavorably.[113] Faced with acute land shortage and government "inaction," Logoli peasants continued cultivating their meager holdings, destroying existing protective work, while paying no attention to their proposed permanence.[114] By the end of 1940, however, deliberate destruction of anti-soil erosion measures were declining. Agricultural officials in NK attributed the reduction in wanton destruction of soil conservation work to the absence of political activity from NKCA.[115]

In 1941, the colonial administration pushed the NK LNC to adopt a resolution imposing a 10–cent cess per bag of maize exported from the district. Proceeds from the maize cess would fund land-reconditioning expenses including purchasing graders, team oxen and payment of wages. Although the resolution did not receive immediate sanction, it nevertheless symbolized the willingness of Luyia leaders to cooperate with the colonial state in funding soil conservation programs.[116]

In most cases, progress in soil conservation work in the densely populated southern locations of Bunyore and Maragoli stemmed from the struggle for survival. Conversely, the presence of a permanent European AO in the northern Bukusu location of Kimilili remained the driving force. In the less populated or not so fertile parts of Buluyia, people remained largely apathetic.[117]

AOs often reported "progress" in order to justify their jobs and demonstrate a degree of achievement in their duties. This explains the porosity between "progress" and "apathy", the two main themes running through archival records.

Sometimes, Luyia opposition to soil conservation measures was a product of inadequate information from agricultural officials. In 1943, the NKCA submitted a memorandum to DC Hislop complaining that, "The Agricultural Officer is cutting a line in the *shambas* in Maragoli and Kitosh [Bukusu]. . . ." They demanded its suspension "because we want to be shown the true reason of this method as people were opposed to the line. . . ."[118] On the PC's instruction, DC Hislop held a public meeting in Mbale and Vihiga locations of Maragoli on August 20, 1943. Meeting attendees included Chief Agoi of Maragoli, and Andrea Jumba and Lumadede Kisala, President and Vice President of NKCA respectively. While the Mbale meeting focused on the AO's prohibition of cultivation below trenches, the Vihiga meeting criticized the policy requiring Logoli peasants to establish strips between two trenches estimated at 25 feet apart. Even if the strips aimed at providing pasture for tethered livestock, for Logoli peasants already eking out a precarious livelihood on tiny holdings, such excisions clearly endangered Logoli survival. Indeed, DC Hislop abandoned grass strips due to the unstable Logoli subsistence economy.[119]

In the meantime, Chief Agoi attributed Logoli opposition to grass strips to the NKCA. Chief Agoi asserted that the introduction of trenches in Maragoli occurred after consultation with the Logoli peasants who appeared to accept them until the association commenced its criticism. The NKCA President, Andrea Jumba, acknowledged seeing the trenches for the first time and expressed his approval. However, Jumba's Vice President, Lumadede Kisala, remained silent. After investigation, Chief Agoi and ex-chief Mnubi concluded that the story was Lumadede's pure *fitina* [intrigue, mischief], rather than a creation of agricultural instructors.[120]

By the end of 1943, therefore, soil conservation efforts had dismal success among the Luyia. As the AO lamented in his annual report:

> People were unwilling to do maintenance work. For instance although slopes have been contour pegged and instructions given as to the work to be done, the people would do absolutely nothing unless an instructor was standing over them until the work was finished. Moreover response to propaganda and teaching on the subject of manuring and better cultural methods had even been more disappointing than the work on soil conservation. In a number of cases, native farmers were not prepared to put in the necessary work, and were 'apparently quite content to sit back and do nothing.'[121]

Despite the lethargy noted above, in 1944, soil conservation efforts picked up in Maragoli, Tiriki, and Bunyore locations. This included the construction of narrow base contour bunds and planting grass at outlets, old roads, and boundaries.[122] Numerous households in the district applied for assistance in terracing land. The long list of applicants made the provincial administration conclude that soil conservation work had attained a measure of popularity in Buluyia.[123] Soon, it became clear to the officials that the increased demand for the services of the terracing units represented peasants who were unwilling to undertake soil conservation work on their own. In order to avoid prosecution for not carrying out the instructions, peasants eagerly signed their names on the "waiting list."[124] In a nutshell, Luyia households were simply encasing themselves against state reprisals and compulsion.

As Bukusu farmers evaded compulsory soil conservation work by guile, AOs shifted attention to southern Buluyia where they introduced continuous contour trenches. This initiative proved a serious setback to routine work on earth bunds and trenches as members of NKCA intensified propaganda condemning the work as superfluous. NKCA urged peasants to demand payment or otherwise cease performing unpaid coerced conservation work. Such activism against unpaid soil conservation work culminated in the destruction of existing anti-soil erosion measures, thereby making it extremely difficult to get the Logoli to carry out conservation work.[125]

Logoli dislike for soil conservation work also manifested itself in careless cultivation. This was particularly evident in *shambas* from Mbale north to the Kaimosi road, through Mago and Mudete. Arguably, the tenacity of Logoli opposition to soil conservation measures reflected their resentment of government interventionist policies in their agricultural activities. Luyia peasants considered the Department of Agriculture an unmitigated nuisance and desired a return to the old days of "non-intervention."[126] As one old man who had destroyed anti- soil erosion work on his holding three times, aptly put it, AOs were at best "a necessary evil with whom they had to put up with. . . ." This remained the dominant attitude in households between Majengo trading center on the main Kisumu-Kakamega road and the Red Ridge. As the area experienced extensive sheet and gully erosion, farm owners refused to do anything to bring it under control. The Logoli remained largely apathetic or distinctly hostile to soil conservation work. As the AO lamented, "(I)t is heartbreaking to see the work ruined by the crass stupidity of the people for whose benefit the work is being done."[127]

Progress was slow in North Maragoli. The NKCA members visited soil erosion sites urging people to discontinue the work and not to turn out for unpaid soil conservation work.[128] In Chavakali area, allegedly a stronghold of the NKCA, soil conservation work lagged behind other areas and ongoing soil conservation work remained either incomplete or had been destroyed.[129]

Members of NKCA persisted in taking the colonial administrators to task over issues relating to soil conservation. In 1944, the association questioned why AOs pulled up standing crops in order to place in soil conservation trenches.[130] At a meeting with the PC Nyanza, on August 21, 1944, the NKCA members criticized agricultural instructors for their inadequate appreciation of soil conservation work, their limited understanding of Luyia land tenure systems, and the use of coleus weed in soil conservation work.[131] While Maragoli, Bunyore, and Tiriki registered some progress in soil conservation work, throughout 1944, Luyia peasants in other parts of the district continued to destroy established narrow base terraces through what colonial agricultural officials termed "careless cultivation."[132]

By 1945, the SAO, Nyanza reported very little progress on permanent lines.[133] In the case of NK, the AO blamed Lumadede Kisala who alleged at a meeting at Hanguru in Maragoli, that agricultural instructors routinely destroyed standing crops when constructing bench terraces. Investigations by agricultural staff, village headmen, and a representative of NCA "uncovered no grounds for Kisala's allegation." The agricultural officials concluded that the NCA was up to its old tricks of sabotaging soil conservation work.[134]

Besides the southern locations, the northern districts of Buluyia witnessed an appreciable amount of resentment against government enforced soil conservation measures. In Bukusuland, peasants destroyed conservation work by moving pegs to conform to their own ideas on ploughing. Peasants in South Bukusu in particular, remained suspicious of government driven agrarian transformation initiatives. At times, peasants exhibited outright hostility to instructions from agricultural officials. This state of affairs grew desperate due to lack of cooperation from local chief and *Olugongo* (headmen).[135] By 1945, Bukusu resentment to government intervention in land use matters had reached a crescendo. Bukusu peasants refused to provide grain for the ox-teams, *posho* (corn meal) for feeding agricultural staff during their field visits, herding the oxen used by the LNC terracing unit, and rejected to perform mandatory unpaid soil conservation work. Under immense pressure, the NK LNC capitulated and withdrew its terracing ox-units from Bukusu.[136]

Fear of losing land to European settlement constituted a central factor in the Luyia anti-soil erosion stance. This deeply rooted suspicion explains why Luyia peasant households gave lip service to soil conservation programs. Usually, peasants destroyed soil conservation work once an agricultural staff left a locality. Such actions sprung from rumors that the colonial state sought to settle Europeans on any restored land in NK. The AO, M. G. Morgan, acknowledged the extreme difficulty of combating such ideas among a people he considered "primitive."[137] The District Agricultural Officer (DAO), Anderson, summarized the situation aptly:

> Soil conservation is undoubtedly one of the most difficult jobs facing the Agricultural staff at present, as hardly any of the leading natives of the district appreciate its importance and are consequently of no assistance as propagandists, although they pay lip service to the theory of soil conservation.[138]

The fear of losing reconditioned land to European settlement had gained tremendous ground by 1945. This fear became deeply entrenched as a result of the longstanding warped colonial land policy in Kenya. The establishment of gold mining and the attendant loss of land in Buluyia solidified the long history of the insecurity of African lands in the colony. The past nurtured these fears. At a NK LNC meeting in November 1945, the PC, Nyanza, reported to have received a protest letter accusing him of planning to introduce Assistant Soil Conservation Officers (ASCO) as a prelude to placing new European settlers on land in Bukusu. The PC dismissed the rumor, emphasizing that the administration had no plans to alter the NK reserve boundary for European settlement. The PC stressed the centrality of agricultural officials in saving and improving the soil for the benefit of the African population.[139]

In 1946, the administration adopted a new plan aimed at winning the support and increasing the power of indigenous elders (*amaguru*) as partners in soil conservation work. DC. L. E. Whitehouse wanted the *amaguru* to assist agricultural officials in soil preservation in their respective areas.[140] This initiative had little impact due to the significantly reduced authority of *amaguru*.

The colonial state had introduced the salaried post of *milango* (singular: *mlango*) headmen in 1926. The system empowered location chiefs to appoint headmen subject to the DC's confirmation. Naturally, the Luyia generally came to assume that *milango* headmen had replaced clan and village elders. Within this context, the Luyia believed that traditional clan authority over land formerly vested in clan elders now rested with the newly institutionalized *milango* headmen. This supposition gained general popularity when *olugongo* sub-headmen replaced *milango* headmen with the adoption of the Native Authority Ordinance in 1937.[141] Clearly, local authorities never guaranteed the success of state driven agrarian change.

As it became increasingly difficult to get Luyia peasants to carry out obligatory soil conservation work, the colonial administration resorted to extra stringent and arbitrary measures. The administration attempted to compel individual farmers to sign a form of Bond, guaranteeing full compliance with instructions regarding the maintenance of terraces on their holdings. Predictably, the colonial state failed to carry Bukusu households with them. This enactment fueled intense political protest in north Buluyia, where peasants believed that signing the Bond was tantamount to handing over their land to the colonial government. This conviction persisted in spite of the numerous explanatory *barazas* and informal talks between AOs and individual farmers. The rumors that soil-reconditioning work was a precursor to European settlement endured throughout 1946.[142] The colonial administration could not quash this rumor.

Early in 1946, DC Hislop unsuccessfully attempted to dampen this rumor. In his farewell speech to the NK LNC in February 1946, Hislop castigated opposition to land improvement and development measures. He cautioned that opposition to soil conservation spelled economic stagnation or retrogression in NK. The DC explained that the rumor about settling Europeans on reconditioned land in the district originated from newspaper reports on the envisaged post-WW II Soldier Settlement Scheme. The scheme planned to settle 500 European WW II veterans in Kenya. Although the newspaper report was accurate, Hislop reassured council members that the White Highlands had the capacity to absorb these veterans.

On the question of soil conservation, Hislop emphasized the expertise of agricultural officers. Hislop compared the devastating effects of soil erosion in Baringo/Tugen areas of the Rift Valley and Machakos (Ukamba) reserves in Kenya to similar calamities in South Africa and North America. The DC urged Luyia leaders to push forward soil conservation measures in order to avert similar adversity. If Luyia households continued opposing soil improvement efforts, based on rumors that soil reconditioning meant preparing land for European settlers, the district's agricultural development would stall. The DC reiterated the security of the Luyia reserve, adding that the advancement of the Luyia community rested squarely on the health of their land.[143] Retrospectively, DC Hislop affirmed the centrality of agriculture in the Luyia economy that surpassed the transformation occasioned by the gold mining industry.

In spite of such eloquent pronouncements, the rumor persisted, often predisposing people to obstruct conservation work. For example, at a PC's *baraza* held to discuss the role of indigenous elders in soil conservation measures, the Vice President of NKCA raised the issue once more. Although the PC categorically denied the rumor, the question of terrace maintenance by individual peasant households, remained a major problem.[144]

Ominously for the colonial state, the establishment of WW II Soldier Settlement Schemes in Trans Nzoia district north of Bukusu in 1946 appeared to "confirm" the rumor. The Soldier Settlement Scheme ignited significant protest among the Bukusu in the religio-political movement, *Dini Ya Msambwa* (DYM) (The Religion of our Ancestors). Besides opposing colonial agrarian policies, DYM sought to expel Europeans from Kenya.[145] DYM also took on the mantle of a "trade union" when the founder, Elijah Masinde, enforced tea breaks for African employees of Asian merchants in Broderick Falls (Webuye). The confrontation between Masinde and the department of agriculture reached a plateau when DYM adherents burned down the residence of the high-handed AAO in Kimilili in 1944.[146] A more overt political response was, however, the transformation of the Bukusu Union into a branch of the Kenya African Union (KAU).[147]

Arguably, this volatile political situation drove the administration into action. In June 1946, the administration arranged a visit of the NK LNC representatives, including "some of the most anti-soil conservation group in the District," to the badly denuded Kiambu and Machakos districts of the colony. The objective was to impress upon these elders the imperative of communal soil conservation work. The administration authorized the *amaguru* (village headmen) (sing. *Liguru*) representing all Luyia locations to publicize their tour experiences in their respective locations on their return. Before the party's departure, the administration invited a SAO, Norman Humphrey to remind the LNC members of the state's land conservation policy and the central role of local elders in the success of the program. Although Humphrey endorsed increased authority for elders in land conservation matters, LNC members disagreed strongly. For example, Pascal Nabwana, a prominent councilor and farmer in Bukusu, objected to the inclusion of illiterate elders who lacked both the skills and knowledge relating to land management. Nabwana also pointed to the diverse customary land use patterns and practices among the various Luyia ethnicities. In councilor Nabwana's world, Buluyia needed less sanctions and more education. Moreover, communal land use patterns, which constituted the benchmark of the pre-conquest Luyia social formations, were rapidly giving way to individualism. Consequently, Nabwana urged the colonial administration to cautiously marry the past and present in confronting land conservation issues.[148]

The visit to Central Province did not prove quite useful to the district administration. Very quickly, the administratively sanctioned public meetings deteriorated into "political agitation," forcing the administration to withdraw its support.[149] When agricultural officials tried to enforce a policy requiring Luyia peasants to purchase manure from distant locations, the Luyia maintained that if the use of manure on individual holdings was a government sanction, then it was the government's responsibility to provide transportation and labor for carting and spreading the manure in the fields.[150] Luyia peasants were not willing to meet the costs of obtaining, transporting and spreading manure in their holdings, using unpaid coerced labor. At the core of the matter, nevertheless, remained the deep-seated sense of grievance brought on by the alienation of land for gold mining.

When the S of S for the Colonies, A. Creech Jones visited Kenya in 1946, various groups including the Luyia presented numerous petitions dealing with land grievances. In its memorandum to the S of S Jones, the Abaluyia Welfare Association (AWA) demanded the return of all lands occupied by European settlers to African reserves to accommodate the growing African population. In reality, the AWA challenged the S of S Jones to abrogate the findings and recommendations of the KLC, which had endorsed a racially segregated land policy in Kenya. Though the S of S promised to consider the petitions, he vigorously affirmed the irrevocability of the KLC. In fact, Jones placed the African dilemma in Kenya at their doorsteps. The S of S maintained that Africans realized neither the importance of proper land use nor the importance of cooperation with the colonial state in its extensive program relating to land rehabilitation and resettlement, which Jones considered "the only practical solution to the problem."[151] By April 1948, the CO had yet to respond to these petitions.

While soil conservation measures proved difficult to achieve, the Luyia continued to intensify their agricultural activities. In 1946, major exports from the district included cotton, maize (corn), *mtama* (sorghum), *wimbi* (millet), simsim (sesame), rice, and cassava (manioc), all valued at £200,000, with maize accounting for £160,000 of this figure. As the production of maize increased, Luyia peasants showed little interest in soil conservation. This trend resulted from the burgeoning significance of cereal crops, particularly maize in the colonial economy. This, coupled with guaranteed high prices, and the growing need for cash to meet the peoples' ever-increasing needs, led to serious over-cropping and the attendant soil deterioration. Although NK district had not attained the "Machakos" stage, discernable signs of soil exhaustion existed.[152]

The anti-soil conservation stance among the Luyia in general, members of NKCA and the Logoli specifically, revolved around other factors. First, Luyia peasants disliked the enforcement of terracing level slopes. Second, Luyia landowners resented the lack of consultation between colonial agricultural officials and individual landholders on sections of their plots to be terraced. Third, the Luyia hated the tendency for agricultural staff to confine terraces on roadside plots. In many cases, these roadside "show cases" ultimately became the focus of government official stopovers during their itineraries. This meant lack of service to holdings located far from main roadways. Fourth, Luyia peasants condemned the dearth of adequately trained agricultural staff in African reserves. Fifth, the Luyia despised the dismal government expenditure on agricultural improvement in reserves. Sixth, the Luyia loathed the absence of a coordinated program for disseminating better farming methods, and disliked the administration's mandate that required agricultural instructors to work in tandem with "untrained, conservative old men."[153] While all these constituted major rallying points for Luyia opposition to colonial agrarian policies, they also caused division in the Luyia body politic.

To the emerging Luyia petty bourgeois, particularly the Logoli, the issue of population growth often elicited some support for soil conservation measures. Ex-serviceman David Mulindi Mbwanga at a general meeting of the Maragoli Society in May 1947, best articulated this. Mbwanga underscored the centrality of land in Luyia livelihood and the primacy of adopting soil erosion control measures. Mbwanga rightly noted that while Logoli population was progressing geometrically, soil erosion remained largely unchecked. However, when the chairman of the Society posed the question, "What hinders the Maragoli people from taking care of the soil?", the responses were quite revealing: these included ignorance, laziness, the prevailing rumor of losing conditioned land to European settlement (as suggested by one Lubanga, a member of the NKCA), and disagreement between administrators and *amaguru*.

Solomon Adagala, a prominent farmer and trader and member of NKCA posited that rumors such as Mr. Lubanga's emerged from the established colonial policy in which the state annexed land in Kikuyu, Maasai, Ukambani, and Bukusu for European settlers. The alienation of land in Western Kenya for the gold mining industry aggravated the situation. Thus, colonial agrarian initiatives in African reserves only served to intensify land insecurity.

A second view maintained that the rumor regarding the possible annexation of Maragoli resulted from inaccurate reading of a report compiled and published by a soil expert from South Africa some years back. Notwithstanding, Kisala averred to the guaranteed security of Logoli land ownership evident in two letters in his possession written by two Kenyan governors at different periods in response to NKCA's protest regarding land.[154]

After considerable discussion, the society passed several resolutions. In acknowledging the importance of soil conservation, the society called for cooperation between Logoli peasants with local elders, chiefs and government officials, in soil conservation efforts. It recommended the creation of a Soil Conservation Committee comprising a majority of clan elders and a minority of government officials entrusted with the primary responsibility of overseeing soil conservation measures in Maragoli. The committee was to enjoy full powers to take action against those who failed to comply. To facilitate soil conservation work, the society appealed to the colonial state for qualified agricultural instructors in Maragoli.[155]

This change of heart raises questions about who these people were, whom they represented and what their motives were. Scholars have categorized

Africans as "resistors and conservatives," or "collaborators, progressives, innovators and improvers."[156] While the former symbolized a rejection of Western civilization, the latter represented the agents of "modernization." Using these categories, the Adagalas, the Mulindis and the Kisalas belonged to the emerging petty-bourgeois class. This crop of Luyia progressives was characterized by straddling. Apart from agricultural production, they earned off-farm incomes as civil servants and traders. Their views represented a classic clash between improvers desirous of power, and the "conservatives." This clash of interests explains the failure of the Maragoli Society.

Despite the mixed record, the administration continued to emphasize the importance of Luyia households in maintaining the health of the soil. Officials enjoined elders to ensure that all villages under their jurisdiction observe soil conservation regulations. Similarly, agricultural instructors dispensed essential technical advice to farmers, including measuring contours. Notwithstanding, AO Anderson lamented the extensive neglect of large numbers of existing contour lines in the district. Moreover, due to the inexorable destruction of constructed banks, several LNC members recommended that *olugongo* and *amaguru* accompany agricultural instructors on contour measuring itineraries as witnesses. Yet, LNC member John Adala contended that the anti-soil conservation stance in Buluyia emanated from the failure of *olugongo* to regard soil conservation duties an integral part of their responsibilities. This often led to conflict between overzealous *amaguru* and the less ardent *olugongo*.[157]

In any case, DC C. H. Williams categorically and correctly asserted that the restoration of denuded parts of the district rested squarely with the people and required their full cooperation. In spite of the inherent opposition to communal soil conservation work, progressive peasants eager to improve their agricultural practices, fervently sought the services of agricultural staff. The enthusiasm from the so-called "improving farmers" prompted the administration to institute a nine months course at Bukura agricultural center to train line levelers. Chief Jonathan Baraza and councilor Pascal Nabwana of Bukusu wanted the government to prioritize the training of individual "progressives" to serve as model farmers in their respective locations. The two also questioned the practice of agricultural instructors routinely destroying standing crop while measuring contour lines on land under cultivation.[158] AO Anderson fervidly denied this charge arguing that farmers whose holding were measured early enough did nothing about the contours until after planting. Anderson called for the adoption of an LNC resolution to prosecute such farmers.[159] The NK LNC never endorsed the idea.

By May 1947, some NKCA members still objected to European officers taking an active interest in land preservation initiatives. Yet, faced with a rapidly growing population in Buluyia, the colonial state stressed the significance of protecting every inch of land in the district to guarantee sufficient food production. This did little to reduce terrace destruction. In May 1947, the colonial state prosecuted three Logoli cultivators for cultivating on a steep slope against the AO's orders. The three, Gibendi, Indulagi, and Chaminyolo, had lost land to Edzawa Mining Company leases.[160]

In the meantime, the administration was also concerned about the effects of the dominant maize culture in Buluyia. In 1948, the colonial government focused on the diversification of agriculture throughout NN district as an alternative to the wartime exigencies that narrowed farming almost entirely to the cultivation of cereal crops, particularly maize. Consequently, the government encouraged the cultivation of root crops, with new varieties of sweet potatoes and cassava as valuable famine reserve crops. To boost food production, state apparatuses zealously pursued the use of manure and contour ploughing on the large farms in the northern part of the district. Apart from encouraging the use of manure and the diversification of crops, agricultural officials also endorsed the policy of promoting group or cooperative farming in Buluyia. These agricultural innovations received funding from the NK LNC Agricultural Betterment Fund (ABF). Ultimately, in 1948, the first three ABF funded units came into existence. Sadly, the bulk of the ABF applicants comprised the emerging petty-bourgeois class of government servants, teachers, and LNC members.[161]

The brainchild of the SAO, P.C. Chambers, group farming, was adopted in 1947. It entailed the voluntary merging of individual holdings aimed at increasing productivity and raising soil fertility by checking fragmentation and soil erosion.[162] Even though the state applied no coercive measures to push group farming forward, assistance from the ABF or other funds to peasants, depended entirely on the adoption of better agricultural practices on group farms.[163]

Luyia reception to group farming was lukewarm. Three weeks after Chambers' rationalization of the new farming method to the LNC, the NKCA and the *amaguru* sought audience with the DC alleging, "Both the LNC, and local communities still did not understand the need for group farming."[164] Therefore, while group farming reportedly faired well in the maize growing locations of Kimilili and South Bukusu, it dismally failed in the southern locations of Bunyore and Maragoli.[165] At the peak in 1948, the district had twenty-seven group farms with 420 members cultivating about 8,700 acres.[166] By 1949 when the LNC withdrew ABF funding, only two group farms existed. The spectacle of group farming failed in Buluyia because the reportedly "mean-souled" peasants rejected strange modes of cultivation clearly divorced from their households and kinship ties.[167] Notably, group farming failed because it stymied the diverse entrepreneurial activities of Luyia straddlers. As this initiative went under, the agricultural officials rebounded on soil conservation.

In 1948, the district administration noted that while turnout for unpaid communal soil conservation work in Bunyore and Maragoli remained problematic, cooperation and goodwill had replaced the peoples' previous antagonistic posture. In the first five months of 1948, Maragoli produced some 27 miles of new terraces on approximately 250 acres. In some villages, 66 percent of adult males turned out for one day's work each month.[168] By the end of the year, 66,237 men days work had created 365,186 yards of new terrace in Maragoli, Tiriki, Nyang'ori, Bunyore, Idakho, Isukha, and Kisa locations. Seemingly impressive, these figures represented a most depressing reality. For example, during the peak month of November, only one-third of males available completed a day's work on soil conservation. AAO G. D. M. Baily believed that in the absence of direct coercion in enforcing LNC resolutions regarding land use in Buluyia, the completion of terracing NN would have lasted between two to three decades.[169]

Nonetheless, the question of overpopulation in Maragoli and Bunyore coupled with the impracticality of devising a suitable system of farming to support it, continued to haunt the colonial state. While emphasis on cereal production, particularly maize, continued unabated, and contour ploughing nearly became the rule throughout the north; in the southern locations of Buluyia, the AO reported that "practically 100% of the land is under the plow and virtually *no land at all* is being rested."[170] As the southern locations grappled with rapid population growth, the colonial administration steered clear of a state driven population resettlement policy. First, the administration dreaded the predictable hostility toward a state enforced population resettlement. Secondly, a government driven resettlement did not guarantee reduced birth rate among those who did not emigrate.[171] Accordingly, soil control measures remained the most favored approach. Although the long-term policy hoped to relieve the heavy pressure on land through either population relocation or perhaps a rural industrialization program, no such initiatives came to fruition.[172] By the early 1950s, both possibilities remained a growing but illusive urgency.[173]

While Luyia cultivators demonstrated a general awareness of the need for soil conservation by the late 1940s, the urgency of combating the problem and getting each peasant farmer involved remained chimerical. In 1949, the government charged and fined sixteen people in Khwisero, five in Nambale, one in Mumias, one in Kimilili, and six in Lurambi, a total of Shs.1, 614 for contravening the Native Authority Ordinance. Besides the cash penalties, the convicts performed unpaid communal service for periods ranging between one week and two months.[174] Due to such coercive measures, the district recorded over 1,000 miles of terraces, doubling the level of the previous year. Even then, development in Buluyia in no way matched the progress in parts of Central Province where peasants constructed between 3,000 and 5,000 miles of terraces annually. It is important to note that the success in soil conservation measures was an outcome of excessive coercion where colonial chiefs forced Kikuyu women to work in communal terracing, leading to the women's revolt.[175]

Luyia peasant households fiercely resisted such compulsion. Therefore, by 1950, it was evident that colonial conservationism had hardly taken root in Buluyia. P. S. Osborne, DC NN, attributed this failure to a lack of organization and leadership in communal groups. Since clans and clan elders constituted the obvious and only existing leadership in Buluyia, Osborne resolved to make another effort at utilizing them before inventing a new form of organization.[176]

Osborne's attempt to revive indigenous land authorities in NN as apparatus for enforcing communal soil control work, and as arbiters in land disputes, proved quite ineffective. The LNC criticized the DC's novelty of training *amaguru* in leveling and other soil conservation measures. Council members contended that soil conservation work was the responsibility of trained and paid agricultural officials. Consequently, LNC members condemned the double-dilemma the unpaid elderly *amaguru* experienced when enforcing the unpopular soil conservation programs.[177]

In addition, the avid fear of losing reconditioned land to the colonial state and European settlers, continued to preoccupy Luyia minds. Luyia response to the 1950 agricultural census planned for the entire province as part of the World Agriculture Census aptly captures this fact. The envisaged census provided fodder to the rumor mill in a fresh wave of propaganda. Rumors circulated alleging that the colonial state wanted to enumerate cultivated fields and plantations with the ultimate aim of alienating empty or unoccupied land for European settlement. The rumor dominated everyday debate especially in the four locations selected for the sample census namely Malakisi, Marachi, Marama and Tiriki.[178] Once more, the provincial administration, led by PC K. L. Hunter, found itself on the defensive.

The Marama Locational Advisory Council (LAC) refused to support the census, prompting the administration to issue direct orders for the census to proceed despite the strong hostility. Five of the farmers chosen for the "Sample Counts" either obdurately refused to provide the necessary information, or refused to permit AAOs access to their holdings. The government subsequently prosecuted and fined them heavily.[179] Such acts of defiance only served to intensify brute coercion in enforcing unpaid compulsory soil conservation work in Buluyia.

For example, in 1951, agricultural officials adopted grass filter strips for all land under a slope of 8 percent. This practically meant the greater part of the district. This measure aroused instantaneous opposition in the more congested southern locations where peasants routinely encroached on wash stops. To undercut this anti-conservationism in Bunyore and Maragoli, ABF targeted agricultural improvement and for the first time, Luyia women officially joined the soil conservation unpaid workforce. To push the program forward, the administration employed two assistant agricultural propagandists in the locations. The propaganda machine captured women at the newly instituted women's *barazas* at which agricultural staff dispensed instruction on innovative agricultural practices.[180]

As late as 1951, therefore, chiefs continued to harangue people to participate in communal soil conservation efforts as a matter of government policy.[181] In 1952, as the mines closed in NN, the district reaped 462,913 bags of maize and 5,857,749 lbs of AR cotton and 385,589 lbs of BR cotton.[182] This illustrates the continued significance of land and peasant production in the Luyia political economy.

CONCLUSION

The inauguration of the gold mining industry in NK reserve and the threat that the rural industrialization program posed to Luyia landowners, produced diverse responses among local peasants. The Luyia acted and reacted first as individuals, and later the resistance crystallized into militant organized protest in the NKCA. While initial resistance entailed obstructionist activities and physical assaults against prospectors, organized protest resorted to deputations, memoranda, petitions and other official ways of highlighting Luyia grievances against gold mining and the colonial land policies. With time, NKCA fastened its protest around soil conservation measures that led to widespread failure of colonial agrarian initiatives. While the question of population pressure loomed large in Luyia opposition to soil conservation measures, the fear of losing recondition land to the colonial state and European settlers, largely shaped and continued to steer Luyia resistance to these measures. The significance of land to Luyia households, the failure of the colonial state to implement long-term policies regarding the population problem, and the short-lived benefits that the gold mining industry provided for the vast majority of Luyia households, remained central factors in Luyia perception of colonial land policies. Apart from the political effects, the colonial project which was meant to promote a capitalist industrial base in the hitherto, predominantly rural agrarian economy in Buluyia, led to significant economic and social impact on the Luyia.

NOTES

1. NKDAR 1933, KNA: DC/NN/1/14; Shivachi, Oral Interview (OI), February 15, 1998; Seleta, OI, February 15, 1998; Claudio Mbanda, OI, February 28, 1998; Akibaya and Otiende, OI, August 20, 1998; Shibelenje Musine, OI, September 21, 1998; James Bukhala, OI, September 24, 1998.

2. Musine, OI, September 21, 1998; Misaki, OI, July 18, 1998; Amahwa, OI, August 29, 1998.

3. NKDAR 1931, KNA: DC/NN/1/12; NPAR 1931, KNA: PC/NZA/1/26; NKDAR 1932, KNA: DC/NN/1/13.

4. DC NK to acting PC Nyanza, January 15, 1932, KNA: PC/NZA/3/14/374; Also See Francis Bode, "Anti-Colonial Politics within a Tribe: The Case of the Abaluyia of Western Kenya," in *Politics and Leadership in Africa*. eds. Aloo Ojuka and William Ochieng' (Nairobi: East African Literature Bureau (EALB), 1972), 109. She was imprisoned for three months. McGregor Ross to Bottomley, February 22, 1933, PRO: CO 533/429/10.

5. KDAR 1932, KNA: DC/NN/1/13; NKDAR 1933, KNA: DC/NN/1/14; Seleta, OI, February 15, 1998; Mbanda, OI, February 28, 1998; Mavia, OI, March 2, 1998; Shitsukane, OI, March 19, 1998; Amahwa, OI, August 29, 1998.

6. NKDAR 1933, KNA: DC/NN/1/14.

7. Mavia, OI, March 2, 1998; Musine, OI, September 21, 1998; Bukhala, OI, September 24, 1998. For a similar dilemma among Kikuyu chiefs, see Marshall S. Clough, *Fighting Two Sides: Kenyan Chiefs and Politicians, 1918–1940* (Niwot: University Press of Colorado, 1990), especially chapters 5 & 6.

8. Mutsotso Khakali, OI, March 4, 1998; Amagadu and Shikali, OI, March 3, 1998.

9. Khakali, OI, March 4, 1998.

10. Warden of Mines, Confidential Monthly Intelligence Reports (CMIR) July and August 1933, KNA: PC/NZA/3/14/372.

11. NKDAR 1932, KNA: DC/NN/1/13; Haruni Litavakha, OI, March 9, 1998.

12. NKDAR 1932, KNA: DC/NN/1/13; Khakali, OI, March 4, 1998.

13. NKDAR 1931, KNA: DC/NN/1/12; NPAR 1931, KNA: PC/NZA/1/26.

14. NKDAR 1934, KNA: DC/NN/1/15; NPIR July 1934, KNA: PC/NZA/4/5/1.

15. NKDAR 1934, KNA: DC/NN/1/15; NKDAR 1935, KNA: DC/NN/1/16.

16. Warden of Mines, CMIR September 1933, KNA: PC/NZA/3/14/372.

17. NKDAR 1934, KNA: DC/NN/1/15; Warden of Mines, CMIR January 1934, KNA: PC/NZA/3/14/372; NPIR January 1934, KNA: PC/NZA/4/5/1.

18. NKDAR 1933, KNA: DC/NN/1/14; Col. & Prot. of Kenya, *MGDAR 1933*, 6–7; NKDAR 1934, KNA: DC/NN/1/15.

19. Warden of Mines, CMIR August 1933, KNA: PC/NZA/3/14/372; Otiende and Akibaya, OI, 20 August 1998.

20. *MGDAR 1935* (Nairobi: Government Printer, 1936), 21.

21. The North Kavirondo Central Association to Sir Philip Cunliffe-Lister, encl. in North Kavirondo Central Association to Charles Roden Buxton, October 27, 1934, Buxton Papers, RHO, MSS. Brit. Emp. S. 405/6/5.

22. North Kavirondo Local Native Council (NKLNC), Minutes of a Meeting held at Matungu on March 18, 1932, KNA (Kakamega): HW/13/1; Extract from Memos by McGregor Ross dated June 14, 1933, in Buxton Papers, RHO, MSS. Afr. Emp. S. 405/6/5.

23. Warden of Mines, CMIRs February and March 1934, KNA: PC/NZA/3/14/372; NPIR March 1934, KNA: PC/NZA/4/5/1.

24. Warden of Mines, CMIR April 1934, KNA: PC/NZA/3/14/372; NPIR April 1934, KNA: PC/NZA/4/5/1.

25. Warden of Mines, CMIR May 1934, KNA: PC/NZA/3/14/372.

26. Warden of Mines, CMIR June 1934, KNA: PC/NZA/3/14/372; NPIR June 1934, KNA: PC/NZA/4/5/1.

27. Surveyor General to Commissioner for Local Government, Lands and Settlement, May 11, 1935, KNA: PC/NZA/2/2/50.

28. DC NK to Senior Inspector of Mines, May 27, 1935, KNA: PC/NZA/2/2/50. Initially called "New Maragoli," the beacon was renamed "Bunyore" on the PC's instructions on May13, 1935 because it was located in Bunyore location.

29. Commissioner for Local Government, Lands and Settlement to Acting PC Nyanza, July 22, 1935, KNA: PC/NZA/2/2/50; NKLNC, Minutes of a Meeting Held at Matungu on August 15 & 16, 1935, KNA (Kakamega): HW/13/2.

30. Acting Commissioner of Mines to PC Nyanza, October 4, 1935, KNA: PC/NZA/2/2/50; NPAR 1936, KNA: PC/NZA/1/31.

31. Assistant Warden of Mines to Commissioner of Mines, December 28, 1939, KNA: DC/KMG/2/18/4.

32. NPIR July 1941, KNA: PC/NZA/4/5/3; NKDAR 1941, KNA: DC/NN/1/23; NKDAR 1942, KNA: DC/NN/1/24.

33. Unpublished MGDAR 1941.

34. Unpublished MGDAR, 1943.

35. NPIR July 1934, KNA: PC/NZA/4/5/1.

36. Sources variously refer to NKCA as the Abaluhya Central Association, Nyanza Central Association or the Central Association. Except in direct quotations, this study will consistently use NKCA.

37. The North Kavirondo Central Association to Charles Roden Buxton, September 27, 1934, Buxton Papers, RHO, MSS. Afr. Emp. S. 405/6/5.

38. NKDAR 1934, KNA: DC/NN/1/15; NKDAR 1935, KNA: DC/NN/1/16; NPIR February 1934, KNA: PC/NZA/4/5/1; NPAR 1935, KNA: PC/NZA/1/30; Wagner, *The Bantu,* Vol. 1., 34–35; Carl G. Rosberg and John Nottingham, *The Myth,* 161–62; Bode, "Anti-Colonial,"108; See also John M. Lonsdale, "Political Associations in Western Kenya," in *Protest and Power in Black Africa,* ed. Robert I. Rotberg and Ali A. Mazrui (New York: Oxford University Press (OUP, 1970), 623–25; B. A.

Ogot. "Mau Mau & Nationhood," in *Mau Mau & Nationhood: Arms, Authority & Narrative*, ed. E. S. Atieno Odhiambo & John Lonsdale (Athens: Ohio University Press, 2003), 3, 73.

39. Warden of Mines, CMIRs February and March 1934, KNA: PC/NZA/3/14/372; NKDAR 1934, KNA: DC/NN/1/15; NKDAR 1935, KNA: DC/NN/1/16; Bode, "Anti-Colonial," 112; Shikali, OI, March 3, 1998; Francis Mwinamo, OI, March 26, 1998; Amahwa, OI, August 29,1998.

40. Clement Muhati, OI, February 13, 1998; Ayiekha, OI, February 15, 1998; Seleta, OI, February 15, 1998; Musalimwa, OI, February 17, 1998; Isenjia, OI, February 24, 1998; Vihembo, OI, March 26, 1998; Enock Mwinamo, OI, March 26,1998; Amahwa, OI, August 29, 1998; Francis Mutsotso, OI, March 4, 1998; Akibaya and Otiende, OI, August 20, 1998. The term agitators in colonial lexicology signified upstarts or trouble makers. In the Luyia context, they were heroes who ardently fought their battles against colonial oppression.

41. Akibaya and Otiende, OI, August 20, 1998; Litavakha, OI, March 9, 1998. The people rejected the NKTWA's suggestion to adopt the name "Abakwe" (the people of the east) or "Abalimi" (the farmers). It was not until 1942, that the NK LNC officially adopted the name "Abaluyia" or "Abaluhyia". This marked the birth of Luyia cultural nationalism and the imagined greater Luyia community. See B. A. Ogot, "Mau Mau & Nationhood," 13.

42. Muhati, OI, February 13, 1998; Francis Mwinamo, OI, March 26, 1998. See also Bode, "Anti-Colonial," 109.

43. Bruce Berman, *Control & Crisis in Colonial Kenya: The Dialectic of Domination* (Athens: Ohio University Press, 1990), 243; Lonsdale, "Political Associations," 626.

44. Warden of Mines, CMIR August 1934, KNA: PC/NZA/3/14/372; NPIR 1934, KNA: PC/NZA/4/5/1; see also Rosberg and Nottingham, *The Myth*, 162; Lonsdale, "Political Associations," 625.

45. Warden of Mines, CMIR August 1934, KNA: PC/NZA/3/14/372; Bode, "Anti-Colonial," 115–16; Otiende, OI, August 20, 1998; Musine, OI, September 21, 1998.

46. Eric Aseka, "Political Economy of Buluyia: 1900–1964," (Ph. D. dissertation, Kenyatta University, 1989), 361.

47. Bruce Berman, "Ethnology as Politics, Politics as Ethnology: Kenyatta, Malinowski, and the Making of Facing Mount Kenya," *Canadian Journal of African Studies (CJAS) /RCEA* 30:3 (1996): 323–24.

48. NKDAR 1936, KNA: DC/NN/1/18; NPAR 1936, KNA: PC/NZA/1/31; North Kavirondo Central Association Petitions 1938, PRO: CO 533/496/6; Brooke-Popham to S of S, Despatch, May 17, 1938, Ibid. Some CO officials believed that NKCA was Owen's creation. See W. C. Bottomley, Minute, June 1, 1938. Ibid; Bode, "Anti-Colonial," 112.

49. Warden of Mines, CMIR August 1934, KNA: PC/NZA/3/14/372; NPIR February 1934, KNA: PC/ NZA/4/5/1.

50. *East Bourne Gazette,* 1 January 1938, PRO: CO 533/501/11.

51. *Salisbury and Winchester Journal*, (16 December 1938), PRO: CO533/501/11. CO officials were divided on how to counter Kenyatta's activities. Cowney White

openly considered Kenyatta an unsuitable person to give lectures in government supported courses and warned the WEA against allowing him access to such platforms. J. J. Paskin favored a more discreet approach in the attempt "to 'stymie' Kenyatta's attempts to spread his poison" so as to avoid awkward questions in Parliament if it became known that the CO had taken steps to restrict Kenyatta's opportunities for expressing his views. Paskin later noted the difficulty of combating Kenyatta's oral "fulminations" at obscure meetings, organized by a perfectly reputable organization in quarters where counter blasts in important organs of the press were not likely to have much effect. Most CO officials did not, however, favor the possibility of counter propaganda. See Cowney White to J. J. Paskin, Minute, November 14, 1938; Paskin to L. B. Freeston, Minute, December 17, 1938; Paskin to Major Dale, Minute, February 14, 1939; Freeston, Minute, February 15, 1939 in Ibid. The S of S for the Colonies, Malcolm MacDonald, consulted the governor seeking information as to whether Kenyatta had ever been to Kenya since 1931, the legitimacy of his claim to be secretary of KCA, his source of funding, whether his activities in London were sanctioned by the Kikuyu LNC and the authenticity of KCA as an organ of Kikuyu opinion. This information was seen to be useful for countering Kenyatta's propaganda" in House of Commons debate. CO to Governor, Despatch, February 24, 1939. Ibid.

52. Rosberg and Nottingham, *The Myth*, 163–78; Berman, *Control & Crisis*, 243; Lonsdale, "Political Associations," 636; J. M. Lonsdale, "The Depression and the Second World War in the Transformation of Kenya," in *Africa and the Second World War*, ed. David Killingray and Richard Rathbone (New York: St. Martin's Press, 1986), 110; J. F. Munro, *Colonial Rule and the Kamba: Social Change in the Kenya Highlands 1889–1939* (Oxford: OUP, 1975), 214–20; Robert L. Tignor, T*he Colonial Transformation of Kenya: The Kamba, Kikuyu and Maasai From 1900 to 1939* (Princeton: Princeton University Press, 1976), 344; Robert M. Maxon, *East Africa: An Introductory History* (Morgantown: West Virginia University Press, 1986), 203–04.

53. George Oduor Ndege, "History of Pastoralism in Kenya 1895–1980," in *An Economic History of Keny,* ed. W. R. Ochieng' and R. M. Maxon (Nairobi: EAEP), 100.

54. NKDAR 1938, KNA: DC/NN/1/20; NPAR 1938, KNA: PC/NZA/1/33; NPIR September 1938, KNA: PC/NZA/4/5/2; Rosberg and Nottingham, *The Myth*, 163; Lonsdale, "Political Associations," 636; Bode, "Anti-Colonial," 124. Lonsdale acknowledges Mulama's contribution of £12 in spite of Bode's contention that "nothing happened"during the visit. Lonsdale, "Political Associations," Ibid.

55. NKDAR 1938, KNA: DC/NN/1/20; NPIRs September and October 1938, KNA: PC/NZA/4/5/2.

56. Memorandum by Members of the Kikuyu Central Association and the Kavirondo Taxpayers Welfare Association in Kenya to the Secretary of States for the Colonies on the Native Lands Trust Ordinance, 1938, n.d but definitely after November 9, 1938, for they annexed a copy of a petition by the Kikuyu of Kiambu KCA dated so. PRO: CO 533/502/3; Lonsdale, "Political Associations," 636; Bode, "Anti-Colonial," 124. Of the total £50,000, £17,000 was to be spent on actual purchases of

land for addition to reserve, £14,000 for compensation as recommended by the Commission, £7,000 for some special items of compensation, and about £7,000 on the cost of the Commission. Of the latter, the colonial state transferred £2,500 for the construction of a bridge over the Athi River which was outside NK reserve. J. Chadwick, Minute, September 1939, PRO: CO 533/502/3.

57. Bode, "Anti-Colonial,"124. For example, John Adala, late Secretary of NKCA, decamped in 1938, and, took over the mantle of Secretary of NKTWA. Adala sought to cooperate with the colonial administration following his election to the LNC. See NKDAR 1938, KNA: DC/NN/1/20; NPIR January 1940, KNA: PC/NZA/4/5/3; Bode, "Anti-Colonial," 122. The colonial state often exercised politics of cooption by appointing critics of colonial policies to positions of power. Otiende, OI, August 20, 1998.

58. Col. & Prot. of Kenya, *The Crown Lands (Amendment) Ordinance, 1938*, PRO: CO 533/502/2; Native Lands Trust Ordinance, 1938, PRO: CO 533/513/1.

59. Joseph Kang'ethe, KCA to CO, Telegram, February 28, 1939 cited in CO to Governor, Despatch, March 29, 1939, PRO: CO 533/502/3; Rosberg and Nottingham, *The Myth*, 163. Popham protested to the CO against Africans communicating directly to Whitehall by disregarding the Kenya administration. He urged the CO to ignore such petitions until they passed through the proper channels. In a paternalistic tone, Popham urged Africans to look at the DC as "their father and mother, and their chief source of help." Brooke-Popham to MacDonald, July 26, 1939. Popham Papers, RHO, MSS. Afr. S. 1120 111/4. Brooke-Popham was recalled to serve with the Royal Air Force in the Far East on September 21,1939, Brooke-Popham Papers, RHO, MSS. Afr. S. 1120 111/7/73; *East African Standard*, 26 September 1939.

61. Sir Walter Harragin, Governor's Deputy to MacDonald, Despatch, August 22, 1939, PRO: CO 533/502/3.

60. Ibid.

62. PC Nyanza to Chief Secretary, May 25, 1939; PC Nyanza to the Secretary, KTWA, November 6, 1939, PRO: CO 533/502/3. Fazan believed that KTWA was being "got at" by external agitators who were using misrepresentations to stir resentment. PC Nyanza to DC Central Kavirondo, May 8, 1939.

63. Bode, "Anti-Colonial," 115–56.

64. NPIR August 1934, KNA: PC/NZA/4/5/1. Colonial administrators stereotyped African political activity as *fitina*- (intrigue), and plots in pursuit of self-serving motives. Unfortunately, such sloppy images shaped the administration's reaction to African politicians and political associations. Berman, *Control & Crisis*, 213–14.

65. Warden of Mines, CMIR August 1934, KNA: PC/NZA/3/14/372.

66. NPIR August 1934, KNA: PC/NZA/4/5/1; NKCA to Buxton, September 27, 1934, Buxton Papers, RHO, MSS. Afr. Emp. S. 405/6/5; Bode, "Anti-Colonial," 116.

67. NPIR November 1934, KNA: PC/NZA/4/5/1; Warden of Mines, CMIR November 1934, KNA: PC/NZA/3/14/372.

68. NPIR September 1934, KNA: PC/NZA/4/5/1.

69. NPIR October 1934, KNA: PC/NZA/4/5/1.

70. Warden of Mines, Handing Over Report (HOR), February 18, 1935, KNA: PC/NZA/ 3/14/372.

71. NPIR April 1935, KNA: PC/NZA/4/5/1.
72. NPAR 1935, KNA: PC/NZA/1/30.
73. NKDAR 1935, KNA: DC/NN/1/16; NPAR 1935, KNA: PC/NZA/1/30; NPIR June 1935, KNA: PC/NZA/4/5/1; Rosberg and Nottingham, *The Myth,* 163; Bode, "Anti-Colonial," 119–120, 130.
74. NKDAR 1935, KNA: DC/NN/1/16.
75. NPIR July 1935, KNA: PC/NZA/4/5/1.
76. NKCA to PC Nyanza, January 8, 1936; DC NK to acting PC Nyanza, January 18, 1936, KNA: (ADM. 9/3/1) 3/1974; NKCA, Petition to S of S, January 15, 1936 encl. in Byrne to W. Ormsby-Gore, Despatch, June 22,1936, PRO: CO 533/473.
77. Byrne to Ormsby-Gore, Despatch, June 22, 1936, PRO: CO 533/473. It appears that NKCA was aware of this Circular. This issue was raised in the House of Commons on December 5, 1934 following the submission of a petition by NKCA directly to the S of S for the Colonies. The S of S failed to give consideration to this petition on ground that it was not submitted through recognized and prescribed channels. See *Parliamentary Debates,* 295 H. C Deb. 1934–35, December 5, 1934.
78. Byrne to Ormsby-Gore, Despatch, June 22, 1936, PRO: CO 533/473; Ormsby-Gore to Byrne, Despatch, September 2, 1936, Ibid; C. R., Minute, CO, August 18, 1936, Ibid.
79. NPIR June 1936, KNA: PC/NZA/4/5/1.
80. NKDAR 1937, KNA: DC/NN/1/19; NPAR 1937, KNA: PC/NZA/1/32. Mulama's grievance entailed the colonial state's refusal to appoint him Paramount Chief of Buluyia.
81. NPIR April 1938, KNA: PC/NZA/4/5/2.
82. NPIR May 1938, KNA: PC/NZA/4/5/2.
83. NKDAR 1939, KNA: DC/NN/1/21; NPAR 1939, KNA: PC/NZA/1/34; Bode, "Anti-Colonial," 117–18. The DC's reference to Adala is misleading as Adala had left the association in 1938.
84. NKLNC, Minutes of a Meeting Held at Kakamega on August 12 & 13, 1940, KNA (Kakamega): HW/13/3; NKDAR 1940, KNA: DC/NN/1/22; NKDAR 1941, KNA: DC/NN/1/23; NPAR 1940, KNA: PC/NZA/1/35; NPIR May 1940, KNA: PC/NZA/4/5/3; NPAR 1941, KNA: PC/NZA/1/36; Lonsdale, "Political Associations," 637; Kanogo, *Squatters*,106; Aseka, "Political Economy," 362. Lonsdale notes that NKCA went into voluntary liquidation less than a month after KCA's ban and the PC's denunciation of opposition during the war. The PC, however, maintained that the fate of KCA had nothing to do with NKCA's temporary dissolution. See NPIR June 1940, KNA: PC/NZA/4/5/3.
85. NKDAR 1942, KNA: DC/NN/1/24; NKDAR 1943, KNA: DC/NN/1/25.
86. NPIR January 1942, KNA: PC/NZA/4/5/3. In fact, the administration rejected NKCA's intention to resume activity in December 1941 on account of the war. See NKDAR 1941, KNA: DC/NN/1/23. Similarly, when the issue came up in 1942, the PC advised members to "please themselves" for as long as they refrained from any political antagonism during the war period. Although they officially declared to be in voluntary dissolution, evidence shows their continued activity during the period.

87. NKDAR 1944, KNA: DC/NN/1/26; See also NKDAR 1942, KNA: DC/NN/1/24 and NKDAR 1943, KNA: DC/NN/1/25.

88. NKDAR 1946, KNA: DC/NN/1/28; Aseka, "Political Economy," 362.

89. For details on metropolitan and colonial concern for soil erosion see David Anderson, "Depression, Dust Bowl, Demography and Drought: The Colonial State and Soil Conservation in East Africa During the 1930s," *African Affairs* 83 (1984): 321– 43; Maxon, *Going Their Separate Ways,* 144–45, 208–218.

90. NK Agricultural Annual Report (NKAgAR), 1935, KNA: AK/2/27; Maxon, *Going Their Separate Ways*, 141–51.

91. AO, Safari Diary, KNA: AK/21/24; Khakali, OI, March 4, 1998.

92. AO NK, Monthly Reports, September-October 1935 and November 1935, KNA: AK/2/ 24; NKDIR October 1935, KNA: PC/NZA/4/5/5.

93. AO NK, Safari Diary, January 1936, KNA: AK/21/24; NKAgARs, 1936 and 1937, KNA: AK/2/26. Mnubi was chief of South Maragoli.

94. NKLNC, Minutes of a Meeting Held at Matungu on August 15 & 16, 1935, KNA (Kakamega): HW/13/2.

95. Ibid.

96. AO NK, Monthly Report, December 1935, KNA: AK/2/24; NKLNC, Minutes of a Meeting Held at Kakamega on April 15 & 16, 1936, KNA (Kakamega): HW/13/2; NPAR 1937, KNA: PC/NZA/1/32. Europeans referred to African men as boys irrespective of age.

97. NPAR 1937, KNA: PC/NZA/1/32.

98. DC North Nyanza (NN) to PC Nyanza, February 15, 1950, KNA: PC/NZA/2/20/28; Khakali, OI, March 4, 1998; Isenjia, OI, February 24, 1998.

99. NPAR 1938, KNA: PC/NZA/1/33. Boundary furrows were trenches used to demarcate individual holdings.

100. Rosberg and Nottingham, *The Myth,* 163–64; Khakali, OI, March 4, 1998; Litavakha, OI, March 9, 1998.

101. African Affairs: Race Relations Committee, RHO, MSS. Afr. S. 596 EEMO. 40/3.

102. NPAR 1938, KNA: PC/NZA/1/33; Henry Onzere Chavasu, "British Colonialism and the Making of the Maragoli Diaspora in Western Kenya, 1895–1963," (M. Phil., Moi University, 1997), 190–94; Maxon, *Going Their Separate Ways*, 200–04.

103. NKLNC, Minutes of a Meeting Held at Matungu on July 11 &12, 1938, KNA (Kakamega): HW/13/2; District Development Team Meetings, 1950–54, KNA (Kakamega): AGU/3/3.

104. Ibid. Underlining as in the original.

105. Meeting of the District Commissioners of the Nyanza Province at Kisumu, July 14–17, 1937, KNA: PC/NZA/4/1/3/1.

106. William Pulfrey to PC Nyanza, February 5, 1940, KNA: PC/NZA/2/2/51.

107. Quoted in B. Izard, Commissioner of Mines to Chief Secretary, May 21, 1940, KNA: PC/NZA/2/2/51.

108. AO NK to SAO Nyanza, February 2, 1940, North Kavirondo Monthly Agricultural Report (NKMAgR), January 1940, KNA: AK/2/23.

109. AO NK to SAO Nyanza, March 8, 1940; NKMAgR February 1940, KNA: AK/2/23.
110. Lonsdale, "Political Associations," 637.
111. NKAAgR, 1940, KNA: AK/2/27; NPIR February 1940, KNA: PC/NZA/4/5/3.
112. NKMAgRs for March and April 1940, KNA: AK/2/23; NPIR March 1940, KNA: PC/NZA/4/5/3.
113. NPIR May 1940, KNA: PC/NZA/4/5/3.
114. AO NK to acting SAO, Nyanza, October 1, 1940, KNA: AK/2/23.
115. AO NK to SAO Nyanza, December 31, 1940, KNA: AK/2/23; NPIRs June and December 1940, KNA: PC/NZA/4/5/3.
116. NKLNC, Minutes of a Meeting Held at Matungu on December 2 & 3, 1941, KNA: HW/13/3; NPAR 1941, KNA: PC/NZA/1/36. The initial suggestion was twenty cents but the LNC found it "excessive."
117. NKDAR 1943, KNA: DC/NN/1/25; NPAR 1943, KNA: PC/NZA/1/38; Litavakha, OI, March 9, 1998.
118. DC NK to PC Nyanza, August 21, 1943, KNA: AK/26/20.
119. Ibid.; NKDAR 1943, KNA: DC/NN/1/25. Agoi became chief of a united Maragoli in 1940. See NPAR 1940, KNA: PC/NZA/1/35.
120. DC NK to PC Nyanza, August 21, 1943, KNA: AK/26/20.
121. NKAgAR 1943, KNA: AK/2/27.
122. NKDAR 1944, KNA: DC/NN/1/26; NPAR 1944, KNA: PC/NZA/1/39.
123. NKDAR 1944, KNA: DC/NN/1/26.
124. AO NK to SAO Nyanza, April 6, 1944, NK Quarterly Report, January–March 1944, KNA: AK/2/23.
125. Ibid.
126. AO NK to SAO Nyanza, Safari Diary, November 1943, KNA: AK/21/24.
127. AO NK, Safari Diary, March 9, 1944, KNA: AK/21/25.
128. Ibid.
129. Ibid.
130. DC to Assistant Agricultural Officer (AAO), Kimilili, April 24, 1944, KNA: DC/KMG /1/1/153.
131. DC NK to AO Kakamega, August 21, 1944, KNA: DC/KMG/1/1/153.
132. AO Kakamega to SAO, NK Quarterly Report, July-September 1944, KNA: AK/2/23.
133. NPAR 1945, KNA: PC/NZA/1/40.
134. AO NK, Safari Diary, November 3, 1945, KNA: AK/21/25.
135. AO NK, Safari Diary, December 14, 1945, KNA: AK/21/25.
136. NKAgAR 1945, KNA: AK/2/28.The LNC purchased two teams in 1941. NKAR 1941, KNA: DC/NN/1/23.
137. NKDAR 1945, KNA: DC/NN/1/27; NKAgAR 1945, AK/2/28; Judith M. Abwunza, *Women's Voices, Women's Power Dialogues of Resistance From East Africa* (Toronto, Ontario: Broadview Press, 1997), 17.
138. NKAgAR 1945, KNA: AK/2/28.

139. NKLNC, Minutes of a Meeting Held at Matungu on November 28–30, 1945, KNA (Kakamega): HW/13/4; NKDAR 1946, KNA: DC/NN/1/28.

140. NKLNC, Minutes of a Meeting Held at Matungu on the 28–30 November 1945, KNA (Kakamega): HW/13/4.

141. NKDAR 1926, KNA: DC/NN/1/7; Col. & Prot. of Kenya, *Report of the Committee on Native Land Tenure in the North Kavirondo Reserve (RCNLTNKR)* (Nairobi: Government Printer, 1930), 18; Chavasu, "British Colonialism," 155.

142. NKAgAR 1946, KNA: AK/2/28; NKDAR 1946, KNA: DC/NN/1/28.

143. NKLNC, Minutes of a Meeting Held at Kakamega on February 26–28, 1946, KNA (Kakamega): HW/13/4.

144. Col. & Prot. of Kenya, *Department of Agriculture Annual Report (DofAAR) 1946* (Nairobi: Government Printer, 1948), 39; NKDAR 1946, KNA: DC/NN/1/28; NKAgAR 1946, KNA: AK/2/28.

145. Bode, "Anti-Colonial," 126. Founded by Elijah Masinde among the Bukusu in 1942, DYM articulated myriad grievances against colonial rule in Kenya in the religious, political and economic realms. Exhaustive studies of the movement include G. S. Were, "Politics, Religion and Nationalism in Western Kenya, 1942–62: Dini Ya Msambwa Revisited," in *Politics and Nationalism in Colonial Kenya. Hadith* 4. ed. B. A. Ogot (Nairobi: EAPH, 1972), 85–104; A. Wipper, *Rural Rebels: A Study of Two Protest Movements in Kenya* (Nairobi: OUP, 1977); J. B. Shimanyula, *Elijah Masinde and Dini Ya Msambwa* (Nairobi: TransAfrica, 1978). For the spread of the movement outside Bukusuland, See Col. & Prot. of Kenya, *Commission of Inquiry into the Affray at Kolloa, Baringo* (Nairobi: Government Printer, 1950); B. E. Kipkorir, "Kolloa Affray, 1950," in *TransAfrican Journal of History* 2, no. 2 *(1972)*:114–29; Kipkorir, "Colonial Response to a Crisis: The Kolloa Affray and Colonial Kenya in 1950," in *Kenya Past and Present* 2, no.1 (1973): 22–35. For insights on the centrality of symbolism in DYM's dramatization of the eventual departure of Europeans from Kenya, See Shilaro, "Kabras Culture." 162–97; Shilaro, "Religion, Culture and Political Protest in Western Kenya: The Case of Dini Ya Msambwa, 1943–63," Paper presented at the Senator Rush Holt History Conference, West Virginia University, September 1999.

146. NKDAR 1944: KNA: DC/NN/1/26.

147. Bode, "Anti-Colonial," 126. A brainchild of Pascal Nabwana and others, Bukusu Union came into existence after their earlier organization, the Kitosh Education Society founded in 1935, was proscribed in 1940. Nabwana received a Catholic education in north Bukusu, before training as a teacher at Jeans School, Kabete. In 1943, he became a leveler with the Soil Conservation Service at Kitale. Wagner, *The Bantu,* Vol. II, 38.

148. NKLNC, Minutes of a Meeting Held at Matungu on June 12–14 and November 26–28, 1946, KNA (Kakamega): HW/13/4; NKAgAR 1946, KNA: AK/2/28; NKDAR 1947, KNA: DC/NN/1/29; DofAAR) 1946, 40.

149. NKAgAR 1946, KNA: AK/2/28; NKDAR 1946, KNA: DC/NN/1/28.

150. Ibid.

151. Petitions and Memorials, PRO: CO 533/544/2. Commenting on this delay, one CO official noted, "All this looks pretty bad. I wonder whether European settler petitions would have been delayed so long. . . If we plead for African cooperation

... we ought to listen to and answer their grievances." A. S., Minute, April 9, 1948, Ibid.

152. NKDAR 1946, KNA: DC/NN/1/28.
153. David Mulindi Mbwanga to PC Nyanza, May 9, 1946, KNA: AK/26/20.
154. Minutes of the General Meeting of the Maragoli Society Held at Ivona on May 31, 1947, KNA: DC/KMG/1/1/153.
155. Ibid.
156. Berman, *Control & Crisis*, Chapter 5; Bruce Berman & John Lonsdale, *Unhappy Valley:Conflict in Kenya & Africa. Book One* (Athens: Ohio University Press, 1992), 81–82, 151, 161–62, 195, 197–78; Margaret Jean Hay, "Economic Change in Luoland: Kowe, 1890–1945," (Ph. D. dissertation, University of Wisconsin, 1972); Odhiambo Ndege, "Struggles for the Market," 2–83, 233– 49.
157. NKLNC, Minutes of a Meeting Held at Kakamega, March 10–13, 1947, KNA (Kakamega): HW/13/4.
158. Ibid.
159. Ibid.
160. M. M. Lusiola, African Assistant Agricultural Officer (AAAO), Safari Diary, May 7, 1947, KNA: DC/KMG/1/1/13; Misaki, OI, July 18, 1998.
161. NNDAR 1948, KNA: DC/NN/1/30.
162. *DofAAR 1947* (Nairobi: Government Printer, 1949), 15; NKDAR 1947, KNA: DC/NN/ 1/29; Aseka, "Political Economy," 350.
163. *DofAAR 1947*, 39–40; North Nyanza LNC (NNLNC), Minutes of a Meeting Held at Kakamega on July 6–9, 1948, KNA (Kakamega): HW/13/4.
164. Nyanza Central Association to DC NN, July 31, 1948, KNA: DC/KMG/1/1/90.
165. NNLNC, Minutes of a Meeting Held at Kakamega on March 8–11, 1949, KNA (Kakamega): HW/13/4. The AO NK contended that group farming was "of utmost importance to the future of African agriculture." Ibid.
166. *DofAAR 1947*, 39.
167. Aseka, "Political Economy," 351–52.
168. Report of the DAO for the Year Ending June 1948, KNA: AK/2/28; NNDAR 1948, KNA: DC/NN/1/30.
169. NNAgAR 1948, KNA: AK/2/28.
170. Ibid. Underlining is that of the AO.
171. AOs Meeting Held on June 20, 1943, KNA: AK/6/8; NKAgAR 1948, KNA: AK/2/28.
172. Ibid.
173. NNDAR 1949, KNA: DC/NN/1/31.
174. DC NN to PC Nyanza, February 15, 1950, KNA: PC/NZA/2/20/28.
175. NNDAR 1949, KNA: DC/NN/1/31; See Kanogo. *Squatters*, especially Chapter 4; David W. Throup. *Economic & Social Origins of Mau Mau. 1945–53.* (Athens: Ohio University Press, 1988), especially Chapter 7.
176. NNDAR 1950, KNA: DC/NN/1/32.
177. NNDAR 1951, KNA: DC/NN/1/33; NNADC, Minutes of a Meeting Held at Kakamega on May 14–17, 1951, KNA (Kakamega): HW/13/4.

178. North Nyanza African District Council (NNADC), Minutes of a Meeting Held at Kakamega on March 7–10, 1950, KNA (Kakamega): HW/13/4. ADCs replaced LNCs in 1948.

179. NNDAR 1950, KNA: DC/NN/1/32.

180. NNDAR 1951, KNA: DC/NN/1/33.

181. AAO to DAO NN, July 9, 1951, KNA: AK/4/14. *"Amri ya Serikali inasema kila mtu atakwenda kwa mitaro* (Kiswahili)." Literally, "Government policy says that every person will go to the terraces."

182. Ibid. AR and BR represent cotton grades.

Chapter Five

Rural Industrialization: The Economic Balance Sheet

Both the colonial state and the imperial government in England justified the alienation of land in the NK reserve for gold mining by emphasizing the numerous benefits Luyia households would reap from the industry. These included handsome compensation rates, employment opportunities, and a market for local produce, improved health, and better communication networks. This chapter examines the economic impact of the gold mining industry on the Luyia community, using indices such as mining leases and compensation, African entrepreneurship in the new industrial capitalist endeavor as miners and laborers, mining accidents, and market opportunities. A critical analysis of these economic indicators illustrates the nature of benefits that accrued to the local population because of the gold mining industry. Although Fearn maintains that the advantages of gold mining in Buluyia outweighed the disadvantages,[1] this study unearths new evidence to the contrary.

This chapter shows that Luyia households lost land for meager cash payments that were not commensurate with the enormous economic, social and psychological effects that accompanied the loss of land to gold mining interests. This in turn created such deep resentment toward the gold mining industry that the Luyia in the core mining areas shunned employment in the industry. As Luyia households grappled with the loss of land, colonial mining laws stifled local entrepreneurship by demanding unattainable English standards as a prerequisite for obtaining mining licenses. As a result, many Luyia resorted to illegal gold dealings often ending in the inevitable costly prison and cash penalties. Also, African mine workers constantly struggled against the specter of death in the mines and endured low wages, poor living and working conditions, and declining health due to increased incidence of malaria, pneumonia and cerebro-meningitis brought about by mining activities. Similarly, both humans and livestock faced major environmental problems. The gold mining industry provided short-lived market opportunities for low priced commodities so that very few Luyia men and women translated their meager gains into meaningful economic prosperity.

LAND ALIENATION, MINING LEASES AND COMPENSATION

The inauguration of gold mining and the subsequent adoption of the NLT(A)O of 1932 led to the alienation of land for large mining concerns in Buluyia. The colonial state promised handsome cash compensation to dispossessed Luyia households. This section examines compensation paid for mining leases and evaluates how "generous" these payments were. By juxtaposing archival sources and oral evidence, it is clear that dispossessed Luyia households found cash compensation both irrelevant and inadequate.

Gold prospecting and mining in NK district preceded the NLT(A)O of 1932 which empowered the colonial state to grant mining leases to major companies that were operating in the Kakamega goldfield. European prospectors acquired mining rights through EPLs or mining permits. By 1934, between 62,000 and 65,000 acres of NK reserve were under mining claims.[2] For lack of a *modus operandi*, claim pegging proceeded haphazardly with unpleasant consequences for the affected Luyia households. The NK DC, E. L. B. Anderson and DO F. D. Hislop, captured the scenario in a 1932 memorandum:

> Claims of 600 by 300 feet are being pegged, almost daily, whole hills and ridges are being so covered, and all this is right among native *bomas* [homesteads] and *shambas* [holdings], prospecting trenches down to 40 feet deep are being dug everywhere in these areas . . . the natives affected being paid a few shillings compensation. . . They detest Europeans coming into their private lands, living in their pastures and *shambas*, and causing the damage and loss, compensated it is true, by a system of enforced arbitration. . . . They do not want the Europeans on their ancestral lands, whatever benefits they bring. . . . No compensation in cash can really meet the case of the expropriated natives, who will become landless in a densely populated part of the reserve, . . . Money has no real value to him.[3]

Anderson's and Hislop's observations raise fundamental questions regarding the value of land to Luyia households and the significance of monetary compensation to those dispossessed by mining interests. Can one limit the analysis to the question of the adequacy of cash compensation? Can one peg the value of compensation solely and strictly on the utilitarian view of land espoused by European mining interests and colonial officials? While these questions constitute important starting points, this study adopts a holistic paradigm that extends the analysis to the intangible and unquantifiable psychological consequences of land loss to dispossessed Luyia households.

The passage of the NLT(A)O in 1932 provided for mining leases to European prospectors and miners. On August 1, 1935, the colonial administration granted the first mining lease consisting of 113.5 acres to Kimingini Gold Mining Company.[4] In determining the commuted rent, the AO NK adhered to the principles laid down by the KLC which had asserted, "it is clear to us that the basis of valuation must be the value which the land possesses for natives," by which the Commission meant . . . "the uses [to] which the native[s] put such land today."[5] The AO's rate of Shs. 5 per annum or £5 commuted rent for 21 years for the Kimingini lease, took into consideration the value of the land for agriculture, grazing, firewood, thatching grass, and trees.[6] In an appeal to the governor over the compensation rates, company officials called it "excessive" maintaining that only 30 acres of the leasehold was arable.[7] The Kimingini lease affected sixty landholders, who received a lump sum of Shs. 10,552. Only six of the sixty individuals received a total of Shs. 709 as compensation for loss of property including dwellings and stores, trees, and crops, ranging from Shs. 200 to Shs. 16.[8] Based on the compensation payment list prepared by the DC, thirty-nine of the dispossessed lost substantial acreage.[9] Appendix II is a list of owners, property and their value on the Kimingini lease.

The question of compensation was controversial because of conflicting conceptions of the value of land between the Luyia and the Europeans. Among the Luyia, land had an intrinsic value as one's ancestral home. It also provided subsistence, grazing for livestock, building materials and general flora and fauna that graced the Luyia social milieu. This all-embracing view conflicted with the European notion that restricted the value of land to its arability. The company's attempt to compare the value of land to Luyia households with existing notions that the colonial state conveyed to European immigrants, was untenable.[10] When the company disbursed the lease compensation in December 1935, the process of sharing the spoils became an arena for remarkable squabbling. These wrangles represented displeasure among the dispossessed Idakho.[11]

The second twenty-one year mining lease involved a surface lease of 137.25 acres and a subterranean lease of 515 acres to Rosterman Gold Mining Company on November 5, 1936.[12] Occupied by the Isukha, a large proportion of this land was under intensive cultivation. To circumvent local opposition, the DC NK sought to secure the "consent" of the local population speedily. He wrote, "It seems to me to be likely that the longer we delay the greater the opportunity for subversive propaganda to penetrate to the local inhabitants."[13]

Nevertheless, the Luyia opposed the proposed lease, with the NK LNC articulating the greatest opposition. At an April 1936 LNC meeting held to discuss the lease, members expressed general opposition to it. The local Isukha Chief Milimu refused to give his views, protesting the administration's failure to seek his opinion in the matter from the outset. LNC member Dominikus Osianjo dismissed the issue as a *fait accompli* that required no discussion, while Jeremiah Masitsa harangued LNC members to reject the lease. Ultimately, the LNC "consented" to the lease as a matter of government policy; as aptly articulated in the minutes: "Various members expressed opinion in opposition but eventually the council agreed unanimously to the granting of the lease, not from their own choice but because they realised it was Government's policy."[14] The Central Land Board subsequently approved the lease, paving the way for Rosterman to pay a commuted rent of about £754 at £5 per acre, on July 21, 1936.[15] From the compensation list, at least four households lost approximately 6 acres each.

Once again, the Rosterman lease brought to the fore the dichotomy between the British land proprietorship and the Luyia land tenure system. In the directive to the Secretary of the Central Land Board regarding the distribution of compensation to dispossessed Luyia households, the PC wrote:

> In my opinion private land rights are not so clearly defined as to constitute a system of private right holding in the area. The question of prior rights will be taken into account when the District Commissioner as chairman of the Local Land Board, makes payment . . . out of the sum paid by way of commuted rent.[16]

This context of conflicting ideas explains why out of more than 300 people who applied for compensation, only sixty received lump sum cash offers ranging from Shs. 600 to Shs. 15. Entitlement to compensation depended on the recommendations of an alien Luo interpreter at the DC's Office, an African assessor, Chief Milimu, an ex-headman, and the local headman.[17] Manifestly, ignorance and personal idiosyncrasies influenced the final list submitted to the administration.[18] Table 5.1 shows the distribution of the rent among individuals displaced by the Rosterman mining lease.

Although dispossessed households resented and opposed the loss of land to mining concerns and demonstrated extreme dissatisfaction with monetary compensation, land alienation continued unabated. Coming quickly on the heels of the Rosterman lease was another 21-year lease comprising 94 acres granted to the Kavirondo Gold Mining Syndicate at Sigalagala (then known as Piccadilly Circus) at a commuted rent of £141.[19] The numerous Isukha

Table 5.1. Numbers Dispossessed and Compensation Values for Rosterman Lease

Number of People	Amount (Shs.)	Total (Shs.)
4	600	2,400
8	400	3,200
1	225	225
1	120	120
43	100	4,300
1	80	80
1	20	20
1	15	15
Total 60		10,360

Source: Compensation List for Rosterman Mining Lease encl. in DC NK to Acting PC Nyanza, August 19, 1936; PC Nyanza to DC NK, August 31, 1936, KNA: PC/NZA/3/14/358 A. Also, found in KNA: DC/KMG/2/18/9.

families affected by this lease rejected cash compensation, forcing the administration to remove them by force.[20] In 1944, Rosterman received a special mining license covering 1,807 acres south of west Kakamega.[21] The PC K. L. Hunter ensured high secrecy while negotiating for the granting of this lease. PC Hunter wanted to circumvent local opposition to the proposed lease by presenting the Isukha with another *fait accompli*. Toward this end, Hunter advised the DC to, "Please conduct these inquiries in such a manner that the natives do not suspect that any negotiations are proceeding, as I wish to complete all details vefore [sic] placing the matter before them."[22]

As the secret negotiations continued on the second Rosterman lease, the administration granted two mining leases to Edzawa Ridge Mining Company in 1941, comprising 21-year leases of 29.3 acres in South Maragoli and 31 acres at Tintax in Kisa location.[23] The official value of both leases stood at approximately £3 per acre. Appendix III lists the number of households affected by the Edzawa Ridge lease and the acreage involved.

By August 1, 1941, a deputation of Edzawa Ridge land right holders visited the DC demanding additional compensation for loss of houses, trees and crops.[24] Although the leasehold contained an acre of banana plantation, the SAO excluded the value of the crop in his valuation.[25] Ultimately, the commuted rent approximated about £141. Local right holders received £30 for loss of property, £65 for loss of land, and the NK LNC received £27, of which £19 paid for loss to the location.[26] Though the company deposited the entire sum with the district administration on January 2, 1942,[27] due to a delay in granting the leases, dispossessed Logoli households earmarked for removal from the leasehold found themselves squatting on their former land. These families endured immense anxiety over the possibility of purchasing alternative land in the overcrowded reserve.[28] Indeed, when the planting season commenced in March, the affected households proceeded to plant crops on what had become company land.[29] Such "defiant violation" of "company property" elicited a wave of complaints from company officials to the district administration. Pressure from European capitalist interests compelled the administration to move quickly using local chiefs to warn displaced Logoli households against continued cultivation on the proposed leasehold land.[30] Notwithstanding, dispossessed Logoli households had nowhere to go. The situation became desperate, forcing the DC to intervene. The DC urged the company to permit Logoli households to reap the standing crop before evicting them.[31] Since the company grudgingly obliged to the DC's request, by the end of harvest season, anxious company officials wanted to remove the Logoli from their lease.[32] This dilemma burst out into internal feuds, as disinherited Logoli sent deputations to Chief Paul Agoi, accusing him of receiving and appropriating compensation money.[33]

In the case of the Edzawa Mining Company's lease at Tintax in Kisa location, the complexity of compensation issues created minimal disturbance because the Tintax leasehold contained less property. Appendix IV shows the numbers dispossessed and the acreage involved. Also unlike the Rosterman lease where the PC and the DC eschewed consultation with the Isukha, DO Lambert consulted the Kisa community over the proposed lease on August 13, 1941. Thereafter, the Local Land Board, with the approval of co-opted members, and the LNC approved the demarcation of 31 acres at Tintax, Kisa for a 21-year lease. In the initial valuation, the Tintax lease approximated £105, at £3 per acre for 21-years, and an additional £5 for actual disturbance. The Nyanza PC viewed the latter as a "sweetener to the natives who will actually have to move their *shambas.*"[34]

Nonetheless, by late September, the PC reduced the Tintax commuted rent to £100.[35] The PC also rejected the £5 recommended by the Local Land Board as compensation for "loss of effort in tilling land." The PC contended that displaced Kisa families stood to gain from their new settlements.[36] At the completion of Tintax lease negotiations in December, the Kisa gave up 29.4 acres for 21-years for a paltry commuted rent of £94.[37] The local right owners received £65 and the NK LNC got £29, £18 of which compensated the location.[38] While the company deposited the full amount with the district administration on January 2, 1942,[39] an audit inquiry in 1947, and an official complaint from Chief Agoi of Maragoli in 1948, revealed that the administration never remitted the portions earmarked for the benefit of both locations.[40]

From table 5.2, it is evident that Luyia households lost their surface land rights to mining companies for long durations in return for minimal cash compensation. Mining leases accelerated the process of disinheritance that had begun with the granting of EPLs and mining claims to European prospectors.

Table 5.2. Twenty-One Year Surface Mining Leases Granted in NK District

Company	Acres	Price (£)	Average per Acre (£)
Kimingini	113.5	601.19.00	5.06.00
Rosterman	137.25	754.17.00	5.10.50
K.G.M.S.*	94	141.00.00	1.10.00
Edzawa	29.3	122.00.00	4.03.25
Tintax	29.4	94.05.00	3.04.10

Source: Assistant Warden of Mines to PC Nyanza, November 6, 1939, KNA: PC/NZA/2/2/68. *K.G.M.S stands for Kavirondo Gold Mining Syndicate.

Clearly, the loss of land to gold mining interests produced far-reaching effects on Luyia households. Rural industrialization uprooted Luyia families from their cherished ancestral lands and social relations, compelling individual owners to pull down their homes for minimal monetary compensation. Notably, the administration left these families to re-establish themselves *wherever they could* (emphasis is mine). In reality, displaced families became at worst, expropriated landless households or at best, tenants-at-will of their neighbors and relations. Such forcible disinheritance had long-lasting effects on the Luyia that no amount of easily expendable cash could have abated. This is particularly true of dispossessed households that never regained their lands on the expiration of mining leases. While land on the Edzawa Ridge and Kimingini leases reverted to former owners, the North Nyanza African District Council (NNADC) took over the Rosterman and Kavirondo leaseholds and established a Multi-Purpose Scheme and a Technical School respectively.[41] To this day, these lands remain issues of court litigation and a permanent source of frustration and disillusionment among dispossessed Luyia households.

Furthermore, former mining sites pose major environmental dangers to the local populations. Both humans and livestock constantly face the threat of unfilled shafts, unstable land surfaces, and derelict lands covered with mounds of bare rocks.[42] Robbed of their basic means of production, and catapulted into the new gold mining based industrial economy, the Luyia perhaps looked to entrepreneurial prosperity in the new industry. As discussed below, colonial mining laws successfully curtailed this option.

STIFLED AFRICAN ENTREPRENEURSHIP

Sheria ya Mzungu haikumruhusu Mwafrika kutafuta dhahabu.
[The white man's law did not allow an African to prospect for gold.]
Africans were not allowed to prospect for or handle gold.
Any African found in possession of gold would be arrested and imprisoned.[43]

The above assertions constituted almost instantaneous responses from most of the informants to the question, "Were there African prospectors?" This contention is important for the analysis and assessment of the significance of the gold mining industry in the Luyia political economy. This is central, especially when contextualized within the prism of the official colonial policy that ostensibly espoused racial equality in the acquisition of gold mining rights; it, therefore, becomes an enigma worth unraveling.

From the very inception of the gold mining industry, the Luyia expressed concern over their inclusion in this promising capitalist endeavor. Chief Segero of West Kakamega raised this question with colonial officials three months after the gold discovery. The administration assured Africans that racial restrictions would not apply in the acquisition of prospecting permits or mining licenses. Notwithstanding, the officials categorically warned that only an African with technical skills could "know what to do otherwise many people would waste their money."[44]

Similarly, in May 1932, the Nyanza PC, H. R. Montgomery, expressed comparable sentiments in response to a question from Dominikus Osianjo, a member of the NK LNC. Contending that mining was "not a simple business" as it "required much knowledge," the PC advised Osianjo, "to keep his money for something else."[45]

Nevertheless, both the colonial state and the imperial government in England reiterated the allegedly "non-racial policy" in subsequent statements and policy pronouncements.[46] Shortly after, the purported "all-inclusive" and "color-blind policy" collapsed into sheer rhetoric. This sharply contrasts with the West African Gold Coast (Ghanaian) case where African entrepreneurship in the gold industry triumphed.[47] In the case of Kenya, the Mining Ordinance of 1933 allowed anyone over eighteen years of age to obtain a mining permit or license on payment of a prescribed fee of Shs. 20. Notably, the ordinance required eligible applicants to prove their English proficiency to the Commissioner of Mines to determine their ability to understand the provisions of the Mining Ordinance and Regulations.[48]

The English proficiency requirement in effect constrained the emergence of vibrant African rural industrial entrepreneurs. In the 1930s, western education was not a democratized commodity in Kenya colony. Few Africans had the rudiments of the English language; leave alone the ability to peruse sophisticated legal documents. For instance, out of twelve EPLs and 429 special permits granted by 1934, only two Africans had claims registered and those claims were already abandoned.[49]

Until 1948, the colonial administration criminalized African mining activities. Both archival and oral evidence reveal numerous cases of what colonial officials called "gold thefts," "illegal gold winning/mining" and "illicit gold buying." Gold theft entailed unauthorized taking of gold from licensed miners or mining companies; illicit gold winning/mining, involved the recovery of gold by non-licensed miners; and illicit gold buying referred to trade in gold by un-licensed miners and buyers.

The colonial state's policy of stifling African entrepreneurship in the gold mining industry took many forms. To counter illegal gold dealings, the state tightened its legal apparatus of control. For instance, on April 24, 1933, the colonial state adopted the, Trading in Unwrought Precious Metals Ordinance (TUPMO), ostensibly to protect the gold industry from illegal dealings. Under this Ordinance, those convicted risked a fine of £1,000, or imprisonment for one year or both.[50] In 1939, however, the administration amended the ordinance to permit a stiffer penalty. The resultant Trading in Unwrought Precious Metals (Amendment) Ordinance (TUPM(A)O) of 1940 empowered police officers to make immediate arrests without warrant and to search property, premises and anyone suspected of illicit possession of precious metal.[51] Those convicted of illegal gold buying risked ten years in prison or a fine of £1,000, or both.[52]

Moreover, African company employees suffered humiliating body searches involving stripping naked, as mining officials looked for stolen gold. The entire gold production and delivery process involved close police supervision.[53] Generally, Europeans perceived Africans as natural thieves, hence movement to and from Labor camps came under strict police scrutiny. Usually, police supervisors freely and savagely meted out instant discipline on suspected gold thieves. Mine supervisors usually flogged and ejected Africans found trespassing on company property.[54]

In spite of this, the outbreak of World War II, coupled with the colonial state's aversion to license African miners, provided impetus to illegal mining and gold pilfering. During the war, gold stealing increased, particularly among Africans engaged in milling operations. As European mining staff left for military service, monumental mill supervision difficulties arose, thereby providing increased opportunities for African employees to steal gold.[55] Against this background, the colonial state amended the mining law to stiffen penalties further. When the TUPM(A)O came into force in August 1940, first offenders faced a fine of £1,000 and/or imprisonment for two years, and second offenders risked a similar fine and/or a five-year prison term.[56] In the minds of colonial officials, the prison epitomized the ultimate school for instilling lasting discipline in Africans. Yet, illegal gold dealings continued unabated as is evident from arrests in Kakamega thereafter.[57] The reigning "high" wartime gold prices made the business worth the risk.

For instance, between July and September 1940, the colonial administration convicted four Africans for stealing gold. Two of the convicts faced a

Shs. 3, 000 fines or nine months in prison; a third received a Shs. 1, 100 fine or nine months in jail, and the fourth got a Shs. 300 fine or three months in jail. Apart from the fourth convict, the first three failed to raise the fines and served prison term with hard labor.[58] From two prosecutions in the last quarter of 1941, illicit gold mining cases increased to ten in 1942 and thirty-one in 1943.[59]

Illicit gold dealing developed into a lucrative economic undertaking. By 1943, Luyia illicit miners responded creatively to the lure of a brisk business in gold. One Atsangu obtained a "bogus permit" from a Goan to hand sluice in the Gori Gori River. However, Atsangu's illicit alluvial dealings ended abruptly when Mrs. Stitts, a prominent alluvial miner in Kakamega, arrested him with an assortment of mining paraphernalia and handed him to the police.[60] By September 1943, mining officials considered gold stealing a "real and grave danger to the mining industry." This led to intense police surveillance in the mines, resulting in numerous arrests and convictions. For example, on September 14, Chief Milimu's police arrested three Idakho men "caught in the act." They were convicted and imprisoned for two months.[61]

Thus, the undercurrent of illicit gold buying and illegal mining continued to plague the industry. Mining officials attributed the phenomenal growth and recrudescence of these activities to several factors: First, officials argued that the attractive prices paid for illicit gold provided an irresistible impetus. Available evidence does not support the profit motive thesis. Rather, it is evident that Africans often disposed of their gold to Asian profiteers at extremely low prices. For example, illicit African gold dealers sometimes received £1 for gold worth £100 on the Asian dominated black market.[62] Clearly, Asian traders garnered most profits arising from the purchase of illicit gold. Second, officials blamed illegal gold dealings on ineffective supervision of mining operations, and, the colonial state's reluctance to release more personnel to revamp mine supervision work. Third, officials pointed to the overwhelming difficulty involved in eliminating gold stealing. An Inspector of Mines, B. Harding, noted: "It is practically impossible 'to beg, borrow or buy a padlock that is worthy of the name to lock up gold in safety' as in most cases, 'it is not even impossible to steal one as the *watu* [people] have done that.'"[63]

Perhaps the realization of the arduous task of combating illegal gold dealings provided the impetus for a reconsideration of gold mining policy in the post-WWII Nyanza Province Five-Year Plan. In this plan, the Nyanza PC called for "simple regulations to allow Africans to pan alluvial, and perhaps to mine to a depth of not more than a dozen feet." The PC's desire for licensing African miners emanated from his conviction that, "even today the

Africans are jealous of the fact that the mining industry within their reserve is entirely in the hands of non-natives. This fact, together with a desire to preserve their tribal lands, makes it difficult to extend the industry in the native reserves."[64] Predictably, the PC's proposal failed to capture the imagination of mining officials and the central bureaucracy. In the case of the Kimberley diamonds in South Africa, miners opposed licensing Africans because they sought to eliminate competition and secure control over labor.[65] In Kakamega, such legal exclusion only served to solidify African ingenuity.

Illicit gold miners utilized threats against licensed miners. For example, Logoli illicit miners armed with machetes pursued two licensed Asian prospectors namely, Curchuran Singh and Sauta Singh. Similarly, angry Logoli villagers threatened and stoned the house of a licensed Asian prospector near the Maragoli Closed Area. Indeed, threats to foreign prospectors had no racial boundaries. In 1948, for instance, Leokadius Ndede, the only licensed African with a claim near Luanda in Bunyore, and prospecting rights on a tributary of River Awach on the Luyia-Luo border, endured constant threats from local Nyore illicit miners. In a written complaint to the Inspector of Mines, Ndede claimed that, "on any day of the week, between 20 and 30 natives would be found working in that area."[66] Concisely, licensed Europeans, Asians and Africans alike struggled against the threat of illicit African miners.

Admittedly, it is difficult to determine accurately the number of Luyia involved in illegal gold mining; it is apparent that by 1948, illegal gold mining had become a common economic activity in Buluyia. In a letter to the Assistant Commissioner of Mines, the Inspector of Mines, P. Westerberg, lamented, "*Illicit gold mining with the consequential illicit disposal of gold won has now become quite a major native industry in some parts.*"[67] Even though mining officials recognized the loss of revenue to the colonial state due to illicit dealers, they radically differed on remedial strategies. Whereas the Inspector of Mines, Westerberg advocated the legalization of African mining, the Commissioner of Mines, D. Harverson preferred the creation of an institutionalized mechanism that would make illicit gold mining risky. Harverson contended that stringent measures, pursued zealously by the police and the administration, could induce Africans to mine legally and to sell their gold through recognized channels. Harverson considered both illegal gold mining and buying to be a drain on the colony's revenue and its gold resources. He contended that traffic in illicit gold often fortified the illicit gold buying network, evident in the presence of many bold illicit miners in the Kakamega goldfield and, manifested in their ability to interfere with legitimate prospecting and mining.[68]

Similar conflicting views characterized the local administration. While the DC believed in intensifying police patrols in mining areas, the PC insisted on opening up the industry to Africans. The latter wrote to the Chief Secretary:

> Personally I have held the view for a considerable time that we are unlikely to succeed in our efforts to combat this illicit gold production until some form of mining is open to them [Africans]. It is true that there is no valid reason why they [Africans] should not take out prospecting licences, but in fact, the law when examined makes this impossible.[69]

The Nyanza Provincial Superintendent of Police, W. R. B. Pugh, echoed similar sentiments:

> I entirely agree that the time has now come when steps should be taken to legalize gold mining by Africans, and thus bring the gold won by them into the legitimate market to the benefit of Govt. [sic] . . . The area concerned is scattered, and to achieve the aims of the Mines Department a colossal force out of all proportion to the economic value of the mining industry would be necessary.[70]

In the absence of consensus or compromise among colonial officials, the colonial state resolved to tighten its oppressive instruments of control. In 1948, the administration convicted three African mill employees at Kavirondo Gold Mines at Sigalagala for illegal possession of gold amalgam. Thereafter, a mining official, E. G. St. C. Tisdall, appealed to the Commissioner of Mines to grant employers the powers to search and detain suspected illicit gold dealers.[71] The 1952 Trading in Unwrought Precious Metals (Amendment) Ordinance (TUPM(A)O) provided for rewards to persons giving information that led to conviction.[72] The contradictions between the state's willingness to maximize the penalty for illicit gold dealing and its reluctance to license the Luyia is quite remarkable.

Several Luyia had unsuccessfully applied for mining licences. In 1943, the colonial state refused to grant a prospecting right to Esawa Eshahidza of Maragoli location, in spite of his willingness to make the requisite deposit of Shs. 500.[73] More Luyia inquired about mining licences in 1945 with similar outcomes. All along, mining officials harped on the irrelevancy of racial consideration, but clearly, this meaningless verbiage rested on the understanding that most Africans could not meet the requisite English standards. As the Commissioner of Mines stated in his response to the PC's inquiry:

> I consider that no racial discrimination should be made in the mining legislation. At the present time there is nothing in the Kenya Mining laws to prevent a native from obtaining a prospecting right and subsequently Mining Title, provided

that he can satisfy me that he can understand the provisions of the Mining Ordinance and Regulations, and is capable of carrying out such provisions.[74]

As posited above, the emphasis attached to one's ability to understand the mining laws stemmed from the general knowledge that most Kenya Africans had limited or no command of the English language. Based on this, mining officials argued that prospective African miners were incapable of executing reef mining and the attendant mining safety regulations. Consequently, when Joseph Simekha of Ebusubi Mundikiri of Bunyore location applied for a prospecting license in 1945, mining officials rejected his application, "as he could not understand the law." Apart from Simekha's weak mastery of the English language, officials noted that he lacked both the requisite technical training and capital resources to undertake gold mining.[75]

In 1946, PC Hunter abandoned the idea of licensing Africans due to lack of support from mining officials.[76] However, the colonial Chief Secretary at the Secretariat in Nairobi took up this issue. Alluding to the scarcity of foreign exchange in the colony, the Chief Secretary called for a "simple set of instructions translated into Kiswahili," to enable a limited number of Africans to participate in mining. This meant the elimination of the high standards of English required of all prospectors. At the core of this initiative lay the significance of gold resources to the depressed post-war British economy. As the Chief Secretary succinctly put it, "every ounce of gold left in the ground which could be won and marketed means less food for the inhabitants of the United Kingdom.," as "the winning of gold was useless unless such gold is sold to authorised dealers and to authorised dealers only."[77]

Although a meeting between the PC, the Commissioner of Mines, the Inspector of Mines, and the Superintendent of Police held in Kisumu on August 16, 1946 rejected racially specific legislation, the officials agreed that prospective African prospectors deposit Shs. 250 for liabilities such as wages, claims compensation and filling in excavations. Moreover, although the officials agreed to license a maximum of thirty Africans,[78] the administration abandoned the proposal in September.[79] Mining officials contended that no adequate safeguards existed to curtail illegal gold buying. Consequently, the administration introduced laws to "reward and provide some security for Africans who made and reported to government discoveries of economic mineral wealth."[80] Such vacillation in official policy formulation, truncated the licensing of African prospectors.

Indeed, records refer to a legal African prospector in 1948. Dubbed "the only intelligent African producer of legally produced gold," Leokadius Ndede came to the scene in a police report chronicling his arrest by plain-clothes police for suspected illicit gold dealing. The site is the shop of an Asian goldsmith where Ndede had taken his gold for weighing. The arrest followed Ndede's refusal to accept Shs. 90 offered by the Asian goldsmith for the weight of a ten-cent piece of gold. In his defense, Ndede stated:

> Shortly after this a plain clothes policeman picked me up and I showed them [sic] the gold I had produced and went to the police station. I showed them my Prospecting Right Registration and they released me. I suspect the Indian goldsmith bore me malice for not selling him the gold and told the plain clothes man about me. Nobody knew I was carrying my gold to-day until I showed the goldsmith.[81]

Ndede's experience demonstrates the prevailing dominant perceptions among immigrant races that Africans were natural thieves. Indeed, between 1949 and 1951, only five Africans qualified as "legitimate miners" in Kenya colony. Mining officials described them as "literate African miners of 'small worker' type, working isolated workings on scattered claims using hand methods only." While records considered their operations successful, their combined gold exports in 1949 aggregated a paltry 34.74 oz of bullion.[82] This figure rose to 113½ oz in 1950, and then plunged to 31 oz in 1951, less than one third of the 1950 export figure. Their workings consisted "almost entirely of small open-casts in rubble or reef." Mining officials attributed the dismal performance to the uneconomic nature of their properties, coupled with a lack of both capital resources and technical know-how.[83] When the industry folded up in 1952, only three licensed African prospectors remained active.[84] Table 5.3 shows licensed Africans, their areas of origin and gold production figures for 1949.

The foregoing discussion reveals several discernable features. The colonial state's emphatic stance on proficiency in English as a prerequisite for granting mining licences excluded Africans generally and the Luyia specifically

Table 5.3. Licensed African Producers, Localities and Gold Production Figures, 1949

Name	Address	Amount of Gold (oz)
Gari Hansen Elijah	Koyo Nandi, PO Kapsabet	47 in seven months
Leokadius O. Ndede	C/o Catholic Mission, Mbaga, PO Yala	30 in eleven months
H. Benjamin Ojango	Luanda, PO Maseno	18 in four months
W. Zadok Okumu	C/o CMS Ngiya School, PO Yala	18 in ten months
Reverend Mathayo Owino	PO Masena	½ in one month Total 113½

Source: *MGDAR 1950* (Nairobi: Government Printer, 1951), 4. A sixth African, Mathias Kipkorir arap Mitei, licensed for reporting a gold find on his farm in Nandi district, produced nil. The government subsequently revoked his license.

from direct participation in the gold mining industry except as low wage laborers. The few African licensees emerge in the dying hours of the industry. Notably, apart from H. Benjamin Ojango from Luanda in Bunyore, the rest belonged to other African ethnicities outside the central mining area.

In light of the names and addresses provided in Table 5.3, Gari Hansen Elijah was a Kalenjin from Nandi district; Leokadius O. Ndede, W. Zadok Okumu, and Reverend Mathayo Owino were Nilotic Luo. These African entrepreneurs lacked technical skills, experienced insurmountable capital constraints, and worked on unprofitable abandoned rubbles. Capital and technological constraints guaranteed insignificant gold production. The colonial state's reluctance to simplify mining legislation and its emphasis on using oppressive apparatuses of control, stifled local entrepreneurship, relegating the Luyia to low waged labor in the new rural industrial edifice. Certainly, the white man's law provided no entrepreneurial possibilities for the vast majority of Africans in the gold mining industry.

GOLD MINING AND LABOR

Gold mining provided opportunities for a new form of industrial labor in Buluyia. Yet, the local Isukha and Idakho communities were not quick in taking advantage of the job market at their doorsteps. Insufficient numbers turned up for work in the mines, while European mine owners considered those who came to the mines "inefficient." The DC's report of 1932 aptly articulated the official view:

> ... it must be recorded with regret that the natives of the two Kakamega locations have made a very poor response. Their irregularity and reluctance to perform a stated task makes employers go elsewhere for their labor wherever possible. It seems to be very generally agreed that the Kakamega [Isukha and Idakho] are the worst laborers in Kenya.[85]

Oral accounts revealed that Idakho and Isukha men loathed underground work for fear of death.[86] Most considered work in mines dangerous, arduous, unfit and dirty. Those with smallholdings and livestock preferred low paying jobs to labor in mines.[87] Some simply refused to be partners in the destruction of their alienated land, which they perceived as more central to their survival than the European gold mining cartel. In addition, a majority resented binding labor contracts.[88]

Labor problems forced miners to pressure the CNC to sanction the importation of labor for the goldfield. However, the CNC found the proposal undesirable. In his view, it was important to win the goodwill of the local population, whose lukewarm response to industrial labor reflected nervousness with the uncertain land situation.[89]

Nevertheless, colonial economic impositions, particularly taxation, brokered the labor situation, pushing some into the mines. In most cases, the Isukha and Idakho workers labored in the mines to secure money to meet their tax obligations as opposed to their general needs.[90] This compelled the district administration to launch *baraza* campaigns to urge Isukha and Idakho men to take advantage of employment opportunities provided by the gold mining industry, or risk forfeiting the opportunity to imported labor from other ethnic groups.[91]

Such efforts yielded little immediate results as Isukha and Idakho men continued to shun work in the mines, making miners eager to recruit outside labor. For example, at a July 1932 meeting with the governor, the Miners' Association challenged the administration to issue permits for importing labor. Miners argued that the local population was unsuited to, and disliked, the heavy manual labor involved in mining operations. Many sought to import their gangs of trained workers accustomed to strenuous manual labor, and, therefore, more suited for work in the mines.[92]

As the Isukha and Idakho remained indifferent to industrial labor in the goldfield, miners turned to labor from other Luyia groups and non-Luyia migrant labor. Bantu Logoli and Nilotic Luo constituted the most favored underground workers while surface workers comprised "any tribe not resident in the immediate vicinity" of the Kakamega goldfield.[93] Many came from nearby Maragoli, which lacked a full-fledged gold mining industry. The rocky terrain, population pressure and land related problems, created surplus labor in Maragoli. In addition, the relatively better wages in mines and the possibility of working near home, thereby avoiding long labor contracts far from home, constituted additional attractions.[94] The Luo, on the other hand, had a long record of mining experience from work in Macalder mines in SK.[95] Table 5.4 shows the various ethnic groups employed in NK in October 1935.

The importation of outside labor into the goldfield was an intricate issue. The influx of large numbers of non-Luyia labor into the district was bound to create great difficulty with the inevitable establishment of labor compounds. Notwithstanding, the PC favored the importation of skilled labor for key mining positions. The PC argued that in order to attract large capital interests to the goldfields, it was paramount to meet the miners' requirements for local unskilled labor.[96]

Indeed, when large companies began to dominate the goldfield in 1934, labor supply increased appreciably. The labor officer's returns for May 1934 showed that well-established and reputable mining companies employed ap-

Table 5.4. Mining (African) Labor Force—October 1935: Tribal (Ethnic) Composition

Ethnicity	Numbers
Not stated	27
Luo	1,815
Luyia	4,515
Gusii	225
Nandi	69
Kipsigis	1
Kikuyu	38
Akamba	6
Tanganyika Territory	73
Uganda	37
Miscellaneous	8
Total	6,814

Source: Labour Return for the Month of October 1935, KNA: PC/NZA/2/20/6.

proximately 4,000 Africans.[97] During the entire year, the NK goldfield had 6,360 African employees.[98]

In the labor report for April 1935, the provincial labor officer, W. P. Shields, estimated that the Kakamega goldfield had an African labor force of between 6,000 and 7,000 persons.[99] By June, the figure had fallen to 4,857.[100] Table 5.4 shows that in October 1935, mining operations in NK engaged 6,814 men. Of these, 4,515 were Luyia, and the rest comprised mostly Luo.[101] Of the Luyia labor force, 3,096 worked on the surface and 1,330 served underground, mostly Logoli.[102] The remainder held few artisan and clerical jobs.[103] Some sources estimate that between 1935 and 1939, about 10,500 Africans were annually employed in the mining industry.[104]

The labor supply in mines improved due to increased use of female labor. By November 1934, one mine was using female labor to crush quartz. The administration stopped this practice and urged all chiefs to curtail the employment of women in mining work.[105] Thus, while the gold mines offered an economic opportunity to some Luyia men, women found themselves excluded from reaping the same benefits. Two years later, Ordinance No. XXXV The Employment of Women, Young Persons and Children (Amendment) Ordinance of 1936 banned the employment of women on underground work in mines in accordance with the regulations of the International Labour Conference of June 1935.[106]

The adoption of this piece of legislation did not significantly alter employment trends in the mines. By 1938, the labor supply increased as most goldfield employers increasingly harnessed children's labor.[107] Both alluvial miners and big companies employed considerable numbers of cheap but agile juveniles.[108] Out of 1,500 juveniles employed in gold mining operations in

Nyanza Province in 1938, NK district accounted for 700 juvenile surface workers.[109] Moreover, private labor recruiters targeted children, inducing them to travel many miles away from the reserve to work on myriad European capitalist enterprises.[110] The use of child labor received strong opposition from some circles. Archdeacon W. E. Owen's press criticism of child labor in Kenya colony aroused public and Parliamentary outcry in England, forcing the CO to order an inquiry into the minimum age, and the terms and conditions governing the use of juvenile labor.[111] Jomo Kenyatta questioned the legality of binding children in contracts, and The London Group on African Affairs (TLGOAA), called for legislation protecting women and children against industrial labor abuse.[112] The Juveniles Employment Committee chaired by CNC E. B. Hosking subsequently recommended a minimum age of twelve for entry into employment and fourteen for work in industrial undertakings.[113] This was in fact legalized child labor.

The outbreak of WWII significantly altered the labor situation in Nyanza Province. In 1940, the province experienced intense private recruiting activities to meet labor demands outside the province. This pressure heightened the pervasive danger of child labor. Consequently, ordered to check this threat, the provincial administration moved quickly to enforce Ordinance No. XXXV of 1936, which regulated the employment of children below sixteen years.[114]

Competition from the parallel Luyia peasant economy exacerbated the labor situation, especially at the start of each year, when Luyia mineworkers consciously chose to return to their holdings in preparation for planting.[115] Indeed, labor turnout for mines rhythmically fluctuated in accordance with the Luyia agricultural calendar and tax obligations of individual households. For example, in April 1935, an acute labor shortage coincided with intense agricultural activities in the reserve; a time when Luyia preferred working in their own holdings. Shortly after, the labor supply improved with the advent of tax collection.[116] Likewise, the settler economy also suffered competition from the peasant economy. For instance, the Muhoroni-Koru farms faced labor shortage in April of 1935, because of mining demands, and, most notably, African labor demands for planting and cultivating their own holdings. In addition, whenever reserves experienced prosperity, the low wages paid by miners and settlers failed to attract the Luyia.[117]

The peasant economy thus continued to influence labor supply to the mines. For example, in 1937, except for Rosterman Company which met its labor requirements, most mining companies suffered labor shortage because of labor returning home "to start cultivating their *shambas* in preparation for the 'Long Rains'."[118] The shortage coincided with the highest crop prices for the decade in the reserve, which reduced the number of people going out for work. To avert serious shortage, mining concerns contemplated offering additional inducements in form of increased wages or larger meat rations. Indeed, labor was responsive to market conditions by selling their labor power to the most competitive employers.[119] Mining concerns had to adjust their working conditions to remain competitive. Although European employers quickly blamed the labor shortage on the cultivation of cotton in reserves, the truth was that the general recovery both in the reserves and on larger estates had pushed up wages, thereby ending the days of *shenzi* [foolish, stupid] laborers.[120] Shortage on smaller mines resulted from reliance on casual labor who worked irregularly.[121]

An economic survey carried out by NK DO E. A. Sweatman in 1937, showed that most Luyia primarily depended on the success or failure of a particular crop. During 1937, the Luyia earned and spent a staggering £92,000 on shop goods. Of this total, between £25,000 and £30,000 came from the mining sector. Notably, Sweatman acknowledged that the amounts expended resulted from better cash crop harvest and the good prices offered especially for cotton and maize. In his view, the more economically sound the Luyia became, the more uncertain the labor supply became. The survey also revealed that the standard of living had changed very little. Most houses had a sleeping mat, an improvised chair and table, or invariably a stool and wooden box.[122] Another survey conducted in 1938 showed that earnings were about £35,000 less and expenditure about £17,000 less than the previous year. Asian traders reported that 60 percent of their sales constituted clothing, 30 percent foodstuffs, and the remainder comprised comparative luxuries such as soap, shoes, sewing machines, ploughs and bicycles. Luyia households spent very little on improved housing and furniture.[123]

Apart from laboring in the mines, numerous Luyia worked as wage-contract laborers in the service of Europeans, particularly on the maize and sisal farms in the neighboring Uasin Gishu and Trans Nzoia districts, as well as in the farming areas of Central and Coast Province, for the Public Works Department (PWD), and in the towns.[124] In 1938, 31,078 Luyia worked for Europeans, 5,000 of whom worked in the gold mining industry. The bulk of these employees opted for a six-month labor contract and returned home at the end of the contract period.[125]

Clearly, employment in the mining industry was not the only source of Luyia incomes during this period. In 1935, wage earnings derived from work within the province amounted to £261,500. These included, mining £93,500 (36 percent); Kericho tea estates £65,000 (25 percent); government departments, excluding the railways £30,000 (12 percent); farms in the Kisumu-Londiani district £40,000 (15 percent); farms in Kericho district other than tea estates £18,000 (7 percent); Kisumu town exclusive of government employees but including the railways £15,000 (6 percent).[126] In 1936, the incomes distribution stood at mines, £85,000 (28 percent); tea estates £70,000 (21 percent); sugar £15,000 (5 percent); sisal, coffee, maize, cattle, mixed estates £35,000 (12 percent); K.U.R £15,000 (5 percent); PWD £24,500 (8 percent); other government departments £24,000 (8 percent); LNC £22,000 (7 percent); fishing £1,500 (1 percent); ginnery £2,000 (1 percent); domestic and shop assistants £9,000 (3 percent); totaling 303,000.[127]

It is difficult to ascertain the amount of money earned by Africans from Nyanza Province from work outside the province, and almost impossible to guess how much of such earnings found its way back into the province. In 1936, contract labor engaged outside the province earned £19,000.[128] In the case of NK district, the mines did not experience any labor difficulties, although the number of Africans going out as contract laborers increased moderately.[129] By September 1936, SK mines faced acute competition from mines in Tanganyika territory, some of which offered Shs. 14 to underground and mill workers.[130]

While only 20 percent of Luyia men worked outside the district in 1932, the figure had risen to 37 percent in 1937, and, doubled in the 1950s.[131] In their 1937 Annual Report, mining officials reported, "It might be thought that the native would prefer to work on the mines nearest to his home. Apparently, however, he prefers to work some distance away and to return for his periods of leave well dressed and with money to burn."[132] This demonstrates that the gold mining industry in no way diminished the significance of labor migrancy in Luyia livelihood. Labor migrancy remained a potent economic option.

Even with the end of WWII, NK district continued to be the largest labor exporting area in Kenya colony. Curiously, the only large gold mining concern in the district, Rosterman Gold Mining Company Ltd., continued to suffer acute labor shortage, particularly for underground work. This shortage compelled the administration to establish labor liaison committees to facilitate labor recruitment for the company and the Trans-Nzoia and Uasin Gishu European farmers.[133] Labor migrancy continued to characterize NK throughout 1947, as men left the district in considerable numbers for work outside the reserve. The reduced opportunities for labor within the district provided the impetus for those seeking better incomes. While Rosterman had experienced an acute labor shortage in 1946, the company was turning away labor in 1947.[134]

Clearly, gold mining provided industrial labor opportunities in Buluyia. Deep-rooted land grievances and the dangers associated with working in mines, however, forced the local Idakho and Isukha men to shun industrial labor. This led to use of migrant Luyia and non-Luyia labor as well as the labor of children and women. Luyia households, nonetheless, continued to rely on the parallel peasant economy for survival, and incomes from other forms of migrant labor besides mine work. Remittances for war service constituted another source of revenue for Luyia households whose members were conscripted into the army. In 1944, for instance, families in NK received £77,976 in family allotments and special remittances. In addition, the DC, F. D. Hislop disbursed £33,152 in cash to African servicemen who were on leave. In the same year, incomes from agricultural activities netted £188,000.[135] Clearly the prosperity of Luyia households in 1945 was principally a result of incomes from the sale of crops, and secondarily from military sources and wages earned outside the reserve.[136] Notably, the loss of land, off-farm labor and coerced wartime labor obligations, contributed to agricultural decline in the core mining locations.

WORLD WAR II AND LABOR CONSCRIPTION

The outbreak of WWII placed tremendous strains on British colonies. In the case of Kenya, the colony was called upon to contribute to the war effort in form of foodstuffs, livestock, money, and most important for this study, labor.[137] Luyia males were recruited in the King's African Rifle (KAR), police, pioneer corps, and a host of other military duties. In addition, demand for African labor emanated from the tremendous labor shortages in the settler economy. Labor demand on European farms in Trans-Nzoia district adversely affected the Luyia. Moreover, an excellent maize and sorghum crop harvest in the reserve enabled most Luyia households to meet their tax obligations without going out to work.[138] The Warden of Mines dejectedly reported, "(T)he natives, having plenty of food and beer, are disinclined to work."[139]

The worrisome labor situation prompted the district administration to intervene. At a meeting with representatives of the Trans-Nzoia flax growers, the DC NK resolved to adopt coercion in pushing up labor supply for European settlers. All chiefs in the adjacent northern locations were required to supply 1,000 men.[140] Such administrative pressure on local authorities temporarily changed the labor supply. For example, between October 1940 and July 1941, 1,497 Luyia men from Kabras, Bukusu, Malakisi, and Elgon locations were dispatched by lorry (trucks) for work on the Trans-Nzoia farms.[141] Although initially reluctant to sanction conscription of labor for civil work, London conceded to such conscription in October 1941.[142] Consequently, the colonial state adopted The Defense (African Labour for Essential Undertakings) Regulations, which enforced the compulsory conscription of Africans.[143]

To push numbers up, the PC Nyanza held numerous propaganda *barazas* in NK district urging the Luyia to come out for industrial labor as part of the war effort. By the end of October 1941, the colonial state adopted a scheme of "administrative assisted recruitment" compelling Africans to work for European farmers.[144] A report by the DC NK to the PC Nyanza, illustrates the approach:

> On arrival at a camp, I always address the labour in this manner. 'You heard what the *Bwana* [sir] Provincial Commissioner said recently at his *barazas*, namely that for the purpose of maintaining our *askaris* [soldiers] in the field it is necessary to keep up production. The Governor, who is our leader, has told us that for the present he has stopped military recruitment as production is lagging, and he therefore wants people to help on the farms etc. If any man refuses to go without good reason, or who goes and slacks, he is a friend of the enemy'....[145]

Between October 20 and November 30, 1941, NK district supplied 1,559 men as assisted labor for sisal, sugar, and coffee farms, fuel and aerodromes. Yet, numerous mines in SK seeking assisted labor, received none.[146] This was due to differences between administrators as to who should receive such labor. The PC favored supplying only industries necessary for better execution of the war. These war-time essential services included sisal and fuel cutting, saw milling, sugar cane cutting, stevedore work, trunk road and aerodrome work, with underground mining coming last. Although the gold industry fell within the purview of essential industries, the PC maintained that assistance, be restricted to viable mines.[147] Thus, mining received little assistance.

Although assisted recruitment proved unpopular shortly after its inauguration, the military imperatives propelled the process. For instance, official records from the Department of Labor showed that in September 1941, 87,139 Nyanza males served in civil employment. This figure rose to 89,448 in October, 91,867 in November, and 93,212 at the end of the year.[148] This administratively enforced recruitment affected labor supply for employers within the province, including the gold industry.[149] In his Annual Report and HOR for 1942, PC Fazan estimated that Nyanza Province had 30,000 men in military service and 110,000 in civil employment outside the reserve.[150] Three years later, 50,738 adult Luyia males were in employment within the district.[151]

As military demands began to decline, production demands rose significantly. Unfortunately, labor demands for European enterprises were surging against a background of heightened resentment among Africans against assisted labor. Consequently, as the administration found it increasingly difficult to fill required labor quotas, it initiated district labor camps in 1943. The general famine of 1943 made an already bad situation worse. The failure of the short rains in 1942 led to famine, which persisted through June of 1944. The more densely populated southern locations of Maragoli and Bunyore, and, the major mining locations of Isukha, Idakho and Tiriki, bore the brunt of this famine.[152] An estimated 30,000 people suffered serious food shortage, with the elderly dying from starvation. The famine brought in its wake a thriving black market in food crops. Profiteers purchased maize from other locations above government controlled prices, averaging between Shs. 30 and Shs.80 per 200 lb bag, compelling the NK LNC to provide Shs. 3,000 for importing 1,000 bags of maize from European farmers in Trans Nzoia district.[153]

The significance of the 1943 famine for this study lies in its impact on the gold mining locations. Mining locations suffered the worst famine situation, which, neither cash compensation from leases nor wage incomes from mine labor satisfied. In fact, the conscription of labor for European settlers, coupled with labor migrancy significantly reduced agricultural production in the reserve. For example, in 1943, Nyanza Province provided 12,844 men for conscripted civil labor.[154] In addition, European farmers increasingly turned to child labor.[155] In 1943, an estimated 55 percent of able-bodied Luyia men were employed outside the district. Thus, the available labor force was insufficient to meet the huge volume of crop production in progress in the reserve.[156] The exodus of males from the reserve put pressure on Luyia women as they grappled with the increased agricultural tasks.[157]

This process continued throughout the war period. According to the labor bulletin for December 31, 1944, 41,834 Luyia males registered in civil employment, the highest ethnic total in the colony.[158] Most of them worked in European settler farming areas and urban centers such as Kericho, Nakuru, Thika, and Mombasa. By December 1945, the number had reached 50,738.

Clearly, WWII placed new demands on Luyia labor for the settler sector and military service. The flow of Luyia labor to these sectors and the gold mines revolved around the demands of the peasant economy. This explains why the colonial state often resorted to coerced labor recruitment for the so-called essential industries, including underground mining. Nevertheless, as the colonial state energetically recruited Africans to service the settler farming community, the gold mining industry received little assistance. The forcible extraction of labor from the Luyia peasant economy affected agricultural production with the outcome that the failure of rain in 1942 occasioned a famine situation in the gold mining locations between 1943 and 1944. Thus, peasant agriculture remained central to the survival of Luyia households. Monetary compensation for mining leases and incomes from mine wages never obliterated the peasant economy.[159]

WORKING CONDITIONS

Conventional knowledge throughout the world attests to the unpleasant nature of work in mining enterprises. A scrutiny of the terms and conditions of work in Kakamega brings to question romantic interpretations of working conditions in the NK goldfield espoused by scholars like Fearn and Roberts, and most official sources. For example, Roberts asserts that juveniles carrying alluvial mud and gravel were "thoroughly enjoying themselves, with a definite will and wish to work."[160] Similarly, the 1937 *NADAR* noted that children performed light duties and enjoyed the protection of labor officers against abuse.[161] However, the observations of the Commissioner of Mines, on a visit to a mining site the same year, revealed that both children and adults endured depressing work conditions:

> A gang of native children were carrying pay-dirt on their heads some 100 yards to the sluice boxes. In the 'paddock' they were struggling up to their knees in thick gluey slush. Ninety trips are made daily on a 'ticket' of 30 days to earn a monthly wage of three shillings. It has been estimated that the weight carried each day is one ton. Older natives were struggling with barrow loads of overburden up a slope, sliding and slipping in the mud to tip the earth not into adjacent pits but on top of a small hill. The European in charge was absent.[162]

A year later, the Committee on the Employment of Juveniles concluded that task work allotted child employees in the mines was "possibly a little heavy." Children under twelve years of age struggled with heavy pans of mud, sorted and broke ore with hammers for five to eight hours a day for a monthly wage of three shillings.[163] African mine workers in Kakamega did not constitute a happy-go-about squad. They performed difficult, monotonous tasks for long hours under hard taskmasters. On arrival at the site, the *nyapara* (African supervisor) stamped the worker's tickets, divided them into groups of ten to fifteen, and escorted them to the underground station in a lift. Equipped with mining paraphernalia and a helmet, workers labored in bending postures in poorly ventilated and lighted conditions, under the piercing eyes of European supervisors.[164] Some informants described the procedure in some detail:

> On arrival at the mine, workers wore a helmet and were given shovels, *jembes* [hoes], axes, steel rods, and *karais* [pans] and transported in a lift in groups of about fifteen to underground stations. Once in the shaft, with the advice of European supervisors, Africans commenced drilling the rocks to reach the ore using boring steel. Explosives would be used to drill the holes- with cold air blowing the hot rocks to cool them. The rocks would then be filled into gallons, taken to the surface by the lift, and carried to the processing site by surface workers.

At times, underground rail cars transported ores to the washing and boiling plant.[165]

A British journalist, Negley Farson recounted similar depressing conditions in a chilling first hand account during a visit to Kakamega in 1947:

> I went down 1,000 ft in one of them [mines] that was still working, and never had I felt such a sense of depression as in that soft, damp heat, lighted by a few lamps, where some black men were clearing a space where a bored white man would drill a hole for one more charge of dynamite into the vanishing reef.[166]

Such observations, representative of the position of most informants, do not depict particularly good working conditions in the NK goldmines. Work in the mines devalued African labor, and as one informant put it, they constituted "a heap of unhappy dusty workers."[167]

Apart from bad working conditions, labor difficulties also resulted from conflict over wages and the length of the working day. Although wages in mines were relatively higher than other wage opportunities, they were generally very low. In 1931, adult laborers received Shs. 8 for less than a full day's work.[168] By 1932, miners were accusing the Luyia of deliberately trying to freeze them out by demanding high wages. They, therefore, demanded standardized tasks and wages.[169] In 1935, most workers earned Shs. 10–14 followed by workers earning under Shs. 10. The Luo had the smaller percentage in the latter category compared to the Luyia. Table 5.5 illustrates the occupations, total earnings for each occupation, and average monthly wages in October 1935.

While in 1933, the President of the Miners Association, Colonel Swinton Howe, alleged that African labor consented to "work for 5 hours daily and is only efficient for 3 hours out of that time."[170] Sources show that in 1934, both European and African employees at Kimingini and Musgraves worked from 6 a.m to 5 p.m, including Saturdays and Sundays.[171] For a majority of the African workers, a normal working day lasted between 6 a.m and 6 p.m. This long working day was a cause of disagreement for local Luyia.

Fearn also applauded mining companies for enhancing the health of their workers by providing improved diets. True, some mining companies provided food rations for workers, except for commuters who lived in their homes. For example, Kimingini, Rosterman and Kavirondo Gold Mines, provided well-cooked meals consisting of porridge with sugar, *posho* (corn meal), beans, *chiroko* (green grams), meat, fresh vegetables and salt.

Surface laborers show the largest total followed by underground workers. In addition to the total of Shs. 141, 858 monthly wages earned there was also a sum of Shs. 23, 341 representing the total value of rations issued which

Table 5.5. Occupations, Total Earnings for Each Occupation, and Monthly Wages, 1935

Occupation	Numbers Employed	Total Earnings (Shs)	Monthly (Shs)
Not stated	106	1,065	10.00
Clerks	45	1,385	30.77
Office assistants	3	75	25.00
Lorry and motor drivers	36	1,156	32.11
Blacksmiths	29	761	26.24
Carpenters	51	1,925	37.74
Masons	15	642	42.80
Other skilled labor	141	2,177	15.44
Police and watchmen	106	1,906	17.98
Headmen	197	3,266	16.58
Laborers (surface)	7,369	60,116	8.16
Laborers (underground)	4,658	51,890	11.12
Hospital attendants	23	590	25.65
Cooks (compound)	27	232	8.59
Laborers (not stated)	607	5,190	8.55
Cooks (domestic)	79	1,955	24.74
Other indoor domestic servants	185	2,772	14.98
Other outdoor domestic servants	32	428	13.38
Fuel cutters	371	3,774	10.16
Sweepers	35	553	15.80
	14,115	141,858	10.05

Source: Labour Return for the Month of October 1935, KNA; PC/NZA/2/20/6.

works out at Shs. 1.65 per individual per month. *Posho* cost about Shs. 1.50 per load of 60 lbs, while other commodities such as fresh vegetables, beans and meat were also cheap.

Yet there existed notable exclusions. Rosterman did not provide food for underground morning shifts. Kavirondo Gold Mines served one and a half pints of hot cocoa to all workers coming off shift, but not day surface workers. Compared to small mines whose rations consisted of maize flour and salt, rations on large mines were reportedly liberal. Aimed at improving the health and capacity of labor, the regularized diet, supposedly contributed to "excellent health, contentment and efficiency among labor."[172] For example, when Kimingini Company was facing an uncertain labor supply in 1937, the management moved swiftly and increased its food rations to four lbs of meat per week, in addition to cooked *posho* and other food. This move quickly paid off as the company attained its labor requirements shortly.[173] When labor officials inquired from large mining companies and the PWD during the same year about their labor supply and its quality, the firms provided mixed responses. Rosterman and Kimingini reported sufficient supplies of "improving" labor, while companies with low wage rates and less attractive conditions of work, on the other hand, reported insufficient and indifferent labor. While blaming the labor dilemma on the reigning high food crop prices in the reserve, most miners nevertheless believed that sufficient inducement including reasonable wages and large meat rations could boost labor turnout.[174] However, by 1941, less than half of the African labor force employed by Kavirondo Gold Mines took advantage of the hot meals in *lieu* of dry rations.[175] This means a small fraction of labor benefited from company administered meals. Indeed, the prevalence of labor unrest among mineworkers points to significant labor discontent in the goldfield.

INDUSTRIAL LABOR PROTEST

Although scholars like Fearn and Roberts, as well as archival sources provided idyllic views of working conditions in the mines, this study shows that while wages offered by larger mining companies were comparatively better than wages obtainable from European settler enterprises in Nyanza Province, worker discontent in the mines was a common phenomenon. The diverse forms of protest mineworkers adopted including strike action, the ultimate form of labor protest, points to serious labor problems in the NK goldfield.

African labor protested unfavorable conditions through absenteeism and desertion. While some scholars perceive absenteeism, desertion, or "absconding" reflexive reaction to bad working conditions, others consider it an embryonic form of labor protest, or a strategy of class action.[176] Initially, absenteeism and desertion constituted the most common exit option from the unfavorable labor market. In 1935, for instance, Kimingini suffered an acute labor shortage with 263 absentee laborers on their books. Contracted labor abandoned work prematurely or simply "went home on receipt of their dues, or over a weekend and 'forget to return' to work," in violation of the requirement to obtain a discharge endorsement on their registration certificates. State apparatuses of control often tracked down such deserters and charged them with breach of contract or forcibly returned them to work.[177] Lacking the power of a union, labor's defense against low wages, extortionate practices such as cutting several days' tickets for a day's delinquency, and poor conditions of employment, was to stay at home and make a more profitable living there.[178]

Although the history of strikes in Kenya dates to the 1900s, the first generation of African strikes occurred in Nairobi in 1922.[179] Major strikes rocked the colonial urban nerve centers of Mombasa and Nairobi in the 1930s, reaching a peak between 1947 and 1952.[180] While no direct evidence links these general strikes to events in the goldfield, their impact reverberated throughout Kenya colony. For example, the Kakamega mineworkers severally resorted to strike action in the 1940s. This study contends that strike action, the temporary withdrawal of labor and the ultimate tool for negotiating better terms and conditions of work, was a manifestation of serious labor discontent. That African mine workers utilized strike action in Kakamega makes idyllic interpretations of the labor experience in NK untenable.

In May 1941, two strikes occurred in quick succession at Rosterman and Kimingini, the largest and, reportedly, the best managed mining concerns in the district. In the case of Rosterman, 250 underground African laborers dominated by Luo, went on strike on the morning of May 19, 1941. They invaded European quarters to gain the support of domestic workers. The DO, Lambert, and another official, A. I. Trent, rushed to the mine where they arrested several workers. Rather than break the strike, the protesters marched to the DC's office in a body, demanding an increase in wages. However, they returned to work when the DC declared their action illegal. Although the proper identity of the strike leaders was never established, the administration dismissed them as troublemakers.[181] Two days after the Rosterman strike, Kimingini (then under the management of Kerebe Mines Ltd.) workers downed their tools. Unlike the Rosterman strike, the Kimingini company management settled the unrest without state intervention. Again, Luo workers dominated the strike.[182]

The Rosterman strike is intriguing if archival reports of welfare conditions of workers at this mine are accurate. A report by the company manager, D. Kerr-Cross, for December 1941 indicated that in addition to wages, all employees received fresh supplies of food rations, averaging fourteen lbs of *posho* (cornmeal), two lbs of *chiroko* (green grams), two lbs of either groundnuts (peanuts) or simsim (sesame), two lbs of meat and half lb of salt. Moreover, some 200 wives with families resident in the labor compound received food rations. Except for unskilled surface workers who earned Shs. 9, minimum monthly wages were reportedly higher than Shs. 10 per month as shown in Table 5.6.[183]

Again, in 1949, all labor at Rosterman went on strike, in spite the efforts of the DC and the labor officer. The management issued an ultimatum to the workers to resume work or take discharges the following day. Although the laborers at first opted for the latter, surprisingly a majority returned to work.

Table 5.6. Labor Categories, Types of Labor and Wages, Rosterman Gold Mining Ltd., 1941

Category	Type of Labor	Wages (Shs.)
Underground	Mucker boys	12.00
	Pump attendant	12.00
	Spanner boys	13.50
	Drill runners	15.00
	Hoist drivers	15.00
Surface Labor	All unskilled	9.00
Power house, Machine shop	Crusher attendant	10.50
	Engine boys	10.50
	Steel shop laborers	10.50
	Fitters mates	10.50
	Drill sharpeners	21.00
Mill	Pump boys	10.50
	Cam boys	12.00
	Stamp boys	12.00
	Agitator boys	12.00
	Headmen	18.00

Source: D. Kerr-Cross, Manager, Rosterman Gold Mines, December 30, 1941, KNA: PC/NZA/2/20/13. Kerr-Cross referred to three "exceptional" cases of skilled workshop "boys" earning Shs. 105, Shs. 58.50 and Shs. 64.50 respectively as turners and welders.

The DC subsequently dismissed the strike as the work of "a small coterie of discontented labourers."[184]

Labor discontent among African employees in the mines emanated from various factors. First, was the issue of the universally low wages paid by mining companies. As early as December 1931, shortly after the gold discovery, mineworkers received Shs. 8 per a thirty-day ticket with very few miners paying Shs. 10.[185] For example, between 1933 and 1940, the same monthly wage regime of approximately Shs. 10 for surface and Shs. 12 for underground work, with *posho* rations reigned in the mines.[186] While skilled African workers, such as masons and carpenters, earned Shs. 42.80 and Shs. 37.74 respectively in 1934, monthly wages for European employees engaged in reef mining in 1934, 1935, and 1936 averaged about Shs. 327, Shs. 441, and Shs. 572 respectively. European alluvial workers earned about Shs. 71, Shs. 215, and Shs. 313 respectively for the same period, while their African counterparts received roughly Shs. 8 in 1934 and Shs. 7 between 1935 and 1936.[187] Such highly differentiated wage levels reflected the deeply entrenched color bar that characterized colonial Kenya.

Second, wage withholding bred labor discontent. Although archival sources are replete with such cases, a few examples will illuminate the magnitude of the problem. As early as December 1931, an investigation into a large employer's complaint over declining labor supply revealed that he had withheld payment for twenty days to workers who had missed their pay date. The labor had to wait thirty days before the next payday.[188] In 1932, three European miners, operating in South Marama, withheld wages and refused to pay compensation.[189] In 1933, another European owed £100 in timber and wages.[190] In February 1934, H. Tooley, was convicted a second time for withholding Shs. 919.45 from 129 laborers.[191] Three serious cases of wage withholding occurred in July 1934.[192] In 1935, one serious case occurred, and the following year, administration recovered Shs. 400 from B. S. Duffy's mining deposit to pay wages owed.[193] The most dramatic case, however, involved F. Davies of Alphega Mining Syndicate in Maragoli who failed to pay Shs. 6,909.53 to 626 laborers and faced prosecution on 150 counts. Davies had also issued Hut tax tickets in payment of wages to 100 laborers but never submitted the money to authorities.[194] In 1939, H. P. Robotham, stationed in South Marama owed wages to the bulk of his labor, which he conveniently laid off, and a Mrs. Jones owed Shs. 1, 700 to ninety-three laborers.[195] As late as 1946, the administration convicted and fined J. N. Deventer Shs. 20, and ordered him to pay Shs. 329 in unpaid wages.[196] These examples illustrate the insecurity of African workers' wages due to non-payment or dismissal following employer's financial vagaries.[197]

Third, African labor constituted a majority in arduous and low cadre jobs in the gold mines due to stiff competition in skilled work from immigrant races. In 1940, assistant Warden of Mines, W. J. Bailey warned about the threat skilled Chinese labor posed to Africans:

> It has just come to light in some quarters that Chinamen [Chinese] are being employed at the Macalder and Bukura Mines. It seems a pity that local native *fundis* [craftsmen] cannot compete with this class of labour and it looks as though the local inhabitants have not very much to look forward to if they are unable to do better.[198]

Indeed, in 1941, the two companies had forty Chinese male *"fundis."*[199] This was also true of Asian labor. In January 1941, for example, Rosterman had forty-nine Asian employees in skilled and comparatively better paying jobs. These included thirteen carpenters, seven fitters, seven timber men, four masons, three hoist drivers, two electricians, two mill fitters, and one each in the following categories; blacksmith, compound clerk, learner mill, molder, pattern maker, pipe fitter, power house, pumping, store keeper, track layer, water supply.[200] In addition, Asians and Goans dominated administrative, technical and construction dockets. Although only three Prisoners of War

(POWs) worked in the goldfield in 1943, by 1945, thirty Italian POWs served as technicians, carpenters, masons and clerks.[201]

Fourth, African labor resented poor housing conditions. Major companies such as Kavirondo, Kimingini, Rosterman, Risks, and Owombo, provided housing for their labor. Except for local labor operating from their homes, all workers resided in company built houses in racially designated and, at times, ethnic-specific compounds. Europeans enjoyed the comfort of elegant permanent buildings, Asians got second best semi-permanent houses, and Africans had to be content with grass thatched mud huts.[202] Sometimes, mining companies erected labor camps in the middle of Luyia homesteads. In November 1933, for example, the administration instructed mining companies to suspend the expansion of camps in the core mining area due to congestion.[203] Moreover, except for Rosterman, tea plantations offered better housing for labor.[204] Notably, labor in alluvial workers' compounds endured the worst housing amenities. For example, in 1937, health officials intervened to get Mrs. Stitt to clean, renovate and whitewash her labor compound.[205] However, Rosterman constructed a Social Hall, and commenced a program of improved African tiled houses in 1943.[206] Yet, save for Rosterman and Kimingini, most companies prohibited workers' families from residing in labor compounds.[207] This destabilized the family institution.

Apart from poor terms and conditions of work, mineworkers endured immense maltreatment from their European employees. Like the Kenya settler community, miners erected an impenetrable racial curtain between them and their African labor, and, Africans in general. In addition to the conventional myth of the "lazy native", Europeans coined and effectively utilized derogatory epithets such as *shenzi* [foolish, stupid], *nyani* [baboon], *nyoka nyeusi* [black snakes], in their day-to-day reference to Africans. Moreover, miners re-organized and regulated every aspect of African lives including dress codes, social interaction, and general movement. As a rule, Africans wore *marikani* [cheap cotton clothes] or khaki shorts. African men could not wear trousers, shoes, and hats in the presence of Europeans. It was mandatory for Africans to give way to European *bwana* [sir] and *memsahib* [madam], and to follow the chain of orders on pain of verbal abuse, threats, flogging, or dismissal.[208] Imbali and Muhati served a month in jail for confiscating a gun from Europeans in a brawl involving failure to give way.[209] Moreover, in March 1934, Rodseth, a compound manager at Risks, gave twenty strokes to an alleged thief, while his *askaris* [policemen] administered thirty strokes to three trespassers.[210] Ironically, a month later, at the Overseas League luncheon, the S of S for the Colonies applauded the alleged excellent racial relations in the goldfield: "I am absolutely convinced that nowhere in the world did better relations exist between Natives and Europeans than in Kakamega."[211]

Predictably, such Machiavellian discipline, and disrespect for Africans snowballed into unrest.[212] In September 1933, local East Isukha Africans attacked non-Luyia workers leaving work at Kenya Consolidated Company. Later, Europeans suffered two similar assaults. The mining fraternity blamed Chief Sore for making no effort to arrest the assailants.[213] The undercurrent of animosity between Risks laborers and the local inhabitants led to assaults on both sides. In one incident, the compound manager intervened in a row, caught an African not belonging to the compound, beat and locked him up for the night. The manager's action created great resentment among Africans after learning that the victim was Chief Milimu's son.[214] Two months later, the administration fined a Dutch miner Shs. 200 and Shs. 50 compensation for severely beating an African.[214] In October 1936, three Cornish miners at Rosterman paid fines for "foul behaviour" and using "filthy language."[216] In August 1936, a European paid Shs. 100 fine for beating a laborer at Owombo, in Tiriki.[217] In September 1936, a disturbance occurred at Edzawa mine due to "tackless [sic] handling of a situation by a European employee."[218] In August 1938, a former-houseboy of one Lendrum, compound manager at Kavirondo Gold Mines, committed suicide because of an assault and wrongful confinement.[219]

The above analysis exposes the fallacy of romanticizing the experience of African labor in the NK goldfield. African mineworkers endured long hours of difficult, dangerous and monotonous tasks for low wages, under strict supervision. Few laborers took advantage of company administered meals and most resented competition from immigrant races in skilled labor. Work in the gold mines devalued African labor. Africans suffered from poor housing conditions and general maltreatment from European employers. This resentment found expression in diverse forms of labor protest, including absenteeism, desertion, and strike action.

GOLD MINES AS DEATH TRAPS

The fear of death in mines contributed to Isukha/Idakho distaste for underground work in the gold mines. Contrary to Fearn and Roberts's contention that the Kakamega goldfield experienced a low rate of accident related deaths,[220] both archival and oral sources indicate the common incidence of fatal mining accidents.

Mining accidents resulted from diverse circumstances: walls caving in; delayed explosives; late entry into the lift or mechanical failure; gas poisoning and suffocation; slipping in shafts; badly filled shafts sinking; falling into open pits; being ensnared in machines; electrocution; drowning in flooded shafts; and motor accidents.[221] Livestock also fell in unfenced pits and died from water pollution.[222]

Informants relived and memorialized major accidents at deep mining sites including Rosterman, Kimingini, Musgraves, Bukura, Piccadilly Circus (Sigalagala), Risks, and Tintax. Table 5.7 shows the number of mining accidents, numbers killed and injured between 1933 and 1952.[223] The table shows that Africans and "non-Africans" sustained more deaths and injuries compared to Europeans. For example, for years providing racial breakdown of those who were killed or injured in mining accidents, Africans were the majority in both cases. In 1934, for instance, ten Africans died in mining accidents. Similarly, between 1936 and 1938, twenty-one Africans perished in mining accidents. During the same period no fatalities were reported for the European population in the goldfield.

The first fatal accident in Maragoli during 1932 inaugurated the "era of death in the mines."[224] In this accident, falling earth entrapped five African workers, resulting in one death. In its aftermath, other workers "lost their heads" and angrily brandished their tools in the faces of the Europeans. According to Archdeacon Owen, mine owners dispatched the laborers "to work without safeguarding against a fall of earth."[225]

In 1933, five deaths and twenty-five injuries occurred in Nyanza Province. In four of the fatal cases, company management was absolved, and only one criminal action was taken. The administration blamed the accidents on the "treacherous nature of the soil," and lack of experience among miners. Except for two cases, compensation was reportedly paid in accordance with the scale laid down by the principal labor inspector.[226]

Oral narratives vividly recounted several mining accidents. The most commonly cited was a 1934 accident in which three Luo and four Logoli underground workers perished at Musgraves. As the walls caved in and the earth swallowed the workers, one of the Logoli victims, Mudavadi, went down singing:

> *Mukali wange yambolelanga, udzia gi wa Mukhomoli?*
> *Na inzi nammbolera kulihonera gi inzala?*
> *Mukali wange yambolelanga kula maduma.*
> *Na inzi nammbolera inzi nadzia wa Mukhomoli.*
> *Mukali wange yambolelanga kulidza ingombe.*
> *Ninzi nimbola nadzia kukasi wa Mukhomoli.*[227]
> My wife told me why are you going to Mukhomoli's?
> I told her how shall we avoid starvation?
> My wife told me to buy maize.

Table 5.7. Mining Accidents, Numbers Killed and Injured, 1933–52

Year	Number of Accidents		Persons Killed	Persons Injured
1933	25	Non-Europeans	5	21
		Europeans	—	2
1934	18	Africans	10	5
		Europeans	—	4
1935	44*	Non-Europeans	6	5
		Europeans	—	4
1936	35*	Africans	7	47
	50**	Europeans	—	3
		Asians	—	1
1937	21*	Africans	7	14
		Europeans	—	1
		Asians	—	—
1938	17*	Africans	7	11
		Europeans	—	5
		Asians	—	1
1939	41	All races	11	33
1940	60	All races	13	54
1941	60	All races	11	57
1943	36	All races	6	30
1944	31	All races	6	25
1945	27	All races	4	25
1946	26	All races	3	23
1948	4*	Africans	—	2
		Europeans	—	1
		Asians	—	1
1949	4*	Africans	—	6
		Europeans	—	—
		Asians	—	—
		Seychellois	—	1
1952	2*	Africans	—	2
		Europeans	—	—
		Asians	—	—
		Seychellois	—	—
Totals	501		96	384

Sources: *MGDAR 1933*, 10; *MGDAR 1934*, 14; NPIR January 1934, KNA: PC/NZA/4/5/1; *MGDAR 1935* (Nairobi: Government Printer, 1936), 61; NPAR 1936, KNA: PC/NZA/1/31; *MGDAR 1936*, 31; *MGDAR 1937*, 35; *MGDAR 1938* (Nairobi: Government Printer, 1939),24; *MGDAR 1939*, 15; NPIR for June 1939, KNA: PC/NZA/4/5/3; *MGDAR 1940*, 6; Unpublished MGDAR 1941; Unpublished MGDAR 1943; *Land, Mines & Surveys Department Annual Report (LMSDAR) 1946* (Nairobi: Government Printer, 1948), 15; *LMSDAR 1948* (Nairobi: Government Printer 1950), 7; *MGDAR 1949*, 8; *MGDAR 1952*, 8; NKDAR 1932, KNA: DC/NN/1/13; NKDAR 1934; KNA: DC/NN/1/15. *denotes Accidents in Kakamega goldfield and **represents total accidents in Kenya colony.

But I told her that I will go to Mukhomoli's.
My wife told me to sell a cow.
But I told her that I was going to work at Mukhomoli's.

This song captures the agony of a Logoli wife concerned about the dangers of underground work and her unsuccessful attempt to stop her husband from taking up work in mines. Conversely, it also illustrates the determination of a Logoli male head of a household, risking death in order to meet the family's subsistence requirements.

Luyia women took the lead in protesting the deaths of their fathers, husbands, brothers, and sons. Incensed by the company's decision to abandon recovery efforts and the subsequent hurried filling in of the "death trap," at Musgraves, a group of Logoli women who came to claim Mudavadi's body roughed up the manager before local police intervened.[228] To these women, company action symbolized the denigration of African lives for the glitter of gold. Accordingly, the women assumed the new mantle of avenging the deaths of their male relations. In fact, the inquiry into the accident absolved the company of any wrongdoing and ordered it to pay a total of Shs. 1,632 to the relatives of the deceased.[229] Moreover, the administration held *barazas* warning the Luyia against "unruliness or mobbing following a mining accident."[230]

A fatal accident at Rosterman in March 1934 was attributed to "gas acting upon the deceased's weak lungs,"[231] suggesting that miners engaged African labor without medical examinations. Indeed, this accident prompted the district Medical Officer of Health (MOH) to recommend contracts and medical examinations for underground labor. In 1935, Adriano Vihembo, a compressor operator at Kimingini, lost two fingers in an explosion that claimed the lives of two co-workers. Vihembo had assumed this highly technical task after a brief two months' "training".[232] In June 1935 alone, Rosterman recorded six mining accidents, two of them fatal. In the subsequent investigation, the administration absolved both companies from any responsibility.[233] Another accident survivor, Joseph Imbali, spent a month in hospital nursing a serious thigh injury sustained in an accident in Rosterman.[234] Imbali blamed the accident on bhang (marijuana) consumption by the lift operator.

On October 22, 1938, a stope caved in at Kimingini mine, imprisoning two local African laborers and injuring six others. The company management abandoned recovery efforts with the approval of the government Mining Engineer, the Inspector of Mines and the DC.[235] A subsequent inquiry termed the accident "unavoidable," hence, no compensation was paid.[236]

The following February, the sides of a trench at Corrie Ferguson's mine fell upon two African laborers, killing one and severely injuring the other. Once more, a crowd of local women gathered and commenced demonstrations. When several hundred men joined them, the protest flared into a threatening stone throwing demonstration. By evening, when mining officials and the police arrived at the scene, the demonstration had died out, but a few Africans still stood about. The demonstration leaders, both female and male, were subsequently prosecuted and imprisoned.[237]

Another death in the mines brought to life disturbing eyewitness accounts of the demise of Museve Shibonje. Shibonje, an underground laborer at Rosterman Gold Mining Company, was crushed beyond recognition by the shaft lift.[238] Shibonje's sister dramatized the pain and anguish over her brother's gruesome death in a funeral lament:

> *Shibonje mwana weru vamuvoyi mulusasi; lusali lwa mabwoni noho*? (Repeatedly).[239]
>
> My brother Shibonje has been tied in a heap; Is he a heap of mashed potatoes?

This funeral dirge clearly illustrates the horrific forms of death that African workers faced in the gold mines. In 1941, local Isukha physically assaulted the European supervisor at Rosterman following the electrocution of Etolondo Shitambasi and Ashivenda Mkaisi.[240] Muhati recalled that their families received Shs. 400.[241]

Minishi Seleta, who had a long service record in the mining industry, explained the minimal European losses in terms of a "conspiracy theory". Munishi claimed that Europeans dispatched large numbers of Africans for underground work under a single European supervisor.[242] Despite such extreme skepticism, evidence suggests that the underground world occasionally bridged relations between European and African labor:

> Mr. Bangley [a European] was firing the last round of nine holes when one hole exploded prematurely; he was blown some little distance away, and his native assistant was struck by a rock and fell into the gutter with his face in the water, so that he was in danger of drowning. Mr. Bangley, though aware that the remaining fuses would shortly explode, returned in the darkness to search for the native. He eventually found him, and managed to carry him without assistance to a place of safety. The remaining charges then exploded. He carried the injured man for some distance before meeting some native labourers who then conveyed the injured man to the cage, and from there to the surface.[243]

This case illustrates the common dangers that characterized the lives and experiences of mineworkers. Nevertheless, sources show that Africans endured most deaths and injuries in the mines. Yet, as the gold mining industry rendered families destitute, and disabled individual workers, the question of mining accidents and compensation remained complex. As late as 1939, mining companies had not institutionalized a workmen's compensation scheme.[244] Owing to increased incidences of accidents, in 1934, the Warden of Mines requested all leading companies to pay half wages to African labor injured or falling sick as a direct result of their employment until they returned to work. While all major companies agreed to the proposal, Rosterman, which accounted for most of the accidents, declined.[245] Moreover, between 1933 and 1938, only one criminal case went to court, and a second prosecution ended in the acquittal of the shift-boss. In all cases, the court absolved company management.[246] While informants alleged failure of companies to pay compensation or claimed mediocre payments, archival records sweepingly claim that miners "promptly and willingly paid" compensation for injuries and livestock falling into unfenced pits.[247]

In the absence of workmen's compensation and because of the nature of complex legalities involved in claims, accident victims and their families often failed to follow up such claims. The Mining Ordinance empowered a magistrate to inquire into all serious mining accidents. However, in practice the Inspector of Mines undertook prior inquiry on the spot and reported to the magistrate, who in turn apportioned blame and assessed compensation.[248] Since the judicial system considered mining accidents as mere accidents, it is possible that most victims did not receive compensation approval. Most important, cash payments inadequately compensated for the loss of life, debilitating injuries and the socio-psychological impact African laborers and their families endured. As workers died and families became destitute, the colonial state and mining interests proceeded with business as usual.

KAKAMEGA: "THE BOOM TOWN"

The dawn of gold mining led to the tremendous growth of Kakamega Township. Gazetted as a post in 1909, Kakamega became the government station for NK district in 1920. In 1927, it officially became the district headquarters.[249] When gold was discovered in 1931, Kakamega was a relatively quiet district station with "a few Indian *dukas* [shops] and a native market." It was the seat of the NK district office [*Boma*] and contained a government Native hospital, a leper camp, a prison, a Government African School [GAS], the district court of appeal, and a seed farm of the agricultural department.[250] When the Local Land Board approved the September 1930 Kakamega Township boundary survey in August 1931, it displaced thirty-three Luyia households living on land that fell within the township boundaries.[251]

Thereafter, the physical terrain of Kakamega experienced a rapid and unprecedented economic transformation. Kakamega turned into a boomtown: in 1932 alone, the DC issued thirty temporary residential plots including plots for hotels and boarding houses, and forty-eight business plots. Tenure was a temporary lease of one year.[252] By December 1932, Kakamega had forty-six European commercial enterprises with all commercial plots in the old *bazaar* taken up by Europeans and Asians.[253] The town was home to two doctors and a chartered accountant, a general store, two garages, a hotel, and butchery.[254] In 1933, the district administration granted twenty-three leases in the commercial area of Kakamega and twenty in the old Asian *bazaar* zone.[255]

Although Asians held virtual monopoly over the business realm before 1932, many of the new businesses belonged to Europeans. From 1933 to 1934, numerous shops catering to European miners' needs, and hotels including the Eldorado, the Golden Hope Inn, the Corkscrew Inn and the Boma, restaurants, and garages appeared. Harold Pemberton, special correspondent with *The Daily Express* captured the changing scenario in February 1933:

> The town of Kakamega has sprung up like a mushroom in the goldfields. A new 'Eldorado Hotel' has been opened. . . . There are also the Nugget Tea Rooms and the Golden Hope Bar. A stockbroker has already set up an office for selling and buying claims on commission. . . .[256]

A month later, the *Manchester Guardian* reported increased activity:

> The new building housing the Mines Department; the appearance of a couple of European stores, thronged with amateur miners, selling groceries, fresh produce from down-country, and the Nairobi papers on mail days; three times a week; and the amount of building which is proceeding on each side of the hot and dusty main street. A brisk demand for township plots has developed, and is being satisfied by grants of 'temporary occupation licences' valid for one year, ... Already 48 business licences and 30 residential plots had been allocated. . . .[257]

In April, C. A. Marsh, a visitor to the goldfield, described the township as consisting of "one street of corrugated iron buildings for stores of various sorts, petrol pumps, etc. . . . two hotels. . . . [and] a bar . . ."[258] At the end of 1934, Kakamega had twenty-nine leased European enterprises in the commercial area, and twenty-one in the Asian *bazaar*. Revenues accruing from leases financed township improvements such as bush clearing, road surfacing, and the construction of a European sports ground.[259]

Kakamega's boom years witnessed investment in better communications. Located thirty-three miles from the railhead of the Kenya-Uganda Railway [K.U.R] at Kisumu, Kakamega was within twenty miles of a branch line to

Butere, and thirty miles from Broderick Falls [Webuye] on the main Uganda line. Although envisaged extensions from Webuye or Butere failed to materialize, the existing Kakamega-Kisumu, Kakamega-Butere, and the Kaimosi-Maragoli-Edzawa valley roads were much improved. In addition, Kisumu became the Clapham junction of the Imperial Air Service, within six days of London and two hours of Nairobi, with a local aerodrome linking Kakamega to Nairobi, Eldoret, and Kisumu. Additionally, a temporary telegraph line connected Kakamega to the main system via Eldoret.[260] From 1933 to 1934, Kimingini, Risks, and Kakamega Ore Reduction, installed power lines over the area covered by their mills, workshops, shaft heads and camps.[261] While such physical assets transformed Kakamega into a "Western type township," it is significant to underscore the fact that these facilities served European mining interests. Undeniably, the infusion of western capitalist apparatuses of economic penetration, and the concomitant urban growth it unleashed in Kakamega had significant implications for the Luyia. Ultimately, these new forces of capitalist expansion produced short-lived benefits to the local population that remained principally a rural agrarian economy.

As newspapers, observers and government sources wrote glowingly about Kakamega's advance, the conditions of the Luyia near the township remained largely depressing. On April 7, 1933 *The Friend* reported, "Native huts are dotted about everywhere . . .with estimated population densities of between 1,000 and 1,200 persons per square mile. Indeed, nearly all the forest trees have been cleared and all available land is cultivated. . . ."[262]

Yet, the tremendous expansion and significance of Kakamega Township was short-lived. Business fortunes assumed a downward trend as early as 1936, when several businesses closed.[263] In 1937 and 1938, a large number of European businesses shut down, including Messrs U. G. Tank & Metal Works, Penfold's Garage, the Corkscrew Inn, the Boma & Cumberland hotels, Stone Co., Rutherford & Co., George Taylor, and Economy Motors Ltd.[264] By 1939, Kakamega, then the second largest township in Nyanza Province after Kisumu, was a shadow of its former self. Quoting the NK DC, the Nyanza PC reported:

> Ichabod is its second name. From two banks, three hotels, two bars, three European garages, two European shops and a European lawyer two years ago are we descended to one bank, no hotels or bars, two European garages, one of which does little business, and no European Lawyer. . . .[265]

The meteoric rise and fall of the township perhaps explains the short-lived nature of the market opportunities the goldfield made available to Luyia producers.

INDUSTRIAL PENETRATION AND THE MARKET ECONOMY

The inauguration of gold mining and rapid urban growth stimulated trade around Kakamega. By 1933, a brisk market existed for local products such as vegetables, potatoes, eggs, chickens, fruit and milk, besides building materials including poles, grass, string, mats, water jars, and firewood.[266] Women from Idakho, Isukha, Maragoli, Bunyore, Marama, and Butsotso dominated this trade. In addition to fresh supplies, these market women sold cooked food, with a few combining both activities. For example, Popai Indushi from Isukha hawked food and a wide range of fresh supplies at Rosterman Mines.[267] A few men also joined the trade. Ralph Khaminwa, an African Assistant Agricultural Officer (AAAO) from Idakho, raised vegetables for sale to Kimingini mines; Litavakha, an explosive machinist at Rosterman, adopted intensive vegetable farming to supply the mines. In Maragoli, Habil Ndagalu, a member of the North Kavirondo Chamber of Commerce (NKCofC), which emerged in 1933 to maximize African trading interests, specialized in fruit production, while Samuel Imbuye and Benjamin Shikomera from Kabras monopolized large-scale milk deliveries.[268]

Trade in local products stimulated the cultivation of exotic vegetables among the Luyia.[269] In 1933, the NKLNC spent Shs. 400 on vegetable seeds for households throughout the district. Demand for the free vegetable seeds existed in areas both close to and remote from Kakamega. The more common vegetables, such as tomato, kale, leek, onion, and cabbage, suited most parts of the district, while onions did well in Marama and Bukhayo locations. Table 5.8 shows types and quantities of vegetable seed distributed by the NKLNC in 1933.

In 1934, the LNC issued 3,600 packets of vegetable seed, prompting Luyia peasants to cash in on the market in mines, Kakamega Township and nearby

Table 5.8. Types and Quantities of Vegetable Seed Distributed by the NKLNC, 1933

Vegetable	Quantity (lbs)
Cabbage, First and Best	8
Onion, Egyptian	12
Tomato, Large Red	5
Kale, Tall Curly	4
Carrot, English Longhorn	9
Leek, London Flag	6
Spinach, New Zealand	8
Lettuce, Iceberg	½
Total	52½

Source: NKAgAR 1933, KNA: AK/2/27.

Kisumu town. The Logoli and Nyore in particular, reclaimed swamps and ravines for all-year round vegetable production.[270] However, free LNC supplies did not last for long. In the rush for free seed, some peasants often obtained a surplus. Many failed to plant the full quantity acquired. Consequently, the LNC replaced free seed issues with outright sales.

Peasants who desired seed were required to pay ten-cents per Ib.[271] This, however, instantly led to diminished demand. In 1935, for example, peasant households in Malakisi, Bukusu, Samia, Bukhayo, and Maragoli locations purchased 2,892 packets only. Other locations showed no interest in seed purchases. The most popular vegetable included cabbage, carrots, onions and cauliflower, with the tomato the least popular.[272] In spite of reduced seed sales, vegetable sales significantly increased during the year.[273]

Increased vegetable production quickly brought forth monumental marketing problems. Labor in the mines alone required approximately two tons of vegetables per week. Therefore, individual vegetable hawking in mining areas proved quite unsatisfactory. Hawking guaranteed neither a constant supply nor a steady market, and was time consuming. To establish regularity, agricultural officials formulated a scheme for erecting a permanent four-store building at a central market, preferably at Chief Milimu's camp [Khayega] where growers would sell produce to competent buyers in rented stores. Officials anticipated that buyers would obtain vegetable contracts with various mines that required bulk purchases. This arrangement aimed at affording both growers and consumers a certain degree of economic security.[274]

The NKCofC carried out discussions for an LNC funded market at Khayega in 1934 and 1935.[275] In his appeal for LNC funding, DC Anderson emphasized the importance of a central marketing facility in meeting the great demand for vegetables for both European and African consumption in the goldfield.[276] Indeed, the LNC agreed to fund the project.[277] To maintain the supply of vegetables during the dry weather, agricultural officials emphasized the necessity for increased riverside gardening and the utilization of low-lying moist land.[278]

Apart from vegetable, NK received 382 bags of maize seed from the Department of Agriculture funds and the Bukura Agricultural farm. The NK LNC also issued 541 bags of maize seed from its farm at Kakamega.[279] While the Khayega market opened in 1936, the DC, J. Clive, disclosed a year later "for nearly a year, it [the market] has remained unused, other than as a shelter for rain." Traders refused to utilize the facility because of lack of overnight storage for surplus vegetables. Dealers from distant locations in particular, expressed concern over the disposal of daily surplus. To circumvent this problem and attract traders to the market, the DC requested the LNC to approve the use of Shs. 1, 005 of its share of compensation from Rosterman mining lease, earmarked for the benefit of the district, for a lock up store. Once again, the LNC approved the expenditure.[280]

In July 1937, the LNC completed and opened the storage facility to traders. Habil Ndagalu of the NKCofC appealed to Luyia producer enterprise for the success of the store.[281] This project, however, never translated into a successful long-term economic investment. Two months later, the economic viability of Khayega market, and its lock up store, was already in question.[282] Within eight months of opening the market, DC Clive admitted its utter failure due to insufficient demand for the produce offered.[283] By the end of 1938, therefore, efforts toward organized vegetable marketing in Kakamega had come to naught. Luyia producers continued to hawk their produce in Kisumu and other towns and in the gold mines.[284] Although the Nyanza LNCs had sold some 3,500 packets of vegetable seeds to growers throughout the province during the year, agricultural officials held that "there is as yet no need for organized marketing of vegetables in Nyanza Province".[285]

Admittedly, the trade in fresh produce benefited a few women and men who adapted and innovated to meet demands in the mines. Indushi, for example, realized savings that she invested in land, cattle, and a semi-permanent house.[286] Others like Shikomera, Imbuye, and Ndagalu invested in improved farming, thereby joining the emerging Luyia petty-bourgeoisie. Taken in totality, however, these low value perishables fetched marginal prices, mostly a few cents or shillings.[287] This fact is appreciated when viewed against the backdrop of the low retail prices offered for NN vegetables at Kisumu, the only outlet for Luyia growers in 1951, as shown in Table 5.9.

The goldfield also provided a market for maize, the staple food ration in the mines. By 1933, between thirty and forty Logoli owned water mills with a total daily output of over 100 bags, for making *posho* had sprung up. Mill owners anticipated securing *posho* supply contracts with the mines. Despite the optimism, the AO painted a pessimistic picture of the impact of mining on agricultural production when he wrote:

> ... Basically the mines have had, to date, little effect on the agriculture of their area, except for providing a good market for crops for local consumption such as vegetables, fruit, potatoes and milk, and mine timber and firewood; markets that were more beneficial than those of an export nature.[288]

In January of 1934, the district AO estimated that African labor in the gold mines would require some 1,800 bags of maize per month.[289] Later that year, a large *posho* mill with an hourly output of twenty bags of maize meal opened in Kakamega, thereby stimulating local prices of maize.[290] In October, the Butsotso-Kakamega areas either directly or indirectly sold 2,400 bags of

Table 5.9. Prices of Vegetables at Kisumu Market, 1951

Vegetable Type	Cost Per lb. (Cts.)
Green beans	15
Beetroot	12
Bhindi (Ladies fingers)	15
Brinjal	20
Cabbage	10
Cauliflower	16
Cucumber	15
Fenegrek (Meith)	30
Karela	25
Leeks	15
Lettuce	25
Carrots	12
Marrows	12
Spring onions	20
Parsnips	15
Green peas	25
Sweet potatoes	08
Pumpkins	08
Spinach	12
Tomatoes	20
Turnips	12
Potatoes	10
Onions	25

Source: DAO NN to DC NN, September 13, 1951 and Assistant Registrar of Cooperative Societies, Nyanza Province to DC NN, October 2, 1951, KNA: AK/4/14.

maize to the *posho* mill, at a price averaging Shs. 4.75 per bag.[291] Yet, mines absorbed only limited quantities of this local maize.

By March 1935, the maize market was reportedly "lifeless" because of massive maize production that led to a slump in prices. AO W. O. Sunman noted in his Monthly Report for April that "market conditions have been extremely dull, with little except maize, and not much of that, coming onto the market." The Kakamega *posho* mill went through a difficult, depressing period with "considerable quantities of maize unsold and flour." The management stopped buying or grinding any outside maize before clearing its own stock.[292] Two months later the "lifeless" market had knocked the "bottom out of the maize market."[293] The market improved slightly in August, particularly in North Maragoli, due to local mining contracts.[294] Before the end of the year, however, trade in maize and maize meal declined owing to reduction of construction staff by some of the leading mines.[295] While local producers benefited from higher prices offered by the Kakamega mill compared to prices at trading centers throughout the district, it was the low value perishables discussed above that enjoyed sustained demand in mines.[296]

Luyia producers also suffered from restrictive state produce marketing policies, which limited direct sales to mining companies. Asian traders constituted the conduit that conveyed African produced maize to the mines. This, coupled with competition from European produced maize from the Kenya Farmers Association, profoundly reduced the value of maize sales to Luyia households.[297]

In addition, the Friends African Industrial Mission at Kaimosi initially dominated the supply of props and construction labor for mining companies and *posho* for their workers. The Industrial Mission's report for 1935 stated: "We have maintained practically a monopoly of supply of maize meal to local settlers and natives. In return they have furnished our supply of logs and maize."[298] These competing and favored sources of mine supplies severely restricted the Luyia market. That the Luyia realized little benefits from the market opportunities provided by the gold mining industry is borne out by Fearn's own research findings in the 1950s, which revealed that out of 132 traders he interviewed in NN district, only one attributed part of his initial capital to his mining savings.[299] As Luyia households grappled with economic discrimination, competition and exploitation, they also faced new health challenges brought on by gold mining in the locality.

CONCLUSION

The advent of gold mining transformed Kakamega into a "modern Western style" rural industrial township. These new Western accouterments were designed to serve European capitalist interests rather than the interests of the Luyia population. Luyia households lost land to gold mining interests. Yet the dichotomy between the European utilitarian conceptions of land value as opposed to the all-embracing Luyia view in which ancestral lands had intrinsic value in comparison to their immediate agricultural use dissolved the importance of the meager monetary compensation they received. The loss of land, the inconveniences associated with dislocation and loss of subsistence rights, the failed labor opportunities, and the small monetary awards could not bridge the little benefits that accrued to the vast majority of Luyia households from gold mining.

Moreover, aspiring African entrepreneurs faced significant legal impediments. The mining laws placed premium on good command of the English language as a prerequisite for obtaining mining rights. Such legal hurdles guaranteed the exclusion of Africans in general and the Luyia in particular, from direct participation as licensed prospectors.

More important, although gold mining opened new employment opportunities to the Luyia, the local Isukha and Idakho who lost substantial land to mining concerns, failed to take advantage of the new opportunities. The Bantu Logoli and Nilotic Luo, who dominated labor in mines, may have reaped some benefits. However, the dangers of death, incapacitating injuries and the meager compensation offered by mining concerns for accident victims, brought significant economic and immense psychological effects on the Luyia community.

In addition, the provision of housing and food rations to mine workers did not significantly alter Luyia welfare. The prevailing low wages, poor living and working conditions, significantly affected the Luyia people. Likewise, the gold mining industry provided short-lived market opportunities for Luyia peasant households, particularly for fresh vegetables and maize. The low value perishables fetched little economic benefits as few Luyia women and men translated their economic gains into improved living standards.

NOTES

1. Fearn, "The Gold-Mining," 58.
2. Confidential Monthly Intelligence Report (CMIR), April 1934, KNA: PC/NZA/3/14/372; Nyanza Province Intelligence Report (NPIR) April 1934, KNA: PC/NZA/4/ 5/1; Col. & Prot. of Kenya, *MGDAR 1933*, 6; Wagner, *The Bantu*, Vol. I., 10.
3. DC and DO NK to PC Nyanza, Memorandum on the Kakamega Goldfields, February 2, 1932, KNA: PC/NZA/3/14/374.
4. NKDAR 1935, KNA: DC/NN/1/16; NPAR 1935, KNA: PC/NZA/1/30; NPAR 1936, PC/NZA/1/31; DC North Nyanza to P. M. Mahida, Midlands Provision Store, November 6, 1956, KNA: DC/KMG/2/18/10.
5. *KLC* III, Sections 221 and 223 quoted in PC Nyanza to Colonial Secretary (CS), September 12, 1935, KNA: DC/KMG/2/18/10.
6. PC Nyanza to CS, September 12, 1935, KNA: DC/KMG/2/18/10.
7. DC NK to PC Nyanza, September 2, 1935, KNA: DC/KMG/2/18/10; North Kavirondo Local Native Council (NKLNC), Minutes of a Meeting Held at Matungu on February 14, 1935, KNA (Kakamega): HW/13/2; NKDAR 1935, KNA: DC/NN/1/16. The Commissioner of Mines alleged that the land "consisted of a barren unoccupied hill, which in the opinion of the Agricultural Officer had never been cultivated." See Min.89 Central Land Board, May 1, 1935, KNA: PC/NZA/2/2/68.
8. DC NK to PC Nyanza, September 2, 1935, KNA: DC/KMG/2/18/10.
9. DC NK to Acting PC Nyanza, December 20, 1935, KNA: DC/KMG/2/18/10.

10. PC Nyanza to CS, September 12, 1935, KNA: DC/KMG/2/18/10.

11. DC to Acting PC Nyanza, December 12, 1935, KNA: DC/KMG/2/18/10.

12. DC NK to Acting PC Nyanza, April 6, 1936, KNA: DC/KMG/2/18/9; NKDAR 1936, KNA: DC/NN/1/18; NPAR 1936, KNA: PC/NZA/1/31; Commissioner of Mines to PC Nyanza, June 22, 1936, KNA: PC/NZA/2/2/17; Acting PC Nyanza to Hugh Sandys, Rosterman, June 30, 1936, KNA: DC/KMG/2/18/9.

13. DC NK to Acting PC Nyanza, April 6, 1936, KNA: DC/KMG/2/18/9; Valuation by AO NK, April 7, 1936, KNA: PC/NZA/3/14/358 A; NKDAR 1936, KNA: DC/NN/1/18.

14. NKLNC, Minutes of a Meeting Held at Kakamega on April 15 & 16, 1936, KNA (Kakamega): HW/13/2 and KNA: DC/KMG/2/18/9; Acting PC Nyanza to Secretary, Central Land Board, August 6, 1936, KNA: PC/NZA/2/2/17. While the DC received the lease application on January 20, 1936, he informed the LNC of the impending lease at this (April) meeting.

15. DC NK to Acting PC Nyanza, July 25, 1936, KNA; DC/KMG/2/18/9; NPIR July 1936, KNA: PC/NZA/4/5/2; NPAR 1936, KNA: PC/NZA/1/31. It was paid in a check worth Shs. 15, 097.50. No additional compensation was paid for disturbance and loss of property. See PC Nyanza to Secretary, Central Land Board, June 11, 1936, KNA: DC/KMG/2/18/9.

16. Acting PC Nyanza to Secretary, Central Land Board, August 6, 1936, KNA: PC/NZA/2/2/ 17, also found in KNA: DC/KMG/2/18/9.

17. DC NK to Acting PC Nyanza, August 19, 1936, KNA: PC/NZA/2/2/17; also found in KNA: DC/KMG/2/18/9; DC to acting PC Nyanza, August 19, 1936, KNA: PC/NZA/2/2/17. The commuted rent was distributed as follows: the NK LNC £77 for NK district as a whole; £154 on behalf of the location to be devoted for some common purpose; £3 to Chief Milimu as reward for his good office; £2 to Chief Milimu for payment to Africans who had rendered services; and £518 to the sixty individual claimants in proportions laid out in Table 5.6. Acting PC to DC NK, August 17, 1936, and DC NK to Acting PC Nyanza, August 19, 1936, Ibid.; PC Nyanza to DC NK, August 31, 1936, KNA: PC/NZA/3/14/358 A; NPAR 1936, KNA: PC/NZA/1/31.

18. The Rosterman farms have remained contested property in public litigation to date, with some litigants claiming foul play on the part of Chief Milimu. Muhati, (Oral Interview) OI, February 13, 1998; Imbali and Mang'ong'o, OI, February 18, 1998; Mbwabi, OI, February 26, 1998; Bukhala, OI, September 24, 1998.

19. Assistant Warden of Mines to PC Nyanza, November 6, 1939, KNA: PC/NZA/2/2/68.

20. Ayiekha, OI, February 15, 1998; Isenjia, OI, February 24, 1998; Amahwa, OI, August 29, 1998; Musine, OI, September 21, 1998.

21. PC Nyanza to DC NK, July 20, 1944, KNA: DC/KMG/2/18/9.

22. PC Nyanza to DC NK, October 12, 1944, KNA: DC/KMG/2/18/9.

23. Minutes of a Meeting of the NK Local Land Board (NKLLB) Held in the Office of the DC on September 13, 1941, KNA (Kakamega): WD/8/15; PC Nyanza to Managing Director, Edzawa Ridge Company, December 1, 1941, KNA: DC/KMG/2/18/12.

24. DC NK to Edzawa Mines Ltd., August 1, 1941, KNA: DC/KMG/2/18/12; DC NK to PC Nyanza, August 13, 1941, Ibid.
25. DC NK to PC Nyanza, August 13, 1941, KNA: DC/KMG/2/18/12.
26. PC Nyanza to DC NK, December 1, 1941, KNA: DC/KMG/2/18/12. All the figures are rounded up.
27. Secretary to DC NK, January 2, 1942, KNA: DC/KMG/2/18/12; Minutes of a Meeting of NKLLB Held in the DC's Office on November 19, 1942, KNA (Kakamega): WD/8/15.
28. DC NK to PC Nyanza, Telegram, February 23, 1942 and PC Nyanza to Commissioner for Lands and Settlement, February 24, 1942, KNA: DC/KMG/2/18/12.
29. Edzawa Ridge Company to DC NK, March 11, 1942, KNA: DC/KMG/2/18/12.
30. DC NK to the Manager Edzawa Ridge Company, March 12, 1942, KNA: DC/KMG/2/ 18/12.
31. DC NK to the Manager Edzawa Ridge Mine, March 17, 1942, KNA: DC/KMG/2/18/12.
32. DC NK to PC Nyanza, November 18, 1942, KNA: DC/KMG/2/18/12.
33. Minutes of a Meeting of NKLLB Held at the DC's Office, Kakamega on November 19, 1942, KNA (Kakamega): WD/8/15.
34. PC Nyanza to DC NK, September 4, 1941, KNA: DC/KMG/2/18/12.
35. PC Nyanza to Managing Director Edzawa Ridge Mines, September 26, 1941, KNA: DC/KMG/2/18/12.
36. PC Nyanza to DC NK, September 26, 1941, KNA: DC/KMG/2/18/12.
37. PC Nyanza to Managing Director Edzawa Ridge Company, December 1, 1941, KNA: DC/KMG/2/18/12.
38. PC Nyanza to DC NK, December 1, 1941, KNA: DC/KMG/2/18/12.
39. DC NK to PC Nyanza, January 2, 1942, KNA: DC/KMG/2/18/12.
40. DC NK to PC Nyanza, October 2, 1947; PC Nyanza to DC NK, October 6,1947; Chief Paul Agoi to DC NK, March 7, 1948, KNA: DC/KMG/2/18/12.
41. Minutes of Isukha Locational Advisory Council Held at Khayega on April 26, 1950, KNA: DC/KMG/1/1/19; Minutes of a Meeting of the Local Land Board Held at the District Commissioner's Office, Kakamega, June 23, 1950, KNA (Kakamega): WD/8/15; Warden of Mines to PC Nyanza, July 1, 1952; DC North Nyanza to P. M. Mahida, Midlands Provision Store, November 6, 1956, and Chief W. Shivachi to DC NN, August 24 and September 25, 1956, KNA: DC/KMG/2/18/10; Mukabwa, OI, February 12, 1998; Ayiekha, OI, February 15, 1998; Isenjia, OI, February 24, 1998; Mbwabi, OI, February 26, 1998.
42. Aseka, "Political Economy," 139; Kizito L. Muchanga, "Impact of Economic Activities on the Ecology of the Isukha and Idakho Areas of Western Kenya c. 1850 to 1945," (M. A. Thesis, Kenyatta University, 1998), 160–62.
43. Mukabwa, OI, February 12, 1998; Ayiekha and Khasandi, OI, February 15, 1998; Seleta, OI, February 15, 1998; Shivachi, OI, February 15, 1998; Musalimwa, Mwidakho and Ingosi, OI, February 17, 1998; Imbali, Mung'alo and Mutondo, OI, February 18,1998; Isenjia, OI, February 24, 1998; Mbwabi, OI, February 26,1998; Andove, OI, February 28,1998; Mavia, OI, March 2, 1998; Amagadu and Shikali, OI,

March 3, 1998; Mutsotso, OI, March 4, 1998; Litavakha, OI, March 9, 1998; Mutembei OI, March 17, 1998; Vihembo, OI, March 26, 1998; Mwinamo, Andrea Mang'ong'o, Julius Shisia, Lesmas Liyai, Shivachi Mang'ong'o, OI, March 26,1998; Akibaya and Otiende, OI, August 20, 1998; Amahwa, OI, August 29,1998; Musine, OI, September 21, 1998; Bukhala, OI, September 24, 1998.

44. Notes of a Meeting Held at Chief Milimu's Camp on December 22, 1931, KNA: PC/NZA/3/14/374.

45. NKLNC Meeting Held at Matungu on May 30, 1932, KNA (Kakamega): HW/13/1.

46. Bi-Monthly Report to the Eastern African Dependencies Trade & Information Office, April 7, 1933, PRO: CO 533/429/1; *Parliamentary Deb.,* 292. H.C. Deb. 5 S., 1933–34, Col. 576.

47. Raymond E. Dumett, *El Dorado in West Africa: The Gold Mining Frontier, African Labor, and Colonial Capitalism in the Gold Coast, 1875–1900* (Athens: Ohio University Press, 1998), 88–90, 98–109, 281–2, 284–87.

48. Col.& Prot. of Kenya, *Mining Ordinance* No LXI 1933 (Nairobi: Government Printer, 1933), 5–6; *News of the World*, December 4, 1932; Eastern African Dependencies, Trade & Information Office, November 10, 1932, PRO: CO 533/427; Extracts from Memos by W. McGregor Ross, June 14, 1933, Buxton Papers, RHO, MSS. Afr. Emp. S. 405/6/5; Assistant Warden of Mines to PC Nyanza, December 16, 1944, KNA: PC/NZA/3/14/353.

49. *Parliamentary Deb.,* 288. H.C. Deb. 5S, 1933–34, April 18, 1934, Col. 948. Eleven EPLs were held by British subjects or companies and one by two Belgians. Three hundred and seventy-four special permits were held by British subjects, thirty-two by foreigners, twenty-two by Asians and one by an Arab.

50. *MGDAR 1933*, 5.

51. *Official Gazette* No. 32, July 11, 1939, KNA: PC/NZA/2/9/22; *MGDAR 1939* (Nairobi: Government Printer, 1940), 4, 7; NPAR 1939, KNA: PC/NZA/1/34.

52. *An Ordinance to Regulate Trading in Unwrought Precious Metals, 1933 (Amendment),* 2, KNA: PC/NZA/2/9/22. Since Africans sold illicit gold to Asian traders who could easily pay off the £1,000 fine, the heavy prison sentence was meant to serve as the ultimate deterrent. Mine employees risked seven years in prison, and concealing gold in or near a mine fetched five years, all with hard labor. Ibid. 3.

53. Shivachi, OI, February 15, 1998; Seleta, OI, February 15, 1998; Isenjia, OI, February 24, 1998. Similar methods prevailed in the Kimberley Diamond industry in the 1860s and 90s. See William H. Worger, S*outh Africa's City of Diamonds: Mine Workers and Monopoly Capitalism in Kimberley, 1867–1895* (New Haven: Yale University Press, 1987)128–31, 139–40.

54. Isenjia, OI, February 24, 1998; Amahwa, OI, August 29, 1998.

55. Isenjia, OI, February 24, 1998. This is also true of illegal mining; Akibaya and Otiende, OI, August 20, 1998.

56. *MGDAR 1940* (Nairobi: Government Printer, 1941), 2; Unpublished Mining and Geological Department Annual Report 1941.

57. Assistant Warden of Mines, CMIR November 1939, March and April 1940, KNA: PC/NZA/3/14/372; Musalimwa, Mwidakho and Ingosi, OI, February 17, 1998; Mbwabi, OI, February 26, 1998; Amagadu and Shikali, OI, March 3,1998.

58. Commissioner of Mines to PC Nyanza, October 14, 1940, KNA: PC/NZA/2/7/117. Two Asians and a Seychellois convict served nine and six months in prison respectively. Out of fourteen prosecutions during the quarter, twelve ended in convictions. Warden of Mines, CMIR August 1940, KNA: PC/NZA/3/14/372; NPAR 1940, KNA: PC/NZA/1/35; NKDAR 1941, KNA: DC/NN/1/23.

59. First Class Magistrate, Kakamega to Secretary of Chamber of Mines, February 16, 1944, KNA: PC/NZA/2/7/117. In 1942, the Mining Department voted £40 for rewarding informers on gold theft. See Commissioner of Mines to PC Nyanza, Resident Magistrate, Kisumu, Superintendent of Police, and Assistant Warden of Mines, Circular, May 12, 1942, Ibid.

60. Mrs. Stitt to Mines Office, Kisumu, n.d., June 1943, September 10, 1943, KNA: PC/NZA/2/7/117.

61. Lieut. Col. Stitt to Inspector of Mines, September 18, 1943, KNA: PC/NZA/2/7/117; NKDAR 1943, KNA: DC/NN/1/25; Amagadu and Shikali, OI, March 3, 1998; Se also NKDAR 1943, KNA: DC/NN/1/25.

62. *An Ordinance to Regulate, 1933*, 2.

63. Inspector of Mines to PC Nyanza, February 7, 1945, KNA: PC/NZA/2/7/117; NKDAR 1945, KNA: DC/NN/1/27.

64. PC Nyanza to Chief Secretary, n.d., but before May 22, 1945, KNA: PC/NZA/3/4/357.

65. Worger, *South Africa's City of Diamonds*, 118.

66. Leokadius Ndede quoted in Inspector of Mines, Confidential Minute, January 13, 1948, KNA: PC/NZA/2/7/117. Ndede called for "real punishment" as opposed to the "light duty" imposed on illicit gold buyers.

67. Inspector of Mines to Assistant Commissioner of Mines, January 7, 1948, KNA: PC/NZA/2/7/117.

68. Commissioner for Lands, Mines & Survey to Chief Secretary, January 23, 1948, KNA: PC/NZA/2/7/117

69. PC Nyanza to Chief Secretary, March 4, 1948, KNA: PC/NZA/2/7/117. See DC NK to PC Nyanza, February 27, 1948, Ibid.

70. Superintendent of Police, Nyanza to PC Nyanza, March 2, 1948, KNA: PC/NZA/2/7/ 117.

71. Tisdall to Commissioner of Mines, July 29, 1948, KNA: DC/KMG/2/18/3.

72. *MGDAR 1952* (Nairobi: Government Printer, 1954), 4.

73. DC NK to Assistant Warden of Mines, February 5, 1943, KNA: DC/KMG/2/18/13.

74. Commissioner of Mines, June 7, 1945 quoted in PC Nyanza to Chief Secretary, KNA: PC/NZA/3/4/357.

75. Joseph Simekha to DC NK, January 25, 1945; Simekha to PC Nyanza, May 18, 1945; PC Nyanza to Joseph Simekha, May 22, 1945, KNA: PC Nyanza LND.11/2 3/3298; NPAR 1945, KNA: PC/NZA/1/40.

76. PC Nyanza to the Chief Secretary, April 4, 1946, KNA: PC/NZA/3/14/357.

77. Chief Secretary to Commissioner for Lands, Mines & Surveys, July 16, 1946, KNA: PC/NZA/3/14/357. The PC rejected the Chief Secretary's proposal that African chiefs be held responsible for the "proper" disposal of gold won by Africans. See PC Nyanza to Chief Secretary, August 8, 1946, and Assistant Commissioner of Mines to Commissioner for Lands, Mines & Surveys, August 9, 1946, Ibid.

78. Assistant Commissioner of Mines, Memorandum on Meeting Held at Kisumu on August 16, 1946, KNA: PC/NZA/3/14/357. Assistant Commissioner of Mines, D. Harverson, favored restricting the area to "productive" areas while the PC sought to extend it to all mining areas except the Maragoli Closed Area.

79. Acting Chief Secretary to Commissioner of Mines, September 23, 1946, KNA: PC/NZA/3/14/357.

80. Assistant Commissioner of Mines to PC Nyanza, October 24, 1946, KNA: PC/NZA/3/14/357.

81. Quoted in Inspector of Mines to the Superintendent of Police, February 19, 1948, KNA: PC/NZA/2/7/117

82. *MGDAR 1949* (Nairobi: Government Printer, 1950), 3; NPAR 1949, KNA: DC/KSI/1/24.

83. *MGDAR 1951* (Nairobi: Government Printer, 1953), 3–4.

84. *MGDAR 1952*, 20–21. Two of the original five were active in 1951 and only one remained among the three listed in 1953. *MGDAR 1953* (Nairobi: Government Printer, 1955), 19.

85. NKDAR 1932, KNA: DC/NN/1/13; See also NPAR 1933, KNA: PC/NZA/1/28.

86. Muhati, OI, February 13, 1998; Shivachi, OI, February 15, 1998; Ayiekha, OI, February 15,1998; Seleta, OI, February 15, 1998; Imbali, OI, February 18, 1998; Isenjia, OI, February 24, 1998; Mbwabi, OI, February 26,1998; Mavia, OI, March 2, 1998; Amagadu and Shikali, OI, March 3, 1998; Francis Mutsotso, OI, March 4, 1998; Litavakha, OI, March 9, 1998; Shitsukane, OI, March 19, 1998; Mwinamo, OI, March 26,1998; Amahwa, OI, August 29,1998; Musine, OI, September 21,1998; Bukhala, OI, September 24, 1998.

87. Mavia, OI, March 2, 1998; Mwinamo, OI, March 19, 1998; Amagadu and Shikali, OI, March 3, 1998; Mbwabi, OI, March 26, 1998.

88. Isenjia, OI, February 24, 1998; Mavia, OI, March 2, 1998; Francis Mutsotso, OI, March 4,1998.

89. Notes on a Meeting Held at Milimu's Camp on December 21, 1931, KNA: PC/NZA/3/14/374; NKDAR 1932, KNA: DC/NN/1/13.

90. Notes of a Meeting Held at Milimu's Camp on December 21, 1931, KNA: PC/NZA/3/ 14/374; Isenjia, OI, February 24, 1998.

91. DC NK to Acting PC Nyanza, December 31, 1931, KNA: PC/NZA/3/14/374.

92. PC Nyanza to CNC, December 10, 1932, KNA: PC/NZA/2/20/6.

93. Joyce L. Moock, "The Migration Process and Differential Economic Behavior in South Maragoli, Western Kenya," (Ph. D. dissertation, Columbia University, 1976), 57; NKDAR 1932, KNA: DC/NN/1/13; NKDAR 1933, KNA: DC/NN/1/14;

NKDAR 1934, KNA: DC/NN/1/15; Muhati, OI, February 13, 1998; Ayiekha, OI, February 15, 1998; Seleta, OI, February 15, 1998; Musalimwa, Mwidakho, and Ingosi Segero, OI, February 17, 1998; Imbali, OI, February 18, 1998; Mbwabi, OI, February 26,1998; Andove, OI, February 28, 1998; Mavia, OI, March 2, 1998; Amagadu and Shikali, OI, March 3, 1998; Litavakha, OI, March 9, 1998; Akibaya and Otiende, OI, August 20, 1998; Musine, OI, September 21,1998; Bukhala, OI, September 24,1998.

94. Akibaya and Otiende, OI, August 20, 1998; Amahwa, OI, August 29, 1998.

95. Mbwabi, OI, February 26, 1998.

96. PC Nyanza to CNC, December 10, 1932, KNA: PC/NZA/2/20/6; NKDAR 1932, KNA: DC/NN/1/13.

97. NPIR May 1934, KNA: PC/NZA/4/5/1.

98. Wagner, *The Bantu*, Vol.1, 10.

99. Provincial Labour Officer to Acting PC Nyanza, April 25, 1935, KNA: PC/NZA/3/13/27.

100. Labour Officer to the Chief Registrar of Natives, June 29, 1935, KNA: PC/NZA/3/13/27. Many companies and individuals did not submit their returns.

101. Nyanza Mining, Monthly Averages Employed for the Year Ending October 31, 1935, KNA: PC/NZA/3/13/27; Labour Returns for the Month of October 1935, KNA: PC/NZA/2/20/6.

102. Wagner, *The Bantu,* Vol. 1, 38; Nyanza Mining, Monthly Averages Employed for the Year Ending October 31, 1935, KNA: PC/NZA/3/13/27; Labour Return for October 1935, KNA: PC/NZA/2/20/6.

103. Fearn, "Gold-Mining," 50, 52; Fearn, *An African Economy*, 138–39.

104. George Oduor Ndege, *Health, State and Society in Kenya* (Rochester: University of Rochester Press, 2001), 114.

105. NPIR November 1934, KNA: PC/NZA/4/5/1.

106. Col.& Prot. of Kenya, *Annual Report 1936,* 53; NPAR 1936, KNA: PC/NZA/1/31.

107. NPIR February 1938, KNA: PC/NZA/4/5/2.

108. Col.& Prot. of Kenya, *Annual Report on Native Affairs 1937* (Nairobi: Government Printer, 1938), 185; Seleta, OI, February 15, 1998; Mung'alo, OI, February 18, 1998; Imbali, OI, February 18, 1998; Amagadu and Shikali, OI, March 3, 1998; Khakali, OI, March 4, 1998; Mwinamo, OI, March 26, 1998; Akibaya, OI, August 20,1998.

109. NKDAR 1938, KNA: DC/NN/1/20; Col.& Prot. of Kenya, *Annual Report 1938*, 30; Col.& Prot. of Kenya, *Report of the Employment of Juveniles Committee* (Nairobi: Government Printer, 1938), 5, PRO: CO 533/497.

110. NPIR November 1938, KNA: PC/NZA/5/4/2.

111. J. G. Hibbert, Minute, October 24, 1938; Archdeacon Owen Interview with the Secretary of State (S of S) for the Colonies, September 13, 1938, PRO: CO 533/497. For a detailed discussion of Owen's earlier criticism of child labor in Kenya, See Opolot Okia, "In the Interest of Community: Archdeacon Walter Owen and the Issue of Communal Labour in Colonial Kenya, 1921–30," in *Journal of Imperial Commonwealth History* 32, no. 1 (January 2004): 27–37.

112. *Salisbury and Winchester Journal*, 16 December 1938, PRO: CO 533/501/11; Memorandum on Labour in Africa, The London Group on African Affairs, n.d., RHO, MSS. Afr. S. 1427 Box 3/3.

113. Col.& Prot. of Kenya, *Annual Report 1938*, 34; Report of the Employment of Juveniles, 1938, 16–17, 26.

114. NPIR March 1940, KNA: PC/NZA/4/5/3.

115. PC Nyanza to CNC, December 10, 1932, KNA: PC/NZA/2/20/6; Sharon Stichter, *Migrant Labour in Kenya: Capitalism and African Response, 1895–1975* (Harlow: Longman, 1982), 84–89, 164.

116. NPIR April 1932, KNA: PC/NZA/4/5/1.

117. NPIR April 1935, Ibid.

118. NPIR January 1937, KNA: PC/NZA/4/5/2; Col.& Prot. of Kenya. *Native Affairs Department Annual Report (NADAR)* 1937: Nairobi: Government Printer, 1937), 152, PRO: CO 533/479/9; NPIR June 1937, KNA: PC/NZA/4/5/2.

119. NPIR August 1937, KNA: PC/NZA/4/5/2.

120. NPIR July 1937, KNA: PC/NZA/4/5/2.

121. *NADAR 1937*, 196, PRO: CO 533/479/9.

122. E. A. Sweatman, Economic Survey of North Kavirondo During 1937, NKDAR 1937, KNA: DC/NN/1/19.

123. Sweatman, Economic Survey of North Kavirondo During 1938, NKDAR 1938, KNA: DC/NN/1/20.

124. Wagner, *The Bantu*, Vol. I, 38; NKDAR 1938, KNA: DC/NN/1/20; Musalimwa and Mwidakho, OI, February 17, 1998; Isenjia, OI, February 24, 1998; Mbwabi, OI, February 26, 1998; Mavia, OI, March 2, 1998; Khakali, OI, March 4, 1998; Akibaya and Otiende, OI, August 20, 1998; Amahwa, OI, August 29, 1998.

125. Wagner, *The Bantu*, Vol. 1, 38; NKDAR 1938, KNA: DC/NN/1/20.

126. NPIR March 1936, KNA: PC/NZA/4/5/2. The total excluded the Kaimosi farms; Kakamega, Kisii and Yala Townships; employees of trading centers and LNCs, as well as wages and earnings from the fishing industry. Ibid.

127. NPAR 1936, KNA: PC/NZA/1/31; *NADAR* 1937, 156, PRO: CO 533/479/9. In addition, Nyanza labor outside the province earned £176,000. Ibid.157.

128. NPIR March 1936, KNA: PC/NZA/4/5/2.

129. NPIR April 1936, KNA: PC/NZA/4/5/2.

130. NPIR May, September, October and November 1936, KNA: PC/NZA/4/5/2.

131. Aseka, "Political Economy," 330.

132. *MGDAR 1937* (Nairobi: Government Printer, 1938), 13.

133. NKDAR 1946, KNA: DC/NN/1/28.

134. NKDAR 1947, KNA: DC/NN/1/29.

135. NKDAR 1944, KNA: DC/NN/1/1/26. The "maize granary" during this time was the northern locations of the district.

136. NKDAR 1945, KNA: DC/NN/1/27.

137. Tiyambe Zeleza, "The Colonial Labour System in Kenya," in *An Economic History of Kenya*. ed. W. R. Ochieng' and R. M. Maxon (Nairobi: East African Educational Publishers, 1992), 175–76. For details on the impact of WWII on Africa See eds. David Killingray and Richard Rathbone. *Africa and the Second World War* (New

York: St. Martin's Press, 1986). A full discourse on the King's African Rifles (KAR) is detailed in Timothy Parsons, *The African Rank-and-File: Social Implications of Colonial Service in the King's African Rifles, 1902–1964.* Portsmouth: Heinemann, 1999.

138. NPIR June 1940, KNA: PC/NZA/4/5/3.

139. Warden of Mines, CMIR August 1940, KNA: PC/NZA/3/14/372.

140. NPAR 1940, KNA: PC/NZA/1/35; NPIR September 1940, KNA: PC/NZA/4/5/3.

141. NPAR 1940, KNA: PC/NZA/1/35; NPIR October 1940, April and July 1941, KNA: PC/NZA/4/5/3.

142. Killingray, "Labor Mobilization," 84.

143. Nicholas Ekutu Makana, "Changing Patterns of Indigenous Economic Systems: Agrarian Change and Rural Transformation in Bungoma District, 1930–1960," (PhD. dissertation: West Virginia University, 2005), 141–45.

144. NPIR October 1941, KNA: PC/NZA/4/5/3.

145. DC NK to PC Nyanza, November 13, 1941; PC Nyanza to Chief Secretary, November 8, 1941, KNA: PC/NZA/2/20/13.

146. Summary of Men Supplied by Districts, October 20 to November 30, 1941, KNA: PC/NZA/2/20/13.

147. PC Nyanza to Secretary, Labour Inquiry Committee, December 4, 1941, KNA: PC/NZA/2/20/13.

148. NPAR 1941, KNA: PC/NZA/1/36.

149. NPIR October 1942, KNA: PC/NZA/4/5/3.

150. NPAR 1942, KNA: PC/NZA/1/37.

151. NKDAR 1945, KNA: DC/NN/1/27.

152. Maxon, *Going Their Separate Ways*, 164–174; NKDAR 1943, KNA: DC/NN/1/25; NKDAR 1944, DC/NN/1/26; NPAR 1943, KNA: PC/NZA/1/38.

153. Maxon, Ibid.; NKDAR 1943, KNA: DC/NN/1/25; NKDAR 1944, KNA: DC/NN/1/26. By 1944, LNC food relief funds had soared to £1,000 (Shs. 20,000).

154. NPAR 1944, KNA: PC/NZA/1/39.

155. Ibid.

156. NKDAR 1943, KNA: DC/NN/1/25.

157. For a detailed discussion on the impact of the colonial economic policies on the role of women in agriculture, see Ruth Nasimiyu, "Women in the Colonial Economy of Bungoma: Role of Women in Agriculture, 1902–1960." in *Journal of Eastern African Research and Development* 15 (1985): 65–66.

158. NKDAR 1944, KNA: DC/NN/1/26.

159. Robert M. Maxon, "Fantastic Prices' in the Midst of an "Acute Famine Shortage": Market, Environment, and the Colonial State in the 1943 Vihiga Famine, *African Economic History* 28 (2000): 841–870: Ibid, *Going Separate Ways,* 164–74.

160. A. D. Roberts, "The Gold Boom of the 1930s in Eastern Africa," *African Affairs.* Vol. 85 (October 1986): 553.

161. NADAR 1937, 207.

162. NKDAR 1937, KNA: DC/NN/1/19. The mines then employed 7,500 Africans, including children. Col.& Prot. of Kenya, *Annual Report 1937*, 28.

163. *Report of the Employment of Juveniles 1938,* 6–7, 17, 26, PRO: CO 533/497.

164. Muhati, OI, February 13, 1998; Mung'alo, OI, February 18, 1998; Isenjia, OI, 24, February 1998; Mavia, OI, March 2, 1998; Khakali, OI, March 4, 1998; Litavakha, OI, March 9, 1998; Mutembei, OI, March 17, 1998; Ludenyo, OI, March 26, 1998; Musine, OI, September 21,1998. The mining office retained the work tickets until check out time. The tickets were also stamped for rations. Underground work was later divided into day and night shifts.

165. Muhati, OI, February 13, 1998; Khakali, OI, March 4, 1998.

166. Farson, *Last Chance,* 247.

167. Mavia, OI, March 2, 1998.

168. Notes on a Meeting Held at Milimu's Camp in December 1931, KNA: PC/NZA/3/14/ 374; NKDAR 1932, KNA: DC/NN/1/13.

169. DC NK to Acting PC Nyanza, January 15, 1932, KNA: PC/NZA/3/14/374.

170. Col. Swinton Howe, President, Miners Association, Memorandum for Consideration to H. E. Re: Kakamega Goldfields encl. in PC Nyanza to Acting Commissioner of Mines, February 4, 1932, KNA: PC/NZA/3/14/374.

171. Warden of Mines, CMIR August 1934, KNA: PC/NZA/3/14/372.

172. NPAR 1935, KNA: PC/NZA/1/30; NPAR 1936, KNA: PC/NZA/1/31; NADAR 1937, 200–01; NPAR 1939, KNA: PC/NZA/1/34; Mukabwa, OI, February 12, 1998; Muhati, OI, February 13, 1998; Ayiekha, OI, February 15, 1998; Imbali and Mung'alo, OI, February 15, 1998; Shivachi, OI, February 15,1998; Musalimwa and Mwidakho, OI, February 17, 1998; Mbwabi, OI, February 26, 1998; Andove, OI, February 28,1998; Mavia, OI, March 2,1998; Khakali, OI, March 4, 1998; Amahwa, OI, August 29, 1998; Bukhala, OI, September 24,1998. Some informants stated that workers initially resented meat rations believed to be zebra meat. Mbwabi, OI, February 26, 1998.

173. NPIR September 1937, KNA: PC/NZA/4/5/2.

174. NPIR October 1937, KNA: PC/NZA/4/5/2. Rations on tea plantations were limited to maize flour. NPAR 1939, KNA: PC/NZA/1/34.

175. Fearn, *An African Economy*, 148.

176. Charles van Onselen, "Worker Consciousness in Black Mines, Southern Rhodesia, 1900–1920," *Studies in the History of African Mine Labour in Colonial Zimbabwe.* ed. I. R. Phimister and C. van Onselen. (Gwelo: Mambo Press, 1978), 9; Stichter, *Migrant Labour*, 158, 162–65; Zeleza, "The Colonial Labour," 184.

177. NKDAR 1934, KNA: DC/NN/1/15; NKDAR 1935, KNA: DC/NN/1/16; Labor Officer, Kisumu to DC NK, January 9, 1935, KNA: PC/NZA/2/20/6; NKDAR 1942, KNA: DC/NN/1/24; Ayiekha, OI, February 15, 1998; Musalimwa and Mwidakho, OI, February 17, 1998; Khakali, OI, March 4, 1998.

178. DC SK to PC Nyanza, Confidential letter, October 20, 1946, KNA: PC/NZA/2/20/13.

179. Zeleza, "The Colonial Labour," 186.

180. Ibid. 186–92; Stichter, *Migrant Labou*r, 165–68.

181. NPIR May 1941, KNA: PC/NZA/4/5/3; Imbali, Petro Mutondo, Munga'lo, OI, February 18, 1998. One of the strike leaders, Alphonse Lishenga, served time in

jail. Yet, a report for December 1941 claimed labor was contented for the past five years, KNA: PC/NZA/2/20/13.

182. NPIR May 1941, KNA: PC/NZA/4/5/3.

183. D. Kerr-Cross, Manager, Rosterman Gold Mines, December 30, 1941, KNA: PC/NZA/2/20/13.

184. NNDAR 1949, KNA: DC/NN/1/31.

185. DC NK to Acting PC Nyanza, December 31, 1931, KNA: PC/NZA/3/14/374.

186. NKDAR 1933, KNA: DC/NN/1/14; NKDAR 1934, KNA: DC/NN/1/15; *MGDAR 1934* (Nairobi: Government Printer, 1935), 3; NPIR May 1937, KNA: PC/NZA/4/5/2; NKDAR 1935, KNA: DC/NN/1/16; NKDAR 1938, KNA: DC/NN/1/20; NKDAR 1939, KNA: DC/NN/1/21; NKDAR 1940, KNA: DC/NN/1/22; NPAR 1935, KNA: PC/NZA/1/30; NPAR 1936, KNA: PC/NZA/1/31; NPAR 1938, KNA: PC/NZA/1/33; NPAR 1939, KNA: PC/NZA/1/34; NPAR 1940, KNA: PC/NZA/1/35.

187. *MGDAR 1936* (Nairobi: Government Printer, 1937), 14; *MGDAR 1937*, 17. For a discussion of legalized differential remuneration in colonial Kenya See Stichter, *Migrant Labour,* 113–14.

188. DC NK to Acting PC Nyanza, December 31, 1931, KNA: PC/NZA/3/14/374.

189. Notes by Shackleton, DO NK, November 4, 1932, KNA: DC/KMG/1/1/165.

190. Warden of Mines, CMIR July 1933, KNA: PC/NZA/3/14/372.

191. Warden of Mines, CMIR February 1934, KNA: PC/NZA/3/14/372.

192. NPIR July 1934, KNA: PC/NZA/4/5/1.

193. NPAR 1935, KNA: PC/NZA/1/30; CMIR December 1936, KNA: PC/NZA/4/5/2.

194. NKDAR 1937, KNA: DC/NN/1/19; NKLNC, Minutes of a Meeting Held at Kakamega on November 21 & 22, 1939, KNA: HW/13/2; Warden of Mines, CMIR August 1937 and July 1939, KNA: PC/NZA/3/14/372; NPIR July 1937, KNA: PC/NZA/4/5/2; Save for fifty laborers, all had ceased working with Alphega. Davies was subsequently employed for £25 a month and agreed to set a side £15 for payment of his debts. NKDAR 1937, KNA: DC/NN/1/19. However, on February 3, a day before the civil case was heard, Davies filed a bankruptcy petition; hence the court proceedings against him were dropped. NPIR February 1938, KNA: PC/NZA/4/5/2. In July, the Labor Officer had advised a DC's meeting to allow workers to "down their tools" for unpaid wages. Meeting of DC's Held on July 14–17, 1937, KNA: PC/NZA/4/1/3/1. Akibaya explained the burning down of the mining office in Kakamega in August 1934 as a miners' ploy to destroy records in order to have their debts written off. Akibaya, OI, August 20, 1998. Official sources labeled the incident an accident, but acknowledged the destruction of accident records, compensation files and general correspondence. NPIR August 1934, KNA: PC/NZA/4/5/1; Warden of Mines, CMIR August 1934, KNA: PC/NZA/3/14/372.

195. DC NK, Safari Diary, South and North Marama, October 13–14, 1939, KNA: DC/KMG/1/1/165; NPIR February 1939, KNA: PC/NZA/4/5/3.

196. Gautama Advocate to PC Nyanza, October 5, 1946, KNA: PC/NZA/3/14/357.

197. For similar experience among South African labor in Natal See Keletso. E. Atkins, *The Moon is Dead! Give Us Our Money! The Origins of an African Work Ethic, Natal, South Africa, 1843–1900* (Portsmouth: Heinemann, 1993), 95–57.

198. Assistant Warden of Mines, CMIR March 1940, KNA: PC/NZA/3/14/372.

199. NPAR 1941, KNA: PC/NZA/1/36. They were located in South and NK respectively.

200. List of Asiatics Working at Rosterman Gold Mining Co. Ltd. as at January 11, 1941, KNA: DC/KMG/2/18/9.

201. Musalimwa and Mwidakho, OI, February 17, 1998; Isenjia, OI, February 24,1998;

202. NKDAR 1943, KNA: DC/NN/1/25; NKDAR 1945, KNA: DC/NN/1/27; North Kavirondo District Intelligence Report (NKDIR) July 1935, KNA: PC/NZA/4/5/5; NPAR 1935, KNA: PC/NZA/1/30; NPIR July 1935, KNA: PC/NZA/4/5/1; NKDAR 1936, KNA: DC/NN/1/18; NPAR 1936, KNA: PC/NZA/1/31; NPAR 1940, KNA: PC/NZA/1/35; Mukabwa, OI, February 12, 1998; Ayiekha, OI, February 15, 1998; Musalimwa and Mwidakho, OI, February 17, 1998; Imbali and Mung'alo, OI, February 18, 1998; Mbwabi, OI, February 26,1998; Andove, OI, February 28,1998; Mavia, OI, March 2,1998; Khakali, OI, March 4,1998; Litavakha, OI, March 9, 998; Andrew Ludenyo, OI, March 26, 1998; Amahwa, OI, August 29,1998; Bukhala, OI, September 24,1998.

203. Warden of Mines, CMIR November 1933, KNA: PC/NZA/3/14/372.

204. NPAR 1939, KNA: PC/NZA/1/34.

205. NPIR July 1937, KNA: PC/NZA/4/5/2.

206. NPAR 1943, KNA: PC/NZA/1/38.

207. Mukabwa, OI, February 12, 1998; Ayiekha, OI, February 15, 1998; Musalimwa and Mwidakho, OI, February 17, 1998; Imbali and Mung'alo, OI, February 18, 1998; Mbwabi, OI, February 26, 1998; Andove, OI, February 28, 1998; Mutsotso, OI, March 4, 1998; Litavakha, OI, March 9, 1998; Bukhala, OI, September 24,1998. In his study of the Kimberly diamond industry in South Africa, William H. Worger found that closed labor compounds were a "mesh of laws and a maze of controls." William H. Worger, *South Africa's City of Diamonds: Mine Workers and Monopoly Capitalism in Kimberley, 1867–1895* (New Haven and London: Yale University Press, 1987), 141–49.

208. Imbali, OI, February 18, 1998; Andove, OI, February 28, 1998; Mavia, OI, March 2, 1998; Amagadu and Shikali, OI, March 3, 1998; Khakali, OI, March 4, 1998; Litavakha, OI, March 9, 1998; Enock Mwinamo and Mang'ong'o, OI, March 26, 1998; Musine, OI, September 21, 1998. See also the *Manchester Guardian*, 23 January 1933. E. S. Atieno-Odhiambo lucidly illustrates the every day nuisances Kenyan Africans endured as a result of "color bar"- the exclusion of Africans from goods and services enjoyed by Europeans in "The Formative Years 1945–55," in *Decolonization & Independence in Kenya 1940–93*, eds. B. A. Ogot & W. R. Ochieng' (Athens: Ohio University Press, 1995), 32–33.

209. Muhati, OI, February 13, 1998; Imbali, OI, February 18, 1998.

210. Warden of Mines, CMIR March 1934, KNA: PC/NZA/3/14/372.

211. S of S for the Colonies quoted in Warden of Mines, CMIR April 1934, KNA: PC/NZA/ 3/14/372.

212. At Watende Mines Ltd. in SK, Rhodesian miners, who made no attempt to understand Africans or their language and maltreated labor, faced a series of strikes, machine destruction and threats of being blown up with explosives. The drudgery of work in mines drove labor into consuming strong drink and bhang to "generate the required energy for completion of tasks." Labor smuggled drugs underground, and liquor production privileges were granted for small bribes to nyaparas and compound managers. All these led to worker aggression and "inexplicable" accidents to labor and machinery. Inspector of Mines, Kisii to Senior Inspector of Mines, Weekly Report, August 24, 1935, KNA: PC/NZA/2/20/6; Imbali, OI, February 18, 1998. It was the usual thing for the white shift boss in Rhodesian mines "to knock the boys" about considerably. The Watende management connived at the practice but disclaimed it officially.

213. Warden of Mines, CMIR September 1933, KNA: PC/NZA/3/14/372.

214. NKDIR September 1935, KNA: PC/NZA/4/5/5; NPIR September 1935, KNA: PC/ NZA/4/5/1; Amahwa, OI, August 29, 1998.

215. NKDIR November 1935, KNA: PC/NZA/4/5/5.

216. NPIR October 1936, KNA: PC/NZA/4/5/1. Two paid £15 and the third £12 in fines.

217. NKDIR August 1935, PC/NZA/4/5/5.

218. NPIR September 1936, KNA: PC/NZA/4/5/2.

219. NPIR August 1938, KNA: PC/NZA/4/5/2.

220. Fearn, "Gold-Mining," 58; Roberts, "The Gold Boom," 553. Fearn argued that between 1935 and 1953, there were six deaths and thirty-two injuries in thirty-three accidents.

221. Muhati, OI, February13, 1998; Shivachi,OI, February 15, 1998; Ayiekha, OI, February 15, 1998; Seleta, OI, February 15, 1998; Musalimwa and Mwidakho, OI, February 17, 1998; Imbali, Mung'alo, and Mutondo, OI, February 18, 1998; Isenjia, OI, February 24,1998; Andove, OI, February 28, 1998; Mavia, OI, March 2, 1998; Amagadu and Shikali, OI, March 3, 1998; Francis Mutsotso, OI, March 4, 1998; Litavakha, OI, March 9, 1998; Mutembei, OI, March 17, 1998; Mbwabi, OI, March 26, 1998; Shitsukane, OI, March 19, 1998; Vihembo, OI, March 26,1998; Misaki, OI, July 18, 1998; Akibaya and Otiende, OI, August 20, 1998; Popai Indoshi, OI, August 30, 1998; Musine, OI, September 21, 1998; Bukhala, OI, September 24, 1998.

222. Warden of Mines, CMIR February 1934, KNA: PC/NZA/3/14/372; Isenjia, OI, March 24, 1998.

223. These are estimates as some companies failed to report accidents. See Warden of Mines, CMIR for December 1939, KNA: PC/NZA/3/14/372.

224. Otiende and Akibaya, OI, August 20, 1998; NKDAR 1932, KNA: DC/NN/1/13.

225. *Times*, 23 February 1933.

226. NKDAR 1933, KNA: DC/NN/1/14; NPAR 1933, KNA: PC/NZA/1/28.

227. Warden of Mines, CMIR January 1934, KNA: PC/NZA/3/14/372; Seleta, OI, February 15, 1998. Mukhomoli, the nickname given to the manager of Musgraves by the Luyia, meant "one who mocks Africans."

228. To date, Logoli women play such a central part in funeral arrangements that it is often stated "it is impossible to have a funeral without women being present." They not only produce the next generation of ancestors, but they also bury the last one. For details see Judith Abwunza, *Women's Voices, Women's Power Dialogues of Resistance From East Africa* (Toronto, Ontario: Broadview Press, 1997), 113–15.

229. NPIR January 1934, KNA: PC/NZA/4/5/1; Warden of Mines, CMIR January 1934, KNA: PC/NZA/3/14/372. During the same month, Mr. Wilcox of Risks Ltd. was prosecuted for negligence with regard to machinery as a result of a drowning accident in the dudgeon shaft.

230. DO Kakamega, Safari Diary, February 4, 1934, KNA: DC/KMG/1/1/1.

231. Warden of Mines, CMIR March 1934, KNA: PC/NZA/3/14/372; NPIR March 1934, KNA: PC/NZA/4/5/1.

232. Amagadu and Shikali, OI, March 3, 1998; Francis Mutsotso, OI, March 4, 1998; Mutembei, OI, March 17, 1998; Amahwa, OI, August 29,1998; Vihembo, OI, March 26, 1998.

233. NKDIR June 1935, KNA: PC/NZA/4/5/5.

234. Imbali, OI, February 18, 1998. Imbali still walks with a limp and flashes a "crater-like" scar in his thigh for which he received no compensation.

235. NKDAR 1938, KNA: DC/NN/1/20; NPIR October 1938, KNA: PC/NZA/4/5/2; Musalimwa, Mwidakho, Ingosi Segero, OI, February 17, 1998.

236. NPIR for November 1938, KNA: PC/NZA/4/5/2.

237. NKDAR 1939, KNA: DC/NN/1/21; NPAR 1939, KNA: PC/NZA/1/34; NPIR February 1939, KNA: PC/NZA/4/5/3. The administration held *barazas* warning against such conduct, as the newly formed Kenya Miners Association protested to the governor on "lack of discipline in reserves."

238. Muhati, OI, February 13, 1998; Musalimwa and Mwidakho, OI, February 17, 1998.

239. Musalimwa, OI, February 17, 1998.

240. *East African Standard*, 5 April 1941; Postmaster General to Secretary, Rosterman Gold Mines, April 22, 1941, KNA DC/KMG/2/18/15; Muhati, OI, February 13, 1998; Imbali, Mung'alo, and Mutondo, OI, February 18, 1998; Mbwabi, OI, February 26, 1998; Andove, OI, February 28,1998. The company was running the electric plant without a permit, contrary to the Electric Power Ordinance.

241. Muhati, OI, February 13, 1998.

242. Seleta, OI, February 15, 1998.

243. Extract from Minute M/1737/29 of October 13, 1936, Ag. Commissioner of Mines to Colonial Secretary, Brooke-Popham Papers, Rhodes House Oxford (RHO), MSS. Afr. S. 1120 111/7/19. Another version is provided in the *NADAR 1937*, 153, PRO: CO 533/479/9.

244. Akibaya and Otiende, OI, August 20, 1998.

245. Warden of Mines, CMIR February 1934, KNA: PC/NZA/3/14/372; NPIRs for February and March 1934, KNA: PC/NZA/4/5/1.

246. NKDAR 1933, KNA: DC/NN/1/14; NKDAR 1934, KNA: DC/NN/1/15; NKDAR 1935, KNA: DC/NN/1/16; NKDAR 1938, KNA: DC/NN/1/20.

247. NKDAR 1933, KNA: DC/NN/1/14; NKDAR 1934, KNA: DC/NN/1/15; NKDAR 1935, KNA: DC/NN/1/16; NKDAR 1936, KNA: DC/NN/1/18; NKDAR 1938, KNA: DC/NN/ 1/20.
248. Warden of Mines, HOR, February 18, 1935, KNA: PC/NZA/3/14/372.
249. Political Record Book Vol. III., KNA: DC/NN/3/4/2.
250. Wagner, *The Bantu,* Vol. I., 10; T*he Daily Express*, 21 March 1933.
251. NPAR 1930, KNA: PC/NZA/1/25; Minutes of the Second Meeting of the NKLLB Held at Kakamega on April 29, 1932, KNA: PC/NZA/2/2/17; Notes by DC NK, August 16, 1933, KNA: DC/NN/3/4/2. These Tsotso and Isukha households entitled to compensation waited until 1939 before they received monetary compensation of Shs. 50 each. DC NK to Chief Mutsembi of Butsotso, June 9, 1939, KNA: DC/NN/ 3/4/2.
252. NKDAR 1932, KNA: DC/NN/1/13; Akibaya and Otiende, OI, August 20, 1998.
253. NKDAR 1932, KNA: DC/NN/1/13. The bazaar was a form of sub-central business district comprising all forms of stores dominated by Asians, Goans and Arabs. Zeleza, "The Colonial Labour," 155.
254. *Times,* 22 December 1932, Lugard Papers, RHO, MSS. Afr. 77/4.
255. NKDAR 1933, KNA: DC/NN/1/14.
256. *Daily Express,* 27 February 1933.
257. *Manchester Guardian*, 21 March 1933.
258. C. A. Marsh, "A Visit to the Kakamega Gold Field," *The Friend, The Quaker Weekly Journal*, April 7, 1933; Marsh to Parliamentary Secretary & Member of the Anti-Slavery and Aborigines Protection Society, PRO: CO 533/429.
259. NKDAR 1934, KNA: DC/NN/1/15.
260. Fearn, "Gold-Mining," 54; NKDAR 1933, KNA: DC/NN/1/14; *MGDAR 1934*, 8–9. Township roads benefited from a Shs. 6, 400 government grant in 1934. NKDAR 1934, KNA: DC/NN/1/15. In 1935, a £20,000 loan from the Colonial Development Fund was earmarked for the Kisumu-Kakamega road. Report of the Committee of the Central Roads and Traffic Board Appointed to Consider the Construction of Roads in Mining Areas out of the Colonial Development Fund Loan, October 5, 1935, PRO: CO 533/459; Memorandum: CO, January 9, 1936, Ibid. In 1936, Kimingini received £50 from the LNC Road Fund for maintenance of the road to its fuel concession. NKDAR 1936, KNA: DC/NN/1/18.
261. Minutes of the Sixth Meeting of the NKLLB Held at Kakamega on December 16, 1933, KNA: PC/NZA/2/2/17; NKLNC, Minutes of a Meeting Held at Matungu on February 14 & 16, 1934, KNA (Kakamega): HW/13/2.
262. Marsh, *Friend*, April 7, 1933.
263. NKDAR 1936, KNA: DC/NN/1/18.
264. Ibid. 1937, KNA: DC/NN/1/19; NKDAR 1938, KNA: DC/NN/1/20.
265. NPAR 1939, KNA: PC/NZA/1/34; NKDAR 1939, KNA: DC/NN/1/21.
266. NKDAR 1933, KNA: DC/NN/1/14; NPAR 1933, KNA: PC/NZA/1/28; NKDAR 1934, KNA: DC/NN/1/15; Ayiekha, OI, February 15, 1998; Shivachi, OI, February 15, 1998; Musalimwa and Mwidakho, OI, February 17, 1998; Isenjia, OI,

February 24, 1998; Andove, OI, February 28, 1998; Amagadu and Shikali, OI, March 3, 1998; Mutsotso, OI, March 4, 1998; Litavakha, OI, March 9, 1998; Enock Mwinamo, March 26, 1998; Misaki, OI, July 18, 1998; Akibaya and Otiende, OI, August 20, 1998; Amahwa, OI, August 29, 1998; Indushi, OI, OI, August 30, 1998; Musine, OI, September 21, 1998. The Kimberley diamonds industry offered similar opportunities. See Worger, *South Africa's City of Diamonds*, 70–2.

267. Andove, OI, February 28, 1998; Amahwa, OI, August 29, 1998; Indushi, OI, August 30, 1998; See also Roberts, "The Gold Boom," 554.

268. Shivachi, OI, February 15, 1998; Andove, OI, February 28, 1998; Amagadu and Shikali, OI, March 3, 1998; Litavakha, OI, March 9, 1998; Akibaya and Otiende, OI, August 20, 1998; Amahwa, OI, August 29, 1998. See also Roberts, "The Gold Boom," 554.

269. Wagner, *The Bantu, Economic Life*. Vol. II. ed. L. P. Mair (London: OUP, 1956), 37.

270. NKAgAR 1934, KNA: AK/2/27; Col.& Prot. of Kenya, *DofAAR 1934* (Nairobi: Government Printer, 1936) 93.

271. NKAgAR 1935, KNA: AK/2/27; D*ofAAR 1935* (Nairobi: Government Printer, 1936), 101.

272. NKAgAR 1935, KNA: AK/2/27; NKAgMR March 1935, KNA: AK/2/24.

273. NKAgAR 1935, KNA: AK/2/27; Shivachi, OI, February 15, 1998; Andove, OI, February 28, 1998.

274. NKAgMR July 1934, KNA: AK/2/24.

275. Ibid; NPIR July 1934, KNA: PC/NZA/4/5/1.

276. NKLNC, Minutes of a Meeting Held at the GAS Hall, Kakamega, May 7, 1935, KNA (Kakamega): HW/13/2.

277. NKAgMR July 1934, KNA: AK/2/24.

278. Ibid.

279. Col. & Prot. of Kenya, *DAAR 1934*. Vol.1. (Nairobi: Government Printer, 1936), 100–01.

280. DC NK to PC Nyanza, November 9, 1936, and DC NK to PC Nyanza, November 11, 1936, KNA: DC/KMG/2/18/9; NKLNC, Minutes of a Meeting Held at Matungu, March 30 & 31, 1937, KNA (Kakamega): HW/13/2.

281. NKLNC, Minutes of a Meeting Held at Kakamega, July 19 & 20, 1937, KNA (Kakamega): HW/13/2.

282. NKLNC, Minutes of a Meeting Held at Matungu, November 16 & 17, 1937, KNA (Kakamega): HW/13/2; NPAR 1937, KNA: PC/NZA/1/32.

283. Wagner, *The Bantu,* Vol. II., 172, 175; NKLNC, Minutes of a Meeting Held at Matungu on July 11 & 12,1938, KNA (Kakamega): HW/13/2.

284. NPAR 1938, KNA: PC/NZA/1/33; *DofAAR 1938* (Nairobi: Government Printer, 1939), 90.

285. *DofAAR 1938*, 90. Revenue from vegetable sales during 1938 for NK and CK growers amounted to £5,000.

286. Indushi, OI, August 30, 1998.

287. Shivachi, OI, February 15,1998; Andove, OI, February 28,1998; Amagadu and Shikali, OI, March 3,1998; Mwinamo, OI, March 26, 1998; Misaki, OI, July 18,1998.

288. NKAgAR 1933, KNA: AK/2/27; *DofAAR 1933* (Nairobi: Government Printer, 1934), 118.

289. AO NK to AO Nyanza Province, January 25, 1934, KNA: AK/11/44.

290. NKDAR 1934, KNA: DC/NN/1/15; NKAgMR October 1934, KNA: AK/2/24; NKAgAR 1934, KNA: AK/2/27. The NKDAR puts the output figure at forty bags per eight hours.

291. NKAgMR October 1934, KNA: AK/2/27.

292. NKAgMR April 1935, KNA: AK/2/24; NKAgAR 1935, KNA: AK/2/27.

293. NKAgMR June 1935, KNA: AK/2/24.

294. NK Crop Report (NKCR) August 1935, KNA: AK/2/58.

295. NKCR October 1935, KNA: AK/2/58.

296. NKAgAR 1935, KNA: AK/2/27. The mill offered 50–75 cents more per bag.

297. Musalimwa and Mwidakho, OI, February 17, 1998; Mbwabi, OI, February 26, 1998; Akibaya and Otiende, OI, August 20, 1998.

298. Report of the Industrial Mission, 1935, quoted in W. T. S. Gould, "Technical Education and Migration in Tiriki, Western Kenya, 1902–1987," *African Affairs* 88, no. 351 (April 1989), 261.

299. Fearn, *An African Economy*, 149.

Chapter Six

At the Crossroads: Socio-Cultural Transformation

GOLD MINING AND HEALTH

Colonial capitalist industrial penetration of Buluyia heralded significant health challenges and accelerated socio-cultural transformation in Western Kenya. Apart from deaths resulting from mining accidents as detailed in chapter 5, a close examination of records shows increased health problems and deaths among workers, contrary to Fearn and Roberts' conclusion that there was low incidence of disease among workers.[1]

As early as 1931, the medical department was worried about the state of public health due to the influx of large numbers of European prospectors working on alluvial claims in riverbeds around Kakamega. Alluvial mining created conducive conditions for anopheles costalis; the malaria-causing female mosquito. This occurred when prospectors exposed streams to light or interfered with river courses. Very quickly, a considerable danger of malaria in epidemic form emerged among Africans and European prospectors living under camping conditions. To counter this menace, mining and health officials emphasized the importance of prospectors taking precautionary measures such as filling up pits and canalizing streams.[2]

In spite of such appeals by officials to prospectors, the situation continued to deteriorate. By the end of December 1931, the NK MOH held a meeting at Chief Milimu's camp in East Isukha to address this issue. The official reiterated the malarial nature of the district with the anopheles mosquito as the principle enemy. Mining operations were creating conditions favorable to its breeding. The MOH urged prospectors and claim holders to fill up holes, and create mechanisms for draining abandoned excavations. While miners promised to carry out anti-malaria measures, most failed to do so.[3] The contribution of alluvial mining in accelerating mosquito breeding and the prevalence of malaria in the goldfield continued to worry both the medical authorities and the district administration.

This was particularly so during the rainy season when Luyia households in the mining locations bore the brunt of malaria epidemics, compared to European prospectors who often sought the security of their farms in the climatically healthy Kenya highlands.[4]

The increased incidence of malaria among Europeans in the goldfield and Kakamega Township, as well as Luyia populations had become a public concern by May 1933. With the advent of rains, alluvial mining engendered water pooling which, coupled with the opening up of streams by alluvial workings, created ideal conditions for mosquito breeding. In addition, employment opportunities in the goldfield attracted Africans from outside the district that brought along new virulent strains of the sub-tertian parasite to a population largely deficient in immunity. Likewise, the haphazard construction of European camps, labor compounds, and the opening up of more streams to alluvial mining, exacerbated the declining health situation.[5]

A clear conceptualization of the gloomy health situation that gold mining engendered in Buluyia emerges from an examination of colonial health services at the time. As noted elsewhere, the global depression of the 1930s had led to the collapse of the Kenya colonial economy. With declining exports and imports, the colonial state sought to tighten its purse strings through reduced government expenditure and retrenched services. For example, government expenditure on medical services fell from £236,934 in 1930 to £222,897 in 1931. In 1932, the figure dropped to £197,260 and remained below the 1930 level until after 1936. Furthermore, funding for health education, an important disease prevention measure, drastically dropped from £3,150 in 1929 to £60 in 1932. More tragically, at the time when Kakamega was under a virtual malarial siege, the administration put off its plan of hiring a special malaria research officer.[6]

Under these circumstances, malaria continued to recur with alarming frequency, forcing the medical department to institute inspection for mining camps and alluvial workings. Nevertheless, the limited resources earmarked for sanitary inspection rendered this effort futile. With a paltry annual budget of £10 dedicated to sanitary inspection for Kakamega Township and the entire district, inspectors barely managed a single annual visit to the goldfield. Notwithstanding, Kakamega continued to attract both European and African populations in large numbers, resulting in a concentration of African dwellings and markets, and European brick fields on the periphery of the township. The rapid physical expansion of Kakamega made the task of mosquito control insurmountable. Capital constraints had forced the administration to limit mosquito control measures within a mile radius of the township boundary. Most of the meager budget went into purchasing oil supplies for anti-malaria work. Moreover, anti-malaria work was constrained by limited medical staff, both Asian and European, which in 1933, remained at the pre-mining level. Similarly, the number of African medical staff remained insignificant. Unfortunately, the success of the anti-malaria program in the goldfields depended on the provision of increased staff and funding.[7] Indeed, in May 1933, the MOH called for:

> The posting of an additional Sanitary Inspector for work on the gold fields, trading centres and native work; the posting of anti-malarial overseer principally for work in the township, but also for work in the gold field; the posting of a clerk for health office work; and the provision of adequate funds for anti-malarial work generally and for essential services in Kakamega township.[8]

Refusing to implement the MOH's proposal, the administration devoted much time to inspection and instruction within the European mining community.[9] In fact, the administration was unwilling to commit more funds and time to the health of Africans. As DC Anderson bluntly put it:

> Considering the taxes he pays [the African], the medical provision for his needs has been excessive in the past. It is, therefore, just that a considerable portion of the time of officers and of the money doubtless originally intended for the native should be diverted to work in connection with Europeans.[10]

The district boss was willing to sacrifice the health of Africans in the goldfield at the altar of European welfare. The limited resources available were to serve European interests. The outcome was that while the work of the Native Hospital at Kakamega had doubled in the first two years of the gold mining era, African public health services in the entire district continued with less supervision and funding. This was in line with the dominant colonial ideology that polarized settler politics against African interests in Kenya.

Apart from malaria, gold mining and the wet conditions prevalent in the goldfield led to increased cases of pneumonia among African mine employees. For example, in November 1933, nine cases of pneumonia occurred in NK district, eight of them among African workers at Risks Mining Company in West Isukha.[11] Risks continued to lead the district in the incidence of pneumonia throughout the year.[12] Nonetheless, by the close of the year, camp sanitation supervision had improved somewhat as small firms continued to merge into larger companies.[13]

There was, however, some policy change in 1934 when the administration relieved the MOH of his hospital duties for six weeks in order to carry out an extensive malaria survey in the goldfield. His findings confirmed the already established link between alluvial mining, trenching and mosquito breeding. The absence of toilets in both European and African camps in the goldfield compounded the health situation further, and miners generally disregarded to keep the requisite First Aid appliances.[14] Such conditions only served to exacerbate diseases and epidemics in the goldfield.

In January 1934, moreover, of the twenty-eight pneumonia cases admitted to hospital for the whole goldfield, twenty-four were employees of Risks Ltd. The figures rose to sixty-four hospital admissions and three deaths by the end of the year.[15] In addition, an outbreak of cerebral-spinal meningitis occurred at Kimingini labor camp, with single cases at Rosterman and Anglo-Continental mines.[16]

Available sources show a high correlation between deteriorating health conditions in the goldfield and the activities of miners bent on flouting mining laws. Although some mining companies established medical facilities for their employees, the general response to the environmental health challenges posed by the gold mining industry was largely unimpressive. This did not augur well for the rest of the population not employed in the mining industry.

In July 1935, incendiary apathy among miners in restoring health conditions in the mines compelled the acting Commissioner of Mines, the NK DC, and the MOH to hold a meeting to discuss possible solutions. The officials agreed to insert into alluvial EPLs a guaranteed monetary deposit for use by health officials in the event that licensees failed to execute their health obligations. The implementation of this measure, however, had to await the PCs approval.[17]

In the meantime, health officials shifted their attention to experimental anti-malaria spraying using Paris Green diluted with powdered charcoal. Shortly after, it became clear that this initiative was propelled by economic imperatives and not simply the need to check malaria. Health officials discovered that the use of oil in alluvial workings to control mosquito breeding often clogged sluice boxes thereby interfering with the process of recovering gold. Hence, mining officials hoped that Paris Green would prove more suited to alluvial operations.[18]

Notwithstanding, the Paris Green experiment proved ineffective because it was suited to larger mosquito breeding sites such as swamps compared to the confined seepage areas of the township. In spite of this, the experiment continued.[19] As officials experimented with alternative anti-malaria measures with the least impact on gold production, the mosquito continued to wreak havoc in the goldfield.

In August 1935, for example, the incidence of malaria increased rapidly at the Rosterman labor camp. Consequently, the colonial state employed a

European overseer and a permanent group of African laborers on anti-larva control, with periodic visits and advice from the field malaria overseer. Despite such measures, unsatisfactory alluvial working conditions persisted. As the goldfield expanded, so did the incidence of malaria. Admittedly, although legal responsibility rested with individual miners for filling excavations, the dichotomy between legislation and its enforcement remained the hallmark of the Mining Ordinance, especially with regard to abandoned pits.[20]

The DCs meeting held at Kisumu on September 30, and October 1, 1935, focused on the question of amending mining permits. Since existing mining laws had failed to adequately deal with mosquito breeding in the goldfield, the administrators unanimously agreed to push for legislation empowering the MOH to stop alluvial work that was most likely to cause mosquito breeding. This measure aimed at providing legal backing for dealing with existing nuisances.[21] This drastic measure could also control mining operations along riverbanks and the diversion of rivers that created a series of excavations that nurtured extensive mosquito breeding. In the face of an impotent Public Health Ordinance and a mining community that disregarded its legal responsibility to fill up prospecting pits, rules aimed at preventing mosquito breeding were a matter of urgency.[22]

Although no major malaria outbreak occurred during the year, mining camps grappled with sporadic cases of cerebro-spinal fever.[23] Despite inspection by health authorities, alluvial mining proceeded in an unsatisfactory fashion. In the words of the MOH: "Visits were made to a large number of old workings on Yala River. . . . Water was standing in a large number of pits and excavations and anopheles costalis breeding was extensive."[24]

The departure of individual miners, particularly alluvial workers, and the ascendancy of reef mining, slightly improved the general public health. Nevertheless, while reef mining was reportedly satisfactory, alluvial mining was not.[25] By December 1935, two cases of epidemic meningitis occurred at the East Africa Mining Areas and an alluvial camp in Kisa location. The low water levels in Yala and Edzawa Rivers intensified pooling and mosquito breeding, and when the rainy season commenced, mosquito larvae freely washed to land.[26]

To counter this menace, the administration intensified bush clearing in mining areas using African labor. This temporarily improved the health condition of the mining community, although many cases of malaria still occurred.[27] By the middle of 1937, malaria was once again rife in the goldfield.[28]

The question of alluvial mining and mosquito breeding continued to exercise the minds of colonial officials in Western Kenya. Faced with the difficult health situation, mining and health officials contemplated halting the extension of alluvial mining. Indeed, when Mrs. Stitt, O. Fayle, D. H. Stanley, and Alluvial Concessions applied for an EPL to extend alluvial mining on River Yala, the MOH NK warned:

> From a health point of view, there would appear to be strong arguments against the extension of alluvial mining in this district. If alluvial mining is to be permitted the special condition No. 3 should restrict the nuisances to the public health which are likely to arise therefrom [sic].[29]

The MOH's opposition to extending alluvial mining attested to the empirical evidence where miners were flagrantly violating mining laws. Mrs. Stitt had been served with numerous notices for mosquito breeding and unsanitary camps. In one instance, it took a court summons to get the nuisances abated. In November 1937, only written threats of revoking her mining licence moved Stitt to restore abandoned claims. Neither were companies clean of such acts. Alphega Syndicate cleaned up its mess under considerable pressure from medical authorities.[30] This context explains the unpopularity of alluvial workers in the corridors of public health.

Such vigilance by medical officials against unsanitary conditions in the goldfield realized some degree of success in 1938. Through treatment and enforced cleanliness in labor camps, the incidence of malaria that characterized the goldfield declined substantially.[31] Further, the administration disallowed new alluvial mining syndicates in the district because of the danger they posed in creating excellent conditions for mosquito breeding.[32]

While the war situation diverted official focus and resources to the war effort, the post-war era saw a return to the theme of health in mining areas. In his Annual Report in 1945, the NK DC documented the prevalence of unhealthy conditions in Kakamega Township, particularly in the overcrowded bazaar area.[33]

Other factors including inadequate medical services and the negative attitude European medical personnel adopted toward African patients significantly impacted African health in the goldfield. In a protest letter to the Director of Medical Services in 1944, the NKCA castigated the attitude of the European doctor in charge of the Native Hospital at Kakamega, demanding his removal. The association condemned the health official for ignoring his duties at the Native Hospital, while devoting much time to the European dispensary at Rosterman. The doctor spent very little time attending to African patients, and allegedly refused "to put his hand on '*nyani*.'"[34]

Despite such cases of deplorable behavior by individual medical personnel, it is nevertheless evident that some mining companies provided health facilities for their employees. Rosterman had its own hospital where, at one time, twenty-three Africans served as hospital attendants.[35] In 1935, the Roman Catholic Mission at Mukumu built an excellent thirty-bed hospital for Africans, with generous support from Rosterman Mine.[36] By 1938, the hospital was catering to European, Asian and African populations. The founder, Dr. Claude Marshall, was in charge, assisted by a qualified dentist, Dr. Bowles. Mission nuns served as nurses. The hospital also received a grant of £600 from the colonial state and support from Messrs Rosterman and Kavirondo Gold Mines.[37] Colonial officials claimed that several mines had ambulances and the services of private doctors. These dispensaries had better medical supplies compared to most that were run by the government. Where necessary, labor received injections for more common severe diseases.[38] The Luyia, however, relied on the LNC ambulance. Nonetheless, oral accounts confirmed that racial strictures governed the separate social space between black and white in Kenya colony. In the absence of direct evidence as to whether African labor had access to such services, it is plausible that only the privileged European miners benefited from such services. Similarly, Rosterman's child welfare, and pre-and post-natal clinics, which became operational in 1943, were most likely European-only facilities.[39]

As late as 1944, Nyanza Province still experienced high incidences of malaria. This was true of NK district, especially the locations within the central mining area and the district headquarters where special staff were engaged to distribute mepacrine in affected areas, and a nursing orderly detailed to give quinine injections in serious cases.[40]

Clearly, gold mining unleashed monumental health challenges in Buluyia. Mining operations, particularly alluvial mining, created excellent conditions for mosquito breeding, leading to declining health and increased fatalities among the Luyia. Although industrial diseases such as silicosis and asthma were not prevalent, other diseases directly resulting from mining undertakings including malaria, cerebro-meningitis and pneumonia, turned the goldfield into an epidemic zone. The racial dimension surrounding health provision in colonial Kenya accentuated the situation. Although a few companies made some conscious effort to provide medical facilities for their employees, racial considerations largely influenced who got the best treatment. Further, the colonial state's inertia in providing funding and personnel to meet the health needs in NK district deepened the health quagmire. In sum, gold mining significantly worsened health conditions in Buluyia and set in motion significant social transformation in Luyia social relations.

GOLD MINING AND
RURAL SOCIO-CULTURAL TRANSFORMATION

The dawn of the gold mining industry in Buluyia posed significant socio-cultural threats to the Luyia social milieu. The gold mining industry and the attendant urban growth that the industry stimulated led to new social challenges that intensified cultural tensions within the Luyia community and between the Luyia and the Europeans. The Luyia opposed the gold industry due to perceived social concerns. Fundamentally, the Luyia dreaded the social implications of the presence of large numbers of Europeans who converged on Kakamega during the gold rush. Most sensed a keen cultural threat. The gold mining industry posed the immediate danger of undermining Luyia cultural norms by creating conditions prone to corrupt Luyia women.[41] As early as January 1932, European prospectors were already a menace to African women. In a Circular of January 27, 1932, DC Anderson aptly noted:

> Rumours have reached me that certain advance has been made to local girls. I would warn anyone who thinks he has come to such a friendly agreement, that the Native of these parts is an expert at Blackmail. They even blackmail each other on these occasions, the lady denying consent. I need hardly say in what an awkward position a European would be placed if such a charge were brought against him, and how it would be difficult to defend himself. . . .[42]

This issue featured prominently at a KTWA meeting on December 31, 1932. In resolutions that subsequently appeared in the British press, the KTWA wrote:

> The inevitable prostitution of native girls and young married women, happily confined to the vicious elements of the goldminers [sic], is breaking down not only the personal morality but also valuable native custom. By native law a father or husband has a right to proceed at law against one who cohabits with a daughter or wife. European law conflicts with native law. Many Africans feel that this divergence will undermine a most valued native safeguard.[43]

Apart from legal frustrations, liaisons between miners and married women brought significant strain on families that in accordance with Luyia customs had to return dowries. Such "dowry refunds" often affected a chain of marriages contracted with the initial dowry.[44] The consequential destabilization of the marriage institution, provided for intense criticism. On January 6, 1933, Archdeacon Owen took up a case of sexual misconduct-involving police constable E. P. C. Slatter's with the CNC, Wade.[45] Deployed as goldmine police and compensation assessor, Slatter so distinguished himself for cohabiting with Luyia prostitutes that the governor ordered his removal from the province.[46]

Perhaps because of this, on January 11, 1933, Luyia elders expressed concern over increasing cases of immorality in the goldfield at a meeting with the CNC. Indeed, the CNC acknowledged that one man had received a "heavy sentence for a particularly nauseating offence."[47] In a private correspondence to Sir Albert Kitson in February, Murray Hughes of the mining department expressed fear over the influx of Greeks, Italians, and a generally mixed crowd of "undesirable" Europeans into Kakamega and their impact on the Luyia social milieu. He reported that most were drinking with African men and sleeping with their women, causing what he described as "most unpleasant and disquieting incidents. . . ."[48]

Although the disturbing news about emerging social challenges in the goldfield reached the S of S for the Colonies through unofficial sources, it nevertheless alarmed the imperial authorities, prompting the CO to seek an explanation from the Kenya government. The S of S called for tighter legislation to facilitate the expulsion of any "undesirable persons from any area, with heavy penalty for disregard of such order, and for return without permission."[49] While the governor denied official knowledge of the allegations, he promised to make enquiries from the Provincial administration.[50] Two months later, the LegCo passed a bill, legalizing the expulsion of persons "deemed undesirable" from the goldfield.[51] However, the legislation never altered the situation.

Urban growth and the presence of miners in Kakamega promised economic opportunities for women in the informal sector. In most pre-conquest rural societies, prostitution was practically unknown and Kakamega appears to have been no exception. Colonial conquest, urbanization and the development of colonial apparatuses of economic penetration, such as communication networks, facilitated mobility and population concentration, which in turn created conditions conducive to prostitution. Luise White situated the emergence of prostitution within the colonial urban space—Nairobi.[52] As one of the few options open to African women in the racially and gendered colonial labor hierarchy, women engaged in prostitution for several reasons. These included personal reasons such as marital discord at home, death of a spouse, and failure to get a husband.[53] Yet, one cannot rule out purely economic motives. In her study of prostitution in colonial Nairobi, White found that prostitutes were among the city's first petty-bourgeois accumulators, especially in real estate.[54] While evidence of accumulation among prostitutes in the goldfield is scant, the underlying economic drive cannot be totally discounted. The rapid transformation of Kakamega into a cosmopolitan township in the midst of a virtual rural agrarian terrain under the strong control of a patriarchal gerontocratic power base, and its equally rapid decline, may have stifled the prostitutes' capital accumulation.

For example, at an August 1933 LNC meeting, Atanas Habwatira drew a direct link between the general increase in prostitution in the district, the development of gold mining and, the influx of women of different ethnicities into the goldfield. Habwatira maintained that this was corrupting Luyia youths and aiding the spread of venereal diseases. Luyia elders condemned Nandi women for causing much antagonism and "debauching . . . local girls."[55] Although both LNC members and the DC acknowledged the presence of Nandi prostitutes in Kakamega, the greatest dilemma remained the strategy for checking their activities.

While the LNC favored forcible expulsion, the DC harbored some reservations. Expelling alien women in the face of a growing demand for prostitutes among goldfield employees would transfer the heavy burden to local women. Ultimately, the LNC resolved to approach the colonial state to facilitate the expulsion of all "alien" prostitutes from the district.[56] Interestingly, the colonial state officially tolerated and at times encouraged prostitution since it served as a "wage depressant, [and] disincentive for laborers to bring their families to towns."[57] As Tiyambe Zeleza aptly puts it, "prostitution facilitated the daily reproduction of male labor power."[58]

Mineworkers constituted the single most important "market" for prostitutes, partly because few establishments encouraged normal family settings in labor camps. Most employees did not allow their African workers to bring their families to their work destinations. Such restrictions drove workers into temporary liaisons with prostitutes. Moreover, the demographic imbalance of the sexes in the goldfield and the lack of employment opportunities for women, sustained prostitution. This provides the context for the involvement of Luyia women in prostitution. Chimoi Shilisia, Mutangale Shivoga, Amina Lukwaa, and Shibesi Andachila from Isukha, and Imbande Wakukha from Idakho, worked side by side with Nandi, Luo and Ganda prostitutes in Kakamega.[59]

Prostitution was also closely related to increased incidence of theft in mining camps. Offenders included not only local Isukha, but also immigrants from other ethnic groups, as well as the Baganda women from neighboring Uganda territory. The difficulty involved in enclosing labor camps, and the inapplicability of the Vagrancy and Pass Laws in the mining areas, compounded the problem further. Vagrancy laws aimed at curtailing the entry of unemployed Africans into colonial urban areas, while the Pass Laws, which were enforced through the *Kipande* (identity card), regulated African labor and wages as well as their movement outside their respective reserves. Faced with this dilemma, the administration contemplated establishing rules for the control of mining camps, and hiring plain-clothes police.[60]

As the administration grappled with the question of controlling prostitution and theft in the goldfield, local men were transforming the "new trade" (prostitution) into personal gain. In January 1934, for example, Luyia men of social standing were housing Nandi and other foreign prostitutes in the neighborhood of mining camps with the purpose of securing profits from rent and other fees. Later, the administration ejected and prosecuted several prostitutes so housed.[61]

While Luyia "notables" near the mines including one Shilibwa and Sunguti were exploiting women prostitutes, the prostitutes themselves suffered arrest, eviction and prosecution. Despite the crackdown on prostitutes, and not their male clients, the practice continued unabated. In 1934, J. D. Otiende, a Logoli student at Makerere Technical College[62] in Uganda, admitted that he was cautioned by his parents and relatives against having sexual liaisons with local girls in the mining area. The community generally assumed that these girls were prime carriers of venereal diseases. As a teacher at Kaimosi Friends African Mission (FAM) School in 1937, Otiende met a Logoli woman who lived with a European miner in Kaimosi.[63]

Luyia women who engaged in prostitution in Kakamega exhibited other types of socially unacceptable behavior. In March 1944, NKCA complained to the DC about women loitering around shops and along roads all day, thereby contributing to prostitution. Once more, the association singled out women for punishment in a recommendation banning their movement throughout the district without passes from location chiefs.[64] Following this restriction on women's movement, the NK LNC passed a resolution to forcibly repatriate Luyia girls earning a living as prostitutes in Kisumu.[65]

On April 4, 1944, Luyia leaders moved swiftly to take a firm stand against the vice. A delegation comprising the NK AO, Chief Segero of Isukha and a representative of Chief Mulupi of Kabras, visited Rosterman's site urging the closing of the market in Rosterman's labor compound due to "much immorality."[66] Although the accomplishment of the delegation is not clear, it is nevertheless manifest that Luyia male leadership continued to press for measures to control prostitution. In September, the NKCA sought to send a delegation to the DC to discuss *inter alia,* prostitution and the labor camp at Rosterman.[67] Clearly, the gold industry institutionalized prostitution with special centers in the goldfield facilitating sexual exchanges. Popularly known as *majengo* [villages], the brothels were strategically located in close proximity to major mining centers, such as the Kavirondo Gold Mining Company site, the Somali camp and Lutonyi, which served both Kakamega township and Rosterman mines.[68]

To deal with the situation, the LNC passed a resolution in 1945, empowering headmen to order owners or occupiers of dwellings, structures and other premises used as brothels or houses of ill fame, or for other purposes deemed prejudicial to morality and public peace, to close down. Hereafter, it was an offense to refuse, neglect or to fail to comply with such an order.[69] Yet, prostitution remained a reality in Kakamega.

Two years later, the Isukha Locational Advisory Council (ILAC) was still groping for a solution to prostitution. The ILAC crossed gender boundaries, banning all Isukha from running brothels or using their private houses for prostitution. The elders lamented the sense of shame they experienced when young men and women hugged and fondled in streets in the scrutiny of the public eye. The social malaise intensified as those infected with venereal diseases concealed their ailment by refusing to seek medical assistance. To avert a social catastrophe, the ILAC empowered local authorities to prosecute such people.[70]

The stigma associated with contracting venereal diseases, became the greatest threat to the Luyia social fabric. In 1935, gonorrhea was the second largest cause of hospital admissions after ulcers, among labor in mining camps.[71] Oral narratives agree that the wider community routinely excluded victims of sexually transmitted diseases (STDs). Although no scientific evidence exists to show that one can contract such diseases by sharing seats or food, most Luyia cautiously avoided suspected STD carriers. The community often dubbed individuals who sought and received medical treatment *efitsiri* [impotent, eunuchs], and death caused by such ailments consolidated fear and prejudice in the community. Such acts of social exclusion and general ridicule compelled victims to choose dying "silently" instead of seeking medical redress. This was particularly true of male household heads with married daughters. These otherwise "respected" males dreaded the possibility of news about their "shameful sickness filtering to their daughters' marital homes." Faced with such potentially socially destructive occurrences, they commonly resolved to "persevere to death." For example, Shilibwa, who illicitly turned his home at Sigalagala into a prostitution center, died such a miserable but "honorable" death. Eyewitness accounts noted, "Shilibwa's stomach opened" as he succumbed to "*kuli kumali,*" literally "the disease that eats and finishes you."[72] Certainly, as late as 1951, when the treatment of venereal diseases with penicillin was proving a marked success, the Luyia in Kakamega and its environs persisted in eschewing seeking medical treatment.[73]

It is extremely difficult to gauge with certainty the benefits of prostitution to Luyia female practitioners. Nevertheless, the fact that the practice prevailed even among Christian girls who moved to Kakamega or Kisumu as full time prostitutes, perhaps points to its viability as a survival strategy for some

Luyia women. Moreover, while documentary evidence on the remuneration for prostitution is lacking, oral evidence posited both monetary and material gains. For example, Amahwa, a former African assessor in Kakamega, maintained that prostitution expanded the informal space for women. Prostitutes rented houses, and, earned additional incomes from illegal beer brewing.[74] In addition to cash rewards, prostitutes with a European clientele enjoyed better clothes and houses.[75] At any rate, monetary and material gains to individuals in no way matched the threat prostitution posed to Luyia social norms and institutions such as marriage and the family.

CONCLUSION

The threat of malaria occasioned by trenching and drilling, diversion of river courses, and the use of poisonous chemicals such as cyanide in the gold recovery process, had significant environmental and health implications. All these transformed the goldfield into an epidemic zone. Sickness and death from malaria, cerebro-meningitis, and pneumonia, coupled with loss of livestock through water pollution, significantly endangered Luyia existence. Moreover, while a few gold mining firms provided health facilities for their employees, Africans generally got the short side of the stick.

The gold industry also affected the Luyia social fabric and threatened the family and marriage institutions by exposing unmarried and young married Luyia women to prostitution. The attendant incidence of venereal diseases such as gonorrhea and syphilis among mine workers bred significant social tensions in the community. Admittedly, prostitution widened the social and economic space for Luyia women. Yet, the community, local authorities and the over-arching colonial establishment, routinely victimized these women.

NOTES

1. Fearn, "Gold-Mining," 58; Roberts, "The Gold Boom," 553. This conclusion was limited to industrial diseases and pneumonia.

2. NK Medical Department Annual Report 1931, KNA: PC/NZA/2/14/6.

3. Notes of a Meeting Held at Milimu's Camp on December 21, 1931, KNA: PC/NZA/3/ 14/374.

4. DC and DO NK to PC Nyanza, Memorandum on the Kakamega Goldfields, encl. in PC Nyanza to Acting CNC, February 8, 1932, KNA: PC/NZA/3/14/374.

5. MOH NK to Director of Medical and Sanitary Services, May 10, 1933, KNA: DC/KMG/ 1/19/6.

6. George Oduor Ndege, *Health, State, and Society in Kenya* (Rochester: University of Rochester Press, 2001), 112–14.
7. MOH NK to Director of Medical and Sanitary Services, May 10, 1933, KNA: DC/KMG/ 1/19/6.
8. Ibid.
9. NKDAR 1933, KNA: DC/NN/1/14; NPAR 1933, KNA: PC/NZA/1/28.
10. DC NK to PC Nyanza, May 12, 1933, KNA: DC/KMG/1/19/6.
11. Warden of Mines, CMIR November 1933, KNA: PC/NZA/3/14/372.
12. Warden of Mines, CMIR December 1933, KNA: PC/NZA/3/14/372.
13. NKDAR 1933, KNA: DC/NN/1/14.
14. Warden of Mines, CMIR January 1934, KNA: PC/NZA/3/14/372; NKDAR 1935, KNA: DC/NN/1/16.
15. Warden of Mines, CMIR January and March 1934, KNA: PC/NZA/3/14/372.
16. Warden of Mines, CMIR November 1934, KNA: PC/NZA/3/14/372.
17. Acting Commissioner of Mines to PC Nyanza, July 26, 1935, KNA: PC/NZA/2/2/34.
18. NKDIR July 1935, KNA: PC/NZA/4/5/5.
19. Ibid, August 1935, KNA: PC/NZA/4/5/5.
20. Ibid.
21. Minutes of the Meeting of District Commissioners, Nyanza Province Held at Kisumu on September 30 and October 1, 1935, KNA: PC/NZA/4/1/3/1.
22. NKDAR 1935, KNA: DC/NN/1/16; NPAR 1935, KNA: PC/NZA/1/30.
23. NPAR 1935, KNA: PC/NZA/1/30.
24. Ibid.
25. NKDAR 1935, KNA: DC/NN/1/16.
26. NKDIR December 1935, KNA: PC/NZA/4/5/5.
27. NPAR 1935, KNA: PC/NZA/1/30.
28. Ibid. 1937, KNA: PC/NZA/1/32.
29. MOH NK quoted in DC NK to Commissioner of Mines, July 23, 1937, KNA: DC/KMG/2/18/4.
30. *MGDAR 1938*, 8.
31. DC NK to Commissioner of Mines, July 23, 1937, KNA: DC/KMG/2/18/4.
32. DC NK to PC Nyanza, January 12, 1938, KNA: DC/KMG/2/18/4.
33. NKDAR 1945, KNA: DC/NN/1/27. *Bazaars* were urban zones designated for Indians.
34. Nyanza Central Association to Director of Medical Services, June 12, 1944, KNA: DC/KMG/1/1/153. The letter was signed by Andrea Jumba and Lumadede Kisala.
35. Fearn, *An African Economy*, 148–89; Muhati, OI, February 13, 1998; Andove, OI, February 28, 1998; Otiende, OI, August 20, 1998.
36. NPAR 1935, KNA: PC/NZA/1/30; NPAR 1936, KNA: PC/NZA/1/31.
37. NKDAR 1938, KNA: DC/NN/1/20. Mining companies in SK district had no health facilities of their own. Hence, they relied on a European ward at Kisii and the Seventh Day Adventist (SDA) Mission Hospital at Gendia. The European ward was

erected in 1938 with government funds. DC SK to PC Nyanza, May 26, 1938, KNA: 2189/PH/2/1/5.

38. NADAR 1937, 177.
39. NPAR 1943, KNA: PC/NZA/1/38.
40. Ibid.1944, KNA: PC/NZA/1/39. It is not clear whether Luyia were included.
41. Bode, "Anti-Colonial," 109. Prostitution remains a culturally unacceptable activity among the Luyia. See Abwunza, *Women's Voices*, 71–72.
42. DC NK, Confidential Notice to All Prospectors, January 27, 1932, KNA: PC/NZA/2/2 /44.
43. *Manchester Guardian*, 12 January 1933, PRO: CO 533/428/9.
44. Khasandi, OI, February 15, 1998; Seth Lugonzo, OI, February 15, 1998.
45. CNC to PC Nyanza, Very Urgent Letter, January 27, 1933, KNA: PC/NZA/2/2/44.
46. Governor to S of S, January 26, 1933, PRO: CO 533/428; Khasandi, OI, February 15, 1998; Imbali and Mung'alo, OI, February 18, 1998; Lugonzo, OI, February 15, 1998; Isenjia, OI, February 24, 1998; Musine, OI, September 21,1998.
47. *Manchester Guardian*, 26 January 1933.
48. Murray Hughes to Kitson, February 8, 1933, PRO: CO 533/429.
49. S of S to Governor, Secret Telegram, February 23, 1933, PRO: CO 533/429.
50. Governor to S of S, Telegram, March 4, 1933, PRO: CO 533/429.
51. Bi-Monthly Report to His Majesty's Eastern African Dependencies Trade Information Office from Kenya for the Period April 7 to May 27, 1933, PRO: CO 533/429.
52. Luise White, *The Comforts of Home: Prostitution in Colonial Nairobi* (Chicago: The University of Chicago Press, 1990).
53. Zeleza, "The Colonial Labour," 180–81.
54. White, *Comforts of Home*, 1–3, 16–18, 45–46, 50, 55–58, 59–60, 64; Zeleza, "The Colonial Labour," 181
55. NKLNC, Minutes of a Meeting Held at Matungu on August 21 & 22, 1933, KNA (Kakamega): HW/13/1. See also DC NK to PC Nyanza, September 5, 1933, KNA: PC/NZA/2/2 /60; Warden of Mines, CMIR September 1933, KNA: PC/NZA/3/14/372.
56. NKLNC, Minutes of a Meeting Held at Matungu on August 21 & 22, 1933, KNA (Kakamega): HW/13/1.
57. White, *Comforts of Home*, 18–20; Zeleza, "The Colonial Labour," 181.
58. Zeleza, "The Colonial Labour," 180.
59. Khasandi, OI, February 15, 1998; Lugonzo, OI, February 15, 1998; Imbali and Mung'alo, OI, February 18, 1998; Isenjia, OI, February 24,1998; Mavia, OI, March 2, 1998; Litavakha, OI, March 9, 1998.
60. Warden of Mines, CMIR August 1933, KNA: PC/NZA/3/14/372.
61. Warden of Mines, CMIR January 1934, KNA: PC/NZA/3/14/372; Isenjia, OI, February 24, 1998; Amahwa, OI, 29 August 1998.
62. Makerere University began as a technical school in 1922. In 1935, it became a center for higher education in East Africa, <http://www.makerere.ac.ug/> (Jan. 12, 2006).

63. Otiende, OI, August 20, 1998.
64. NKCA to DC NK, March 9, 1944, KNA: DC/KMG/1/1/153.
65. NKLNC, Minutes of a Meeting Held at Kakamega on March 15 & 16, 1944, KNA (Kakamega): HW/13/3.
66. AO NK to SAO Nyanza, Safari Diary, April 24, 1944, KNA: AK/21/25.
67. NCA to DC NK, September 9, 1944, KNA: DC/KMG/1/1/153.
68. Akibaya and Otiende, OI, August 20, 1998; Amahwa, OI, August 29, 1998.
69. NKLNC, Minutes of a Meeting Held at Matungu on June 7 & 8, 1945, KNA (Kakamega): HW/13/4.
70. Minutes of ILAC Held on January 4, 1947, KNA: DC/KMG/1/1/19.
71. NKDAR 1935, KNA: DC/NN/1/16.
72. Amahwa, OI, August 29, 1998; Musine, OI, August 30, 1998; Bukhala, OI, September 24, 1998.
73. Medical Department Report for Consideration at the District Development Team Meeting on December 10, 1951, KNA (Kakamega): AGU/3/3.
74. Amahwa, OI, August 29, 1998.
75. Akibaya and Otiende, OI, August 20, 1998.

Chapter Seven

"A Failed Eldorado"

The two decades between 1931 and 1952 witnessed dramatic transformation in a portion of Western Kenya. The dawn of the gold mining industry in 1931 proved a most transforming experience for the Luyia inhabitants of the then colonial NK district. The gold discovery and the subsequent establishment of a rural colonial industrial capitalist economy in Kakamega impacted the Luyia in diverse ways. The march of colonial industrial capitalism in Buluyia, however, did not efface the dominant Luyia agrarian economy that preceded and outlived it. The centrality of land and the vital role of agricultural production in the Luyia political economy both for domestic consumption and market, allowed rural agriculture to remain an equally powerful competitor to the needs of the emerging rural industrial regime.

All things considered, gold output began to decline during the WWII period. This downward trend persisted until 1952, when the last most important firm, Rosterman Gold Mining Company, ceased operations. Kakamega, the much-anticipated "Eldorado" did not become an East African Rand. This was a classic example of a failed Eldorado.

Before the discovery of gold in Kakamega in 1931, the colonial state consciously pursued a policy that stifled industrial development in Kenya colony, except for primary processing services for European settler products. With the discovery of gold in Kakamega, both the colonial state and the imperial government moved rapidly to exploit the gold resources in the colony. The Luyia inhabitants in the goldfield experienced a new form of European industrial capitalist penetration with significant political, economic, social and cultural consequences.

As detailed in these chapters, the development of a rural industrial base in NK district precipitated the amendment of the NLTO of 1930 which provided security of tenure for African reserves as demarcated in 1926. Unlike the 1920s, when the imperial government intervened in Kenya's affairs, the potential for an East African Eldorado compelled the CO to support the colonial state and the powerful European dominated LegCo in adopting the NLT(A)O of 1932, which allowed for the alienation of the Luyia reserve for cash compensation. Notably, while the Labour S of S for the Colonies, Lord Passfield had imposed the NLTO of 1930 on the Kenya administration in April 1931, neither Governor Grigg nor his successor, Sir J. A. Byrne (1931–1936), was willing to protect this external imposition.

Predictably, the discovery of gold in NK reserve in 1931 marked a turning point in the fortunes of Luyia land rights. Coming at the time of immense economic distress brought on by the Great Depression of the 1930s and the locust devastation to the settler economy in Kenya, the gold discovery was considered a panacea to the colony's and British economic woes of the time. The discovery also coincided with heightened land insecurity among Africans and the investigations of the KLC. The imperial government had appointed the Commission in 1932 to adjudicate the conflicting settler and African interests over land in Kenya colony. However, it is clear that the colonial state, the imperial government and the KLC grossly sacrificed Luyia land rights at the altar of colonial industrial capitalism.

The declining fortunes of Luyia land holders were sealed with the ascendancy of the Conservative Sir Philip Cunliffe-Lister to the CO, whose policies resulted in a dramatic erosion of Luyia land rights. Cunliffe-Lister approved the amendment of the NLTO of 1930 to facilitate lease grants to European industrial interest in the Kenya goldfield. Moreover, this radical policy reversal received overwhelming support from representatives of African interests in Kenya and the KLC, in spite of criticism both at home and in Kenya. This symbolized a classic collision between principle, morality and economic pragmatism. The Kakamega gold industry exemplifies the crucial role economic factors played in pushing aside Luyia land concerns in spite of the moral obligation incumbent upon the imperial government to safeguard Luyia land rights.

A European rush to Kakamega quickly followed the discovery of gold in NK reserve. This led to general nervousness among the Luyia over the fate of their land. When Governor Byrne visited the goldfield in July 1932, and later the KLC in September 1932, the Luyia challenged the governor and the commissioners to partake a solemn oath guaranteeing the security of their land. While the governor provided a verbal assurance, both the colonial state and the KLC emphasized the centrality of the Crown's right to the mineral resources of the colony as provided in the NLTO of 1930. With European prospectors demanding land for gold mining, the colonial state moved swiftly and amended the NLTO of 1930 with the approval of both the imperial government and the KLC.

It was the supposed guardian of African interests in Kenya, the CNC, who introduced and articulately justified the need for change in land policy to facilitate gold mining in Buluyia. Aware that the proposal was unpopular to the Luyia, the CNC asserted the inevitability of amending the ordinance to provide for mining leases to European prospectors and concerns in Kenya without consulting Luyia landholders and the LNC. The resultant NLT(A)O of 1932, thus eliminated Luyia consent and legalized monetary compensation. While the ordinance aroused much opposition both in Kenya and Britain, the Conservative S of S, Sir Cunliffe-Lister, countered the storm of criticism by emphasizing the temporary nature of the measure, adding that Luyia households dispossessed by gold mining would receive generous compensation, suffer minimal physical disturbance, and reap significant economic benefits from the industry. In any case, Cunliffe-Lister squashed all criticism against the amending ordinance by stressing that the KLC, then at work in Kenya, would provide a lasting and favorable solution to the Kenya land question.

Like the colonial state and the imperial government, the all-European Commission did little in resolving the raging land issue in Kenya. The Commission endorsed the amending ordinance, recommended mining leases and accepted cash compensation for land alienated for the gold mining industry. African surface rights could not negate the Crown's rights to mineral resources, endorsed in the existing colonial land and mining laws. Indeed, the colonial state, the CO, and champions of African interests in Kenya and Britain, concurred on the centrality of exploiting gold, but they differed on how to treat dispossessed Luyia households. All these institutions and individuals failed to protect Luyia land rights as the glitter of gold overshadowed the burden of trusteeship.

This legal triumph facilitated gold mining in NK reserve. European miners, prospectors and speculators from all over the world in search of quick economic gain soon crowded Kakamega. Both individual prospectors with little capital, and large companies representing a wide range of interests with substantial capital investments, came to the new "Eldorado." British, Australian, American, French, South African, Rhodesian and local Kenya settler interests converged on the goldfield. During the initial prospecting era (1931–34), small capital investments and cheap gold winning procedures sustained the rush to Kakamega. Individuals and companies obtained mining rights in form of EPLs or claims for a minimum fee and proceeded to peg and prospect for gold. However, by 1934 as reef surpassed alluvial mining, large companies moved in and commenced trenching, diamond drilling and deep level mining. Under the NLT(A)O of 1932, companies applied for mining leases, which accentuated land matters in Buluyia. Liberal land grants in EPLs, claims and leases brought to the fore the incongruence between colonial land and mining laws on one hand, and African surface rights on the other. While the Luyia enjoyed legal surface rights in the reserve, the colonial state had unquestionable rights to the minerals underneath. However, the NLT(A)O of 1932 gave legal power to the Mining Ordinance of 1931 by allowing companies to obtain leaseholds for small cash payments. This in turn ushered in the era of prospecting by large companies, concentrating on deep level mining.

The gold mining industry provided relief to disillusioned Kenya settlers and some of the larger mining concerns, contrary to the conclusions of Fearn and Roberts who argue that the industry constituted an economic drain to the colonial state and European industrial capitalist interests, both local and foreign. Like any other speculative ventures, there were losers and winners. Ex-Kenya settler farmers like Smallwood and O'Brien, formed successful medium size deep level mining companies, which realized handsome dividends between 1936 and 1940. Mr. and Mrs. Stitts, Everett and others were examples of successful alluvial workers in the Kakamega goldfield. The colonial state also reaped significant incomes from gold mining in form of assays, fees, forest timber concessions and a 5 percent royalty from miners. Indeed, between 1937 and 1944, gold was one of the most important exports of the colony. From third place among the colony's major exports in 1937, gold became the first most valuable export in 1941 when it accounted for 17 percent of the total value of the colony's exports. A complex interplay of factors constituted important upsets to the gold mining industry in Buluyia, including exhaustion of valuable ore deposits, immense difficulties with both local and foreign finance capital caused by the global economic depression, the Italian-Ethiopian War (1935–41), and the impact of WWII. In addition to skilled European miners and workers leaving for military service, the industry grappled with significant labor management problems, an unstable and discontented African labor force, shortage of mining materials, and increased production costs.

While Kakamega did not become another Rand, the gold mining industry ignited significant protest from the Luyia landholders. The Luyia took, first, individual initiative to protect their holdings from European prospectors. They obstructed work, physically engaged prospectors and destroyed symbols of European "ownership" of their lands such as pegs and beacons. These individual acts of resistance quickly crystallized into organized political protest under the umbrella of the NKCA. From 1933 onward, NKCA bombarded the colonial state and the imperial government with memoranda, petitions, and letters of protest. They also led delegations to the provincial and district administration, and challenged the administration in public *barazas*. With assistance from missionaries in Kenya, particularly Archdeacon Owen, and political links with the KCA whose secretary, Jomo Kenyatta, was then living in London, the NKCA highlighted the Luyia land grievances to a wide audience. The association condemned the breach of pledge given in the NLTO of 1930, resented the European invasion of their lands, condemned cash compensation, and protested colonial policies such as compulsory reduction of African livestock, and soil conservation measures. Indeed, NKCA kept alive the rumor that anti-soil erosion measures were aimed at preparing the land for new European settlers. Although the rumor may have been just that, it emerged within the context of increased land insecurity in Buluyia brought about by the gold mining experience. In a brilliant study of rumor and history in Africa, Luise White aptly notes that what is important in such stories is how they "describe meaning and power and how people thought and behaved."[1] White adds that these narratives elucidate the "grim and mercenary nature of the colonial state,"[2] and constitute commentaries on the state of African land rights in colonial Kenya. These testimonials in archival sources and oral interviews are a powerful form of protest that detailed the tensions, apprehensions, and distrust among Luyia households toward state-sponsored agrarian policies. The Luyia, like the Kikuyu, the Maasai, and the Kipsigis, had already lost land to European settlers. In the Kakamega gold rush, the Luyia lost more land to European industrial capitalism. The Kakamega experience created, nurtured, and sustained intense distrust for, and deep resentment against colonial agrarian policies, especially in Maragoli and later Bukusu, the strongholds of the NKCA. This political protest, however, did not move the colonial state and the imperial government. Instead, the colonial state ridiculed, intimidated, fined and imprisoned the NKCA members.

Both the colonial state and the imperial government eloquently justified the excision of land in NK district for the gold mining industry in terms of the tremendous economic prosperity Luyia households stood to reap. Did the Luyia achieve the envisaged economic breakthrough? In what ways did the colonial state and the metropolitan government facilitate African entrepreneurship, the development of industrial skills, and an expanded market for African produce?

The analysis of the economic and social balance sheet of the gold mining industry in Buluyia shows that Luyia households witnessed minimal benefits from the colonial rural industrial project centered on the extraction of gold. The Luyia lost substantial land to prospectors and mining companies through EPLs, claims and mining leases, in return for meager cash payments. The conflicting perception of the value of land between the dispossessed Luyia, European miners and colonial officials, explains why cash compensation was a strongly contested issue. In addition to utilitarian value, ancestral lands had intrinsic, intangible, and unquantifiable psychological value to the Luyia. Although both the colonial state and the metropolitan government approved the alienation of land in Buluyia as a "temporary" measure, those whom the gold mining industry dispossessed, had to contend with twenty-one year leases, and most of the leasehold lands never reverted to the original owners as promised. The Luyia had little use for cash, except for meeting colonial impositions such as taxes. Their land constituted their sole source of livelihood. Hence, easily expendable cash did not guarantee basic survival.

As Luyia households lost land to the gold industry, both the imperial government and the colonial state pledged racial equality in reaping the benefits of exploiting gold. Yet, the colonial state subjected the Luyia to myriad legal hurdles that effectively locked them out of the industry as miners. The unrealistic high levels of English standards required of miners thwarted African entrepreneurship. In fact, the colonial state criminalized African mining activities until 1948 when the first African was licensed. All African mining activities fell under the rubric of "gold stealing", "illegal gold mining" or "illicit gold buying." Even after WWII, when the exploitation of gold in Kenya was crucial to the revival of the colonial economy and the imperial exchequer, the inclusion of Africans in the gold mining industry proved a controversial and contradictory discourse. While the CO and the provincial administration in Kenya favored opening up the closed Maragoli Gold Mining Scheme to African prospectors, the Secretariat in Nairobi radically opposed such a policy. Indeed, the controversy surrounding the re-opening of the Maragoli Gold Mining Scheme pointed to the deep divisions that characterized the policy formulation process in the British imperial edifice. The colonialists perfected their apparatuses of oppression and control with the result that between 1933 and 1952, the administration severally amended the Dealing in Unwrought Precious Metals Ordinance to provide for stiffer prison and cash penalties for illegal gold dealing. Moreover, the few Africans who were licensed in the dying years of the industry were not from the locations most affected by land alienation. Deficient in technical skills, facing insurmountable capital constraints, and working on uneconomic abandoned rubbles, they produced insignificant gold for export. The colonial state's reluctance to simplify mining laws and its determination to utilize oppressive legalities effectively checked legal African participation.

Although the policy formulation is at times cast in the mold of London versus Nairobi, Kenyan Africans were not malleable, disinterested subordinate observers. Africans constantly contested unfavorable land policies. While pro-African elements in Kenya and champions of African interests in Britain corroborated and elaborated such African voices, a pro-settler governor in

Kenya and a Conservative S of S for the Colonies at the CO, were least suited to pass a pro-African land policy. For example, while colonial legal strictures arrested African entrepreneurship in the gold mining industry, the Luyia in particular faced significant constraints as laborers in the new industrial enterprise. The local Isukha and Idakho who lost most land to mining interests failed to take advantage of the industrial labor market. Deep-seated land grievances, fear of death in the mines, and the specter of binding labor contracts, constituted crucial factors in shaping Idakho and Isukha response to industrial labor. While miners were compelled to rely on migrant Luyia and non-Luyia labor, colonial economic impositions, especially taxation, pushed many reluctant Luyia into the mines. Moreover, the new industrial labor ushered in several challenges to Luyia households through increased use of women and children's labor power. The demands and needs of the parallel Luyia peasant economy often shaped and controlled the flow of labor to the gold mines. Luyia labor consciously entered and withdrew from the labor market in accordance with their agricultural calendar.

Further, it is clear that the mines never constituted the sole source of Luyia incomes throughout the period under study. The long established tradition of labor migrancy to settler farms, government service and the emerging colonial urban centers continued to infuse significant incomes into the Luyia economy. While only 20 percent of Luyia men worked outside the district in 1932, the figure rose to 37 percent in 1937 and doubled in the 1950s. Labor demands in the settler economy and the colonial state's official policy of coerced labor recruitment during WWII, accelerated this trend. All these affected Luyia peasant production, occasioning a famine situation in 1943 in locations in the epicenter of the Kakamega goldfield.

Contrary to what scholars like Fearn and Roberts would have us believe, the terms and conditions of work in the Kakamega goldfield were deplorable. Children, women and men endured long hours of difficult, dangerous, and monotonous tasks in the mines for minimal wages. Work in the mines devalued African labor. In addition, a minimal percentage of labor benefited from company administered meals and rations, and most contended with poor housing conditions. These conditions intensified African resentment, which they expressed in diverse forms of labor protest.

African mine workers actively contested bad working conditions. Besides absenteeism and desertion, labor protest peaked with a series of strikes at Rosterman and Kimingini mines in the 1940s. Labor action resulted from universally low wages and unrealistic racially derived wage differentials between Europeans, Asians and Africans. Other grievances included wage withholding, the reservation of low paying arduous jobs for African labor due to competition from skilled Chinese, Asian, Seychellois and Italian Prisoners of

War. African workers also resented racially engineered everyday maltreatment at the hands of their European employers and supervisors.

In addition to work-related grievances, Africans, and the Luyia in particular, suffered the numbing effects of mining accidents. Between 1933 and 1952, at least 501 recorded accidents claimed ninety-six lives and caused 384 injuries throughout the colony. Most of these accidents occurred in the Kakamega goldfield and Africans suffered higher fatalities and injuries. Stripped of their usufruct land rights and dependent on their male relatives, Luyia women took the lead in protesting and contesting the deaths of their husbands, fathers, brothers and sons, through demonstrations, physical assaults against miners, and funeral dirges. Although official sources attributed only one death to "gas acting on the weak lungs" of the victim, all accidents were directly mining-related. Death and injury from such accidents had a huge impact on the Luyia community.

The significance of these mining disasters becomes important when cast against the issue of accident compensation. The absence of a Workmen's Compensation plan for mineworkers up to 1939 and the complex claim procedures involved often resulted in no compensation. Dispossessed and destitute households grappled with untold economic and socio-psychological repercussions.

Land loss, a stifled local entrepreneurship, failed job opportunities, intolerable working conditions and the devastating accidents were in no way relieved by the transformation of Kakamega into a Western-type urban space. Kakamega experienced meteoric growth as a mining rural-urban outpost. The physical infrastructural transformation served to maximize profits for European local and foreign capital interests in the goldfield. The Luyia benefited secondarily, and the LNC funded some of these infrastructures. The rapid decline of Kakamega Township between 1936 and 1939, therefore, left little permanent evidence of the benefits of the industry on the Luyia economic landscape. The industry provided only short-lived market opportunities for low priced Luyia produce. Nevertheless, Luyia producers, traders, and hawkers had to contend with restrictive state marketing policies, cutthroat competition from Asian middlemen, and the largely protected European produced maize, so that few Luyia women and men graduated into a petty-bourgeois class. To date, the vast majority of Luyia do not look to the gold mining era with nostalgia.

Luyia woes were further evident in the monumental health challenges brought on by gold mining activities in their locale. The major culprit, alluvial mining, coupled with the miners' flagrant disregard for mining laws, accentuated mosquito breeding and transformed the goldfield into an epidemic zone. Moreover, the influx of migrant labor into Kakamega introduced new virulent strains of malaria to a population lacking immunity. Malaria, cerebro-meningitis, and pneumonia devastated both mineworkers and Luyia not employed in the mines. In spite of this, a few mining companies made tepid and short-lived efforts at providing health care for workers as well as supplementing mission health endeavors. The colonial state's unwillingness to fund African health care and the unfavorable disposition of European medical personnel toward their African patients compounded this dilemma.

Mining also created important socio-cultural tensions within the Luyia body politic and between the Luyia and the European miners. The convergence of large numbers of European males on Kakamega and the presence of African male workers created, nurtured and sustained new forms of sexual relations with African women. As one of the few options open to women in the racially and highly gendered colonial labor hierarchy, prostitution promised opportunities for women's economic advancement. Yet, the community and local colonial functionaries criminalized prostitution. Both Luyia and non-Luyia prostitutes in Kakamega were victims of maltreatment and exploitation. Prostitutes endured restricted movement, arrest and forcible evictions. Since the community stigmatized prostitutes as carriers and transmitters of venereal diseases, this often led to social fissures within the Luyia community. Although prostitution and beer brewing widened the social space for Luyia women and provided opportunities for economic gain, it is clear that the meteoric rise and decline of the urban phenomenon in Kakamega, coupled with the deeply entrenched gerontocratic system, curtailed the prostitutes' ability to accumulate real wealth.

This study of Western Kenya's developmental patterns has significant lessons for present-day Kenya's development trajectories. It raises questions about development based on the exploitation of non-renewable industrial resources as opposed to making agricultural development the basis of Kenya's economic growth. Colonial rural industrialization policies centered on gold mining, led to significant conflicts between the competing interests of Luyia peasants and European mining interests. Although European capitalist interests carried the day, gold extraction brought limited benefits to Luyia households while at the same time demonstrating the centrality of peasant agriculture in Buluyia. The exhaustion of gold ores and the demise of gold mining left no lasting evidence of economic transformation in Kakamega. This shows that Kenya's development policies should emphasize the mobilization of agricultural resources as the basis for industrial growth;—and not the reverse.

Next is the issue of policy formulation. The study has demonstrated the failure of the classic colonial "top down" policy of development which independent Kenya bureaucrats continue to emphasize to date. Throughout this study, it is clear that the exclusion of local populations from the process of formulating agrarian policies was largely counterproductive. The Luyia opposed and rejected external impositions concerning land alienation, gold mining and soil improvement measures, specifically because of the deep-seated distrust for the unfavorable colonial land policies. Independent Kenya policy makers, therefore, should make deliberate attempts to integrate local populations in the development agenda through consultation and consensus. They should endeavor to understand the needs of local populations in formulating people-friendly policies to facilitate grassroots acceptance. The Nairobi bureaucrats should stop assuming that "they know what the people want" and instead they should foster conditions for the flowering and maturation of local initiative.

In retrospect, it is clear that the failure of state agrarian initiatives in Buluyia down to the 1950s was the most lasting legacy of the Kakamega gold rush. The gold rush, the violation of imperial trusteeship and the alienation of land in Buluyia, occasioned significant distrust and hostility toward state agricultural initiatives that outlived British colonialism.

NOTES

1. Luise White, *Speaking With Vampires: Rumor and History in Colonial Africa* (Berkeley: University of California University Press, 2000), 89.
2. Ibid.

Appendix One

The NLT(A)O of 1932: Letters of Protest in England

The Georgian Society to Rt. Hon. R. MacDonald, Leader of the Late Labour Government, December 16, 1932.

Isabel Ross (wife of McGregor Ross) to the Editor, Golders Green Gazette, January 3, 1933.

Mrs. Barger, wife of Professor Barger of Edinburgh University to the Prime Minister, January 3, 1933.

Women's International League: Golders Green and Hampstead Garden Suburb Branch to Secretary of State, January 4, 1933.

Archbishop of Canterbury to Sir Philip Cunliffe-Lister, Confidential, January 4, 1933.

H. O. Daniels to Secretary of State, January 5, 1933.

Executive Committee of the Society of Friends (The Meeting for Sufferings) to The RT. Hon. J. Ramsay Macdonald [sic], Prime Minister, January 6, 1933.

Society of Friends to Mr. Baldwin, January 6, 1933.

The Penn Club to Secretary of State, January 8, 1933.

Jordan's Preparative Meeting of the Religious Society of Friends to Secretary of State, January 8, 1933.

The Central Committee of the Catholic Crusade to the Prime Minister, January 9, 1933.

Society of Friends to Major A. J. Muirhead, MP, Oxford, January 10, 1933.

The Stirling and Dunblane Presbytery of the Church of Scotland to Prime Minister, S of S, and Members of Parliament for the Presbytery area, January 10, 1933.

Women's International League: British Section of the Women's International League for Peace and Freedom to The Rt. Hon. J. Ramsay MacDonald, MP, January 11, 1933.

Amalgamated Society of Woodworkers to the Colonial Secretary, January 11, 1933.

E. R. Nickol to The Rt. Hon. Ramsay MacDonald, Prime Minister, 12 January 1933.

D. J. Jeffery Williams to Sir Philip Cunliffe-Lister, January 12, 1933.

The Anti-Slavery and Aborigines Protection Society to The Rt. Hon. Sir Philip Cunliffe-Lister, January 13, 1933; January 18, 1933; January 20, 1933.

Archbishop of Canterbury to Sir Philip Cunliffe-Lister, Private and Confidential, January 20, 1933.

Source: Goldfields in the Kavirondo District, PRO: CO 533/429

Appendix Two

Kimingini Gold Mines Ltd., Commuted Rent

Names	Shs.
Mubiru Mabayaka	160
Miheso Tunganji	160
Miheso Amwaga	600
Edoshi Makamala	160
Mathayo Bitalo	160
Robeni Shitsamani	160
Miheso Khasiani	160
Mirajisinzu Mwabishi	160
Eragula Ashiano	160
Andabe Lisualizu	160
Sunguti Asunza	210
Lishanga Muyuka]	
Miheso Muyuka]	600
Mbukha Sikonye	160
Mahiyela Ikhasi	210
Andambe Muanje	210
Nzaka Shahiya	210
Muchiri Virondo	210
Mushibochi Shitujani	210
Musindahi Malialo	210
Andalo Ambale	210
Lijodi Shigali	210
Petero Indahi	210
Shiyononyinya Khabugate	475
Lumati Liyai	210

Khayimba Khangazi	210
Agala Mamba	210
Libando Makinyi	210
Ligane Asunza	210
Lungabia Shisaka	210
Miheso Chitambe	210
Miheso Huganu	210
Muhangi Shisundi} Amwaga Muhangi}	210
Muhaya Khayumbi	210
Shivachi Miheso	700
Ahubisie	575
Lisamula Ambayachi	100
Muzoli Shitsugane	575
Shitsame Lisuza	575
Chief Milimu (sic)	50
Shivachi (ex-chief)	50
Yacobo Mwidakho (Mlango)	20
Ashiono	30
Shimonia	15
Lumati	15
Miheso Shivachi	15
Likari Asutsa	15
Musidia Khasiani	46
Shilayila Khasiani	46
Shilayila Shiabala	46
Bigatsi Shiabala	46
Shisanza Maburi	28
Igaida Bisaho	46
Adegwi Khaweli	28
Musoko Sheheni	28
Ilagula Masigi	28
Shigori Imbuga	46
Total	10,552

Source: Kimingini Lease 1935-59, KNA: DC/KMG/2/18/10

Compensation for Loss of Property

Names	Shs.
Amwaga Nuhangi	30
Magaka Musinia	50
Musinia Magaka	48
Akhusheyekha Meraho	55
Amwaga Muhangi	125
Magaka Musini	200
Akhusheyukha Meraho	125
Sunguti Asunza	16
Lijodi Shigali	60
Total	709

Source: Kimingini Lease 1935-59, KNA: DC/KMG/2/18/10

Appendix Three

List of Owners, Property and their Value on the Kimingini Lease

Names	Property	Shs.	Value Cts.
Amwaga Muhangi	1 hut	30	00
Magaka Musinia	1 hut, 2 grain stores	50	00
Musinia Magaka	1 hut	30	00
	2 *shambas*	17	88
Akhushekukha Meraho	Banana plantation	55	00
Amwaga Muhangi} Magaka Musinia} Akhushekukha Mereh}	Joint holding of bananas, gum trees and a *shamba*	450	00*
Sunguti Asunza	33 Bananas	16	00
Lijodi Shigali	224 Bananas	60	00
	Total	708	88 (approx. 709)

Source: KNA: DC/KMG/2/18/10. *Divided as follows: Amwaga Shs. 125; Magaka Shs. 200; Akhusheyukha Shs.125.

Appendix Four

List of Those Affected by the Edzawa Ridge Mine Lease (1941) and the Amount of Land Alienated

Names	Yards	Square Yards
Lusina Akunaba	65X30	1950
Nzobo Omulamba	65X38	2470
Ambeba Indumwa	103X34	3502
Chunguli Lumbasio	76X25	1900
Odali Amutsa	37X28	1036
Asena Mahonga	222X55	12210
Obwanga Inyangala	53X30	1590
Ombale Antiimili	53X30	1590
Mabinda Namanywa	19X34	646
Kitenye Tsisaka	239X102	24378
Nyenyele Lumwagi	123X60	7380
Ukwiya Alibitsa	30X8	240
Elekilitsu Akabahagi	118X48	5664
Ogada Kibibi	335X20	6700
Oboyo Kimikuli	250X20	5000
Ajega Mahoga	182X59	10738
Isendi Lusangi	350X50	17500
Sikuku Amugune	167X75	12525
Ilayiha Kisimba	75X28	2100
Asindabisa Ilahiya	210X32	6720
Alitsa Katsira	65X24	1560

Source: List by Assessor Samson Amahwa on Edzawa Ridge Mine, July 29, 1941; List by C. F. Parry, DC NK to Chief Paul Agoi, December 23, 1942, KNA: DC/KMG/2/18/12.

Appendix Five

List of Those Affected by the Edzawa Ridge Mine Lease, Tintax-Kisa Location and the Amount of Land Alienated

Names	Yards	Total
Idele Osale	135X55	7,315
Ikhumwa Irambo	298X50	14,900
Otembo Olwala	68X61	4,148
Muyokhwe Moyi	200X30	6,000
Mbukha Mukholo	84X44	3,699
Luseno Ingingi	174X50	8,700
Owaka Andenga	160X24	3,840
Mukuyia Kochwa	152X60	9,120
Onyango Mukatu	97X40	3,880
Alukwe Mbukha	57X50	2,850
Alinda Mbukha	145X55	7,975
Muka Akhu	60X30	1,800
Chief Jeremiah Shipiri	150X55	8,250
Akhalira Anyanye	93X10	930
Angumo Muka	59X32	1,888
Alukhone Onalo	36X30	121
Ikhumwa Moyi	158X34	5,374
Total	85,284.5 Yards (approx. 18 acres)	

Source: List by Assessor Samson Amahwa, August 7, 1941, KNA: DC/ KKMG/2/18/12.

Bibliography

PRIMARY SOURCES

Newspapers and Internet Sources

Daily Express, 31 December 1932–21 March 1933
Daily Herald, 16 January 1930.
Daily Telegraph, n.d.
East Africa, 12 January 1933.
East African Standard, 12 April 1930–5 April 1941
East African Standard: Uganda Argus, 21 December 1929; 26 September 1939
East Bourne Gazette, 1 January 1938
East Coast Herald, 22 December 1929.
Economist, 16 November 1935
Habari (News), *Juni* (June) 1931.
Manchester Guardian, 12 January 1933–5 September 1933
Mombasa Times, 22 December 1929.
News of the World, 4 December 1932
Spectator, 12 December 1932–19 January 1933
*Scotsma*n, 19 January 1933.
Tanganyika Standard, n.d.
Times of East Africa, 21 December 1929.
Times, 20 January 1930–19 July 1934.
http://www.makerere.ac.ug/ > (Jan. 12, 2006).

Manuscripts

Great Britain, Public Record Office London, Colonial Office Archive

CO 533/367.
CO 533/375.

CO 533/395.
CO 533/396.
CO 533/416.
CO 533/427.
CO 533/428.
CO 533/429.
CO 533/434.
CO 533/436.
CO 533/442.
CO 533/459.
CO 533/473.
CO 533/476.
CO 533/479.
CO 533/496.
CO 533/497.
CO 533/501.
CO 533/502.
CO 533/513.
CO 533/544.

Kenya, Kenya National Archives, Kakamega

ADM./9/3/1 3/1974 NKCA, n.d.
AGU/3/3. District Development Team Meetings, 1950–54.
HW/13/1. North Kavirondo Local Native Council (NKLNC) Minutes, 1929–33.
HW/13/2. NKLNC Minutes, 1934–39.
HW/13/3. NKLNC Minutes, 1940–45.
HW/13/4. NKLNC Minutes, 1945–52.
WD/8/15. North Kavirondo/North Nyanza Local Land Board, 1939–54.

Kenya, Kenya National Archives, Nairobi

Department of Agriculture

AK/2/23. North Kavirondo Agricultural Monthly Reports (NKMAgRs), 1938–44.
AK/2/24. NKMAgRs, 1934–37.
AK/2/26. NKAgAR, 1937.
AK/2/27. NKAgARs, 1934–45.
AK/2/28. NKAgARs 1946–50.
AK/2/58. North Nyanza (NN) Crop Reports, 1935–52.
AK/4/1. South Kavirondo Future Development, n. d.
AK/4/14. NN Development and Policy, 1951–53.
AK/6/8. Agricultural Staff Conferences, n. d.
AK/11/44. Maize General, n. d.
AK/21/24. AO, NK Safari Diary, 1935.
AK/21/25. Diary: North Nyanza, 1944–53.

District Commissioner, Kakamega/North Nyanza

DC/EN/3/3/2. Abaluyia Land Law and Custom, 1930.
DC/KMG/1/1/1. Bidakho, Safari Diary, 1929–39.
DC/KMG/1/1/19. Minutes of Meeting Isukha Location Advisory Council, 1946–55.
DC/KMG/1/1/22. Meeting of Idakho Advisory Council, 1946–55.
DC/KMG/1/1/90. Native Associations and Agitators—Nyanza Central Association, 1944–49.
DC/KMG/1/1/145. Meeting of Provincial Commissioners, 1944–52.
DC/KMG/1/1/153. Native Associations, 1943–59.
DC/KMG/1/1/165. Marama South, 1929–42.
DC/KMG/1/19/6. Medical Department, Medical Safari Programme, 1931–55.
DC/KMG/2/18/3. Permit to Export Gold, 1936–50.
DC/KMG/2/18/4. Permits and Prospecting Licences, 1937–65.
DC/KMG/2/18/9. Rosterman Gold Mines, 1935–55.
DC/KMG/2/18/10. Kimingini Lease, 1935–59.
DC/KMG/2/18/11. Mining Lease, Messrs Risks Limited, 1936–59.
DC/KMG/2/18/12. Setting Apart for Land Lease, 1941–52.
DC/KMG/2/18/13. Mining Claims and Permits, 1933–43.
DC/KMG/2/18/15. Electric Power, 1933–41.
DC/NN/1/11–29. North Kavirondo District Annual Reports (NKDAR) 1930–47.

DC/NN/1/30–34. North Nyanza District Annual Reports (NNDAR) 1948–52.
DC/NN/3/4/2. Political Record Book Vol. III.

Provincial Commissioner, Nyanza

PC/NZA/1/25–38. Nyanza Province Annual Report (NPAR) 1930–43.
DC/KSI/1/24. NPAR 1944–54.
PC/NZA/2/2/17. Native Land Trust Boards- North Kavirondo, 1931–39.
PC/NZA/2/2/34. Native Reserves Prospecting and Mining- Nyanza, 1932–34.
PC/NZA/2/2/44. Native Reserves. Boards, 1932–46.
PC/NZA/2/2/50. Survey, 1932–50.
PC/NZA/2/2/60. Mining and Prospecting, 1933–41.
PC/NZA/2/2/68. Native Reserves, 1934–45.
PC/NZA/2/7/54. Fire, 1934–47.
PC/NZA/2/7/117. Illicit Gold Buying, 1940–48.
PC/NZA/2/14/6. Public Health, 1930–42.
PC/NZA/2/14/40. Sanitation, 1939–50.
PC/NZA/2/20/6. Mining and General: Labour, 1932–46.
PC/NZA/2/20/13. Assisted Labour, 1941–43.
PC/NZA/3/13/27. Labour, 1931–42.
PC/NZA/3/14/351, Enquiries & Reports, 1932.
PC/NZA/3/14/353. Prospecting of Mines, 1933–42.
PC/NZA/3/14/357. Gold Production, 1946–51.

PC/NZA/3/14/355. Permits to Prospect and /or Mine, 1947–49.
PC/NZA/3/14/358 A. Rosterman Mining Syndicate, 1935–37.
PC/NZA/3/14/372. Mining, 1933–41.
PC/NZA/3/14/374. Mining, 1931–32.
PC/NZA/4/1/2/1. Visit of His Excellency The Governor, 1939–44.
PC/NZA/4/1/3/1. District Commissioners' Meeting: Agenda and Minutes, 1935–38.
PC/NZA/4/5/1. Nyanza Province Intelligence Reports (NPIRs) 1934–36.
PC/NZA/4/5/2. NPIRs 1936–39.
PC/NZA/4/5/3. NPIRs 1939–42.
PC/NZA/4/5/5. North Kavirondo District Intelligence Reports (NKDIRs), 1934–36.
PC/NZA/4/5/6. NKDIRs 1936–39.
PC Nyanza LND. 11/2 3/3298. Prospecting for Mines, 1933–45.*

*Refers to KNA, Syracuse source not reconciled with KNA, Nairobi Reference

Oxford, Rhodes House, Oxford University.
African Affairs: Race Relations Committee, MSS. Afr. S. 596.
Buxton Papers, MSS. Afr. Emp. S.405/6/5.
Clive John, MSS. Afr. S. 678.
Fabian Colonial Bureau, MSS. Brit. Emp. S. 285.
MSS. Afr. S. 1281 (i)
The London Group on African Affairs, 1936–38, MSS. Afr. S. 1427.
Lugard Papers, 1934–41, MSS. Afr. 77/7.
Popham Papers, MSS. Afr. S.1120.
Walker Papers, MSS. Afr. S.717.

Unpublished Government Sources

Colony and Protectorate of Kenya

Hitchen, Stansfield C. "Report on the Underground Geology of Kimingini Mine, Kakamega, July 25, 1938. Eldoret/20." Mines and Geological Department Library, Nairobi.
Mines and Geological Department. "Tintax Property Final Report on the Work Carried Out By New Consolidated Gold Fields Limited on Tintax Property During 1956–58. Eldoret/63." Mining and Geological Department Library, Nairobi.
Mining and Geological Department Annual Reports, 1941, 1943 and 1944.

Government Publications

Colony and Protectorate of Kenya

An Ordinance to Regulate Trading in Unwrought Precious Metals, 1933. Nairobi: Government Printer, 1933.
Commission of Inquiry into the Affray at Kolloa, Baringo. Nairobi: Government Printer, 1950.

The Crown Lands (Amendment) Ordinance 1938. Nairobi: Government Printer, 1938.
Department of Agriculture Annual Reports (DofAARs) 1933–1939, 1945, 1948, Nairobi: Government Printer, 1934–1940, 1946, 1949.
Huddleston, A. *Geology of the Kakamega District.* Nairobi: Government Printer, 1952.
Humphrey, Norman. *The Liguru and the Land: Sociological Aspects of Some Agricultural Problems of North Kavirondo.* Nairobi: Government Printer, 1947.
Mining and Geological Department Annual Report (MGDAR) 1933–1940; 1949–1953. Nairobi: Government Printer, 1934–1941, 1950–1955.
Mining Ordinance No. LXV 1933. Nairobi: Government Printer, 1933.
Land, Mines & Surveys Department Annual Report (LMSDAR) 1946–1948. Nairobi: Government Printer, 1948–1950.
Legislative Council Debates, 1929–1932. Nairobi: Government Printer, 1930–1932.
Legislative Council Speech by H. E. Governor Sir Edward Grigg, 18 May 1928. Nairobi: Government Printer, 1928.
Native Lands Trust Bill. Confidential Passfield to Grigg 6 March 1930. Nairobi: Government Printer, 1930.
The Native Land Trust (Amendment) Ordinance, 1932. Nairobi: Government Printer, 1932.
Native Affairs Department Annual Report 1937. Nairobi: Government Printer, 1937.
The Native Lands Trust Ordinance, 1930. Nairobi: Government Printer, 1930.
Official Gazette, July 11, 1939.
Report of Committee on Native Land Tenure in the North Kavirondo Reserve. Nairobi: Government Printer, 1931.
Report of the Kenya Land Commission. Nairobi: Government Printer, 1933.
Report by Sir Albert Kitson. Nairobi: Government Printer, 1933.
Report of the Employment of Juveniles Committee. Nairobi: Government Printer, 1938.
Staff Lists 1934–37, 1939, 1941–3; 1947–55. Nairobi: Government Printer, 1934–7, 1939, 1942–43; 1947–55.

Republic of Kenya

DuBois, C. G. B. *Geological Survey of Kenya: Minerals of Kenya.* Nairobi: Government Printer, 1966.
Kakamega District Development Plan 1984–1988. Nairobi: Ministry of Finance and Planning, 1984.
Laws of Kenya: The Mining Act Chapter 306. Nairobi: Government Printer, 1987.

Great Britain

Annual Report of the Social and Economic Progress of the People of the Kenya Colony and Protectorate, 1932. London: His Majesty's Stationery Office (HMSO), 1934.
Colonial Annual Reports, Kenya 1946–1949. London: HMSO, 1947–1950.
Colonial Reports, Kenya 1950. London: HMSO, 1951.

Colonial Reports, Kenya 1951-1952. London: Her Majesty's Stationery Office, 1952-1953.
Hansard Parliamentary Debates, Commons, 5th Series (1931-37).
Hansard Parliamentary Debates, House of Lords, 5th Series (1931-33).
Joint Select Committee on Closer Union in East Africa. Vol.1. London: HMSO, 1931.
Kenya Land Commission: Evidence and Memoranda. Vol. III. London: HMSO, 1934.
Kenya Land Commission Report: Summary of Conclusions Reached by His Majesty's Government, May 1934. London: HMSO, 1934.
Kenya Colony & Protectorate Annual Report, 1933-1938. London: HMSO, 1934 - 1939.
Memorandum on Native Policy in East Africa Cmd. 3573. London: HMSO, 1930.
Mercer, William H. *Colonial Office Lists 1925-1931*. London: Waterlow & Sons, 1925-1931.

Oral Sources

Oral Interviews

Sakwa Mwela, OI, January 18, 1990.
Teka, Laban Ikhale, January 26, 1990.
Abwao, Priscilla Ingasiani, September 11, 1998.
Akibaya, Jairo Eloga, OI, August 20 and September 3, 1998.
Amagadu, Josphat Ingosi, March 3, 1998.
Amahwa, Samson Shikanga, August 29 & 30, 1998.
Andove, Zacharia Muganzi, February 28, 1998.
Ayiekha Erasto, February 15, 1998.
Bukhala, James A., September 24, 1998.
Mbanda, Claudio Edward, February 12, 1998.
Imbali, Joseph, February 18, 1998.
Indushi, Popai Khayumbi, August 30, 1998.
Isenjia, Beti Joseph, February 24, 1998.
Khasandi, Ruth Ayiekha, February 15, 1998.
Kiradi, Ezina Agufana, July 22, 1998.
Litavakha, Haruni Ashiluka, March 9, 1998.
Liyai, Lesmas Charles, March 26, 1998.
Ludenyo, Andrew, March 26, 1998.
Lugonzo, Seth, OI, February 15, 1998.
Mang'ong'o, Alfred Shivachi, March 26, 1998.
Mang'ong'o, Andrea Mamakaka, March 26, 1998.
Mavia, Christopher Khayumbi, March 2, 1998.
Mbwabi, Likhaya Peter, February 26, 1998.
Misaki, Priscilla, July 18, 1998.
Muhati, Clementi Miheso, February 13, 1998.
Mukabwa, Murunga Andrea, February 12, 1998.

Mulindi, Mulama, March 26, 1998.
Mung'alo, Kaitani, February 18, 1998.
Musalimwa, Gabriel, February 17, 1998.
Musine, Shibelenje Paulo, September 21, 1998.
Mutembei, Frederick Shivachi, March 17, 1998.
Mutondo, Mboi Petro, February 18, 1998.
Mutsotso, Francis Khakali, March 4, 1998.
Mwidakho, Herman M., February 17, 1998.
Mwinamo, Enock Shimechero, March 26, 1998.
Mwinamo Francis Likhanga, March 26, 1998.
Otiende, Joseph Daniel, August 29, 1998.
Segero, Ingosi Elijah, OI, February 17, 1998
Seleta, Minishi Lugonzo, February 15, 1998.
Shikali, Zablon Etambo, March 3, 1998.
Shisiali, Julius, March 26, 1998.
Shitsukane, Peter, March 19, 1998.
Shivachi, Gabriel, February 15, 1998.
Vihembo, Adriano, March 26, 1998.

SECONDARY SOURCES

Unpublished Theses and Dissertations

Aseka, Eric M. "Political Economy of Buluyia: 1900–1964." Ph. D. dissertation, Kenyatta University, 1989.

Breen, Rita Mary. "The Politics of Land: The Kenya Land Commission (1932–33) And Its Effects on Land Policy in Kenya." PhD. dissertation: Michigan State University, 1976.

Burt, Eugene C. "Toward an Art History of the Baluyia of Western Kenya," PhD. dissertation, University of Washington, 1980.

Chavasu, Henry Onzere, "British Colonialism and the Making of the Maragoli Diaspora, Western Kenya, 1895–1963," M. Phil. Thesis, Moi University, 1997.

Hay, Margaret Jean. "Economic Change in Luoland: Kowe, 1890–1945." PhD. dissertation, University of Wisconsin, 1972.

Makana, Nicholas. E. "An Economic History of the Batsotso of Western Kenya: 1800–1945." M. Phil. Thesis, Moi University, 1994.

——."Changing Patterns of Indigenous Economic Systems: Agrarian Change and Rural Transformation in Bungoma District, 1930–1960," PhD. dissertation: West Virginia University, 2005.

Moock, Joyce L. "The Migration Process and Differential Economic Behavior in South Maragoli, Western Kenya." Ph. D. dissertation: Columbia University, 1976.

Muchanga, Kizito. L. "The Impact of Economic Activities on the Ecology of Isukha and Idakho Areas of Western Kenya c. 1850–1945." M. A. Thesis, Kenyatta University, 1998.

Ndege, Peter Odhiambo. "Struggles for the Market: The Political Economy of Commodity Production and Trade in Western Kenya, 1929–1939." PhD dissertation, West Virginia University, 1993.

Shilaro, Priscilla M. "Kabras Culture Under Colonial Rule: A Study of the Impact of Christianity and Western Education." M. A. Thesis, Kenyatta University, 1993.

Wafula, Sibiko. "Colonial Land Policy and the North Kavirondo African Reserve to 1940." B. A. Thesis, University of Nairobi, 1981.

Wako, Dan M. "Luyia Principles and Procedures of Conflict Resolution and Settlement of Disputes." M. A. Thesis, Makerere University, 1978.

Unpublished Conference and Seminar Papers

Shilaro, Priscilla M. "British Trusteeship, African Land Rights in Kenya: Could Trusteeship be Trusted?" paper presented at the African Studies Association (ASA) Annual Meeting, San Francisco, CA (November 1996).

——. "British Trusteeship, the Kakamega Gold Rush and the Kenya (Carter) Land Commission 1932–34," paper presented at the ASA Annual Meeting, Columbus, OH (November 1997).

——. "'A House Divided Unto Itself'": The Colonial State and the Stillborn Maragoli Gold Mining Industry, 1936–55," paper presented at the Historical Association of Kenya Annual Conference. Eldoret, Kenya (July 1998).

——. "British Trusteeship, Luyia Land Rights and the Kakamega Gold Rush, 1930–52, From the Idea to the Present," paper presented at the Institute of Research and Postgraduate Studies, Maseno University College (September 1998).

——. "Religion, Culture and Political Protest in Western Kenya: The Case of *Dini Ya Msambwa*, 1943–63," paper presented at the Senator Rush Holt History Conference. West Virginia University (September 1999).

——. "The Politics of Land Among the Luyia of Western Kenya, 1930–52," paper presented at the ASA Annual Meeting, Philadelphia, PA (November 1999).

Turner, Simon. "The Black Monica: Rumour Mongering and Ideology in a Refugee Camp in Tanzania," Paper presented at the ASA Annual Meeting, Philadelphia, PA (November, 1999).

BOOKS

Abwunza, Judith M. *Women's Voices, Women's Power: Dialogues of Resistance From East Africa.* Toronto, Ontario: Broadview Press, 1997.

Atkins, Keletso E. *The Moon is Dead! Give Us our Money! The Origins of An African Work Ethic, Natal, South Africa, 1843–1900.* Portsmouth: Heinemann, 1993.

Barnes, John and David Nicholson, ed. *The Leo Amery Diaries, 1896–1929.* Vol.1 London, Auckland, Johannesburg: Hutchinson, 1980.

Bates, Robert H. *Markets and States in Tropical Africa: The Political Basis of Agricultural Policies.* Berkeley and Los Angeles: University of California Press, 1981.

Bennett, George. *Kenya, A Political History: The Colonial Period.* London, Ibadan, Accra: Oxford University Press (OUP), 1963.
Berman, Bruce. *Control & Crisis in Colonial Kenya: The Dialectic of Domination.* London: Athens: Ohio University Press, 1990.
Berman, Bruce & John Lonsdale. *Unhappy Valley: Conflict in Kenya & Africa.* Book One. Athens: Ohio University Press, 1992.
——. *Unhappy Valley: Conflict in Kenya & Afric*a. Book Two. Athens: Ohio University Press, 1992.
Clough, Marshall S. *Fighting Two Sides: Kenyan Chiefs and Politicians, 1918–1940.* Niwot, Colorado: Colorado University Press, 1990.
Constantine, Steven. *The Making of British Colonial Development Policy, 1914–1940.* London: Frank Cass, 1984.
Cook, Chris & John Stevenson. *The Longman Handbook of Modern British History, 1714–1995.* 3d ed. London, New York: Longman, 1996.
Cooper, Frederick, Allen Isaacman, et. al. *Confronting Historical Paradigms: Peasants, Labor and the Capitalist World System in Africa and Latin America.* Madison: University of Wisconsin, 1993.
Dilley, Marjorie Ruth. *British Policy in Kenya Colony.* New York: Thomas Nelson and Sons, 1973.
Dumett, Raymond E. *El Dorado in West Africa: The Gold-Mining Frontier, African Labor, and Colonial Capitalism in the Gold Coast, 1875–1900.* Athens: Ohio University Press, 1998.
Ellis, F. *Peasant Economics: Farm Households and Agrarian Development* (Cambridge: Cambridge University Press, 1988), 9–13.
Farson, Negley. *Last Chance in Africa.* New York: Harcourt Brace and Company, 1950.
Fearn, Hugh. *An African Economy: A Study of the Economic Development of the Nyanza Province of Kenya, 1903–1953.* London, New York, Nairobi: OUP, 1961.
Gregory, Robert. G. *Sidney Webb and East Africa: Labour's Experiment with the Doctrine of Native Paramountcy.* Berkeley and Los Angeles, University of California Press, 1962.
Harlow, Vincent and E. M. Chilver, ed. *History of East Africa* Vol. II. Oxford: At the Clarendon Press, 1961.
Havighurst, Alfred F. *Twentieth-Century Britain.* 2d ed. New York: Harper & Row Publishers, 1962.
Hodges, Geoffrey. *The Carrier Corps: Military Labor in the East African Campaign, 1914–1918.* New York: Greenwood Press, 1986.
Huxley, Elspeth J. G. *No Easy Way. A History of the Kenya Farmers' Association and Unga Ltd.* Nairobi: East African Standard Ltd., 1957.
——. *White Man's Country: Lord Delamere and the Making of Kenya.* New Edition. New York: Praeger, c. 1967. First Published in 1935.
Kamoche, Jidlaph G. *Imperial Trusteeship and Political Evolution in Kenya, 1923–1963: A Study of Official Views and the Road to Decolonization.* Washington D. C.: University Press of America, 1981.

Kanogo, Tabitha. *Squatters & the Roots of Mau Mau, 1905–63*. Athens: Ohio University Press, 1987.

Kaplinsky, Raphael, ed. *Readings in Multinational Corporation in Kenya*. Nairobi, New York, Oxford: OUP, 1978.

Kenyatta, Jomo. *Kenya: The Land of Conflict*. London: African International Service Bureau, 1945.

Killingray, David and Richard Rathbone, ed. *Africa and the Second World War*. New York: St. Martin's Press, 1986.

Leo, Christopher, *Land and Class in Kenya*. Toronto, Buffalo: University of Toronto Press, 1984.

Leys, Norman. *Kenya*. 3d ed. London: Hogarth Press, 1926.

Mair, L. P. *Native Policies in Africa*. New York: Negro Universities Press, 1936. Reprinted. 1969.

Maxon, Robert M. and Thomas P. Ofcansky. *Historical Dictionary of Kenya*. 2d ed. Lanham, Md. & London: The Scarecrow Press, Inc., 2000.

Maxon, Robert M. *John Ainsworth and the Making of Kenya*. Boston: University Press of America, 1980.

———. *Struggle for Kenya: The Loss and Reassertion of Imperial Initiative, 1912–1923*. Rutherford, Madison, Teaneck: Farleigh Dickinson University Press, 1993.

———. *East Africa: An Introductory History*. 2d Revised ed. Morgantown: West Virginia University Press, 1994.

———. Going *Their Separate Ways: Agrarian Transformation in Kenya, 1930–1950* (Madison. Teaneck: Farleigh Dicknson University Press, 2003.

Mungeam, G. H. *British Rule in Kenya*. London: Frank Cass and Co. Ltd., 1968. (First Published in 1966).

Munro, J. F. *Colonial Rule and the Akamba: Social Change in the Kenya Highlands, 1889–1939*. Oxford: OUP, 1975.

Murphy, Mary. *Mining Cultures: Men, Women, and Leisure in Butte, 1914–41*. Urbana and Chicago: University of Illinois Press, 1997.

Ndege, George O. *Health, State, and Society in Kenya*. Rochester: University of Rochester Press, 2001.

Ochieng', W. R., ed. *A Modern History of Kenya*. Nairobi: Evans Brothers, 1989.

Ochieng', W. R. and Robert M. Maxon, ed. *An Economic History of Kenya*. Nairobi: East African Educational Publishers (EAEP), 1992.

Ogot, B. A., ed. *Politics and Nationalism in Colonial Kenya. Hadith 4*. Nairobi: EAPH, 1972.

Ogot, B. A. and W. R. Ochieng', ed. *Decolonization & Independence in Kenya, 1940–1993*. Athens: Ohio University Press, 1995.

Ojuka, Aloo and William Ochieng', ed. *Politics and Leadership in Africa*. Nairobi: East African Literature Bureau (EALB), 1975.

Otiende, J. D. *Habari Za Abaluyia*. Nairobi: The Eagle Press, 1949.

Parsons, Timothy, *The African Rank-and-File: Social Implications of Colonial Service in the King's African Rifles, 1902–1964*. Portsmouth: Heinemann, 1999.

Pearson, Scott *et. al.* ed. *Agricultural Policy in Kenya: Applications of the Policy Analysis Matrix*. Ithaca: Cornell University Press, 1995.

Phimister, I. R. and C. Van Onselen, ed. *Studies in the History of African Mine Labour in Colonial Zimbabwe.* Gwelo: Mambo Press, 1978.
Rosberg, Carl Gustav and John Nottingham. *The Myth of "Mau Mau", Nationalism in Kenya.* Stanford, New York: Praeger, 1966.
Robinson, Ronald and John Gallagher, with Alice Denny. *Africa and the Victorians: The Official Mind of Imperialism.* London: Macmillan, 1961.
Ross, McGregor. *Kenya From Within: A Short Political History.* New Impression. London: Frank Cass, 1968. [First Published in 1927].
Rotberg, I. and Ali A. Mazrui, ed. *Protest and Power in Black Africa.* New York: OUP,1970.
Sangree, Walter H. *Age, Prayer and Politics in Tiriki, Kenya.* New York: OUP, 1966.
Shilaro, Priscilla M. "Colonial Land Policies: The Kenya Land Commission and the Kakamega Gold Rush, 1932–4." Pp.110–128 in *Historical Studies and Social Change in Western Kenya: Essays in Memory of Professor Gideon S. Were*, edited by William R. Ochieng'. Nairobi, Kampala, Dar es Salaam: East African Educational Publishers, 2002.
Shimanyula, J. B. *Elijah Masinde and Dini Ya Msambwa.* Nairobi: TransAfrica, 1978.
Sifuna, D. N. *Vocational Education in Schools: A Historical Survey of Kenya and Tanzania.* Kampala, Dar es Salaam: EALB, 1976.
Sorrenson, M. P. K. *Land Reform in Kikuyu Country: A Study in Government Policy.* Nairobi: London: OUP, 1967.
——. *Origins of European Settlement in Kenya.* Nairobi: OUP, 1968.
Stichter, Sharon. *Migrant Labour in Kenya: Capitalism and African Response, 1895–1975.* Harlow: Longman, 1982.
Swainson, Nicola. *The Development of Corporate Capitalism in Kenya, 1918–1977.* Berkeley and Los Angeles: University of California Press, 1980.
Tignor, Robert L. *Colonial Transformation of Kenya: The Kamba, Kikuyu and Maasai From 1900–1939.* Princeton: Princeton University Press, 1976.
Trench, Charles Chenevix. *Men Who Ruled Kenya: The Kenya Administration, 1892–1963.* London, New York: The Radcliffe Press, 1993.
Van Zwannenberg, R. M. A. with Anne King. *An Economic History of Kenya and Uganda, 1800–1970.* New Jersey: Humanities Press, 1975.
Wa-Githumo, Mwangi. *The Impact of Land Expropriation and Land Grievances Upon the Rise and Development of Nationalist Movements in Kenya, 1885–1939.* Washington D. C.: University Press of America, 1981.
Wagner, Gunter. *The Bantu of North Kavirondo.* Vol. I. New York: OUP, 1949.
——. *The Bantu of North Kavirondo: Economic Life.* Vol. II. New York: OUP, 1956.
Watkins, Elizabeth. *Oscar From Africa: The Biography of Oscar Ferris Watkins, 1877–1943.* New York: The Radcliffe Press, 1995.
Were, Gideon S. *History of the Abaluyia of Western Kenya c. 1500–1930.* Nairobi: East African Publishing House, 1967.
White, Luise. *The Comforts of Home: Prostitution in Colonial Nairobi.* Chicago: The University of Chicago Press, 1990.
——. *Speaking with Vampires: Rumor and History in Colonial Africa* (Berkeley: University of California Press, 2000.

Wipper, A. *Rural Rebels: A Study of Two Protest Movements in Kenya.* Nairobi: OUP, 1977.
Worger, William H. *South Africa's City of Diamonds: Mine Workers and Monopoly Capitalism in Kimberley, 1867–1895.* New Haven: Yale University Press, 1987.

ARTICLES AND PERIODICALS

Anderson, David. "Depression, Dust Bowl, Demography and Drought: The Colonial State and Soil Conservation in East Africa During the 1930s." *African Affairs* 83, no. 332 (1984): 321–44.
Ashforth, Adam. "Reckoning Schemes of Legitimation: On Commissions of Inquiry as Power Knowledge Forms." *Journal of Historical Sociology* 3, no.1 (1990): 1–22.
Bruce, Berman. "Ethnography as Politics, Politics as Ethnography: Kenyatta, Malinowski, and the Making of *Facing Mount* Kenya." *Canadian Journal of African Studies (CJAS) /RCEA* 30, no. 3 (1996): 313–44.
Crush, Jonathan, "Swazi Migrant Workers and the Witwatersrand Gold Mines, 1886–1920," *Journal of Historical Geography* 12 (1986): 27–40.
Duder C. J. and Simpson, G. L. "Land and Murder in Colonial Kenya: The Leroghi Land Dispute and the Powys 'Murder' Case." *Journal of Imperial and Commonwealth History* 25, no. 3 (1997): 440–65.
Falola, Toyin. "'An Ounce is Enough' ": The Gold Industry and the Politics of Control in Colonial Western Nigeria." *African Economic History* 20 (1992): 27–50.
Fearn, Hugh. "The Gold-Mining Era in Kenya Colony." *Journal of Tropical Geography* XI (1958): 44–58.
"Gold Mining Expansion in Kenya Colony." *The Rhodesian Mining Journal* (1935): 67–70.
Gool, Selim. "Mining Capitalism and Black Labour in the Early Industrial Period in South Africa." Book Review. Jeanne Penvenne. *The International Journal of African Historical Studies* 18, no.1 (1985): 109–38.
Gould, W. T. S. "Technical Education and Migration in Tiriki, Western Kenya, 1902–1987." *African Affairs* 88, no. 351 (1989): 253–71.
Kipkorir, B. E. "Kolloa Affray, 1950." *TransAfrican Journal of History* 2, no. 2 (1972): 114–29.
———. "Colonial Response to a Crisis: The Kolloa Affray and Colonial Kenya in 1950." *Kenya Past and Present* 2, no. 1 (1973): 22–35.
Lonsdale, John and Berman, Bruce. "Coping With the Contradictions: The Development of the Colonial State in Kenya, 1895–1914," *Journal of African History* 20, no. 4 (1979): 487–505.
Lonsdale, John. "States and Social Processes in Africa: A Historiographical Survey." *African Studies Review* XXIV, nos. 2/3 (1981): 139–225.
Lucas, Robert E. B. "Emigration to South Africa's Mines." *The American Economic Review* 77 (1987): 313–30.

Marsh, C. A. "A Visit to the Kakamega Gold Field." *The Quaker Weekly Journal* (1933): 1–2.

Maxon, Robert M. "Judgement on a Colonial Governor: Sir Percy Girouard in Kenya." *TransAfrican Journal of History* 18 (1989): 90–100.

———. "Up in Smoke: Peasants, Capital and the Colonial State in the Tobacco Industry in Western Kenya, 1931–1939." *African Economic History* 22 (1994): 111–39.

Menkin, A. A. "Gold Mining in Kenya: A Review of 1935–36." *East African Annual* (1936–37): 123–26.

Ogot, Bethwell A. "Mau Mau and Nationhood," in *Mau Mau & Nationhood: Arms, Authority & Narrative,* E. S. Atieno Odhiambo & John Lonsdale eds., (Athens, Ohio University Press, 2003), 8–36.

Okia, Opolot. "In the Interest of Community: Archdeacon Walter Owen and the Issue of Communal Labour in Colonial Kenya, 1921–30." *Journal of Imperial Commonwealth History* 32, no.1 (January 2004): 27–37.

Pycroft, Christopher and Munslow, Barry. "Black Mine Workers in South Africa: Strategies for Co-option and Resistance." *Journal of Asian and African Studies* 23 (1988): 156–79.

Roberts, Andrew Dunlop. "The Gold Boom of the 1930s in Eastern Africa," *African Affairs* 85 (1986): 545–62.

Ross, McGregor. "Gold in Kenya and Native Reserves." *Nature* 131, no. 3298 (1933): 37–38.

Salisbury and Winchester Journal, 16 December 1938.

Shilaro, Priscilla M. "Colonial Land Policies: The Kenya Land Commission and the Kakamega Gold Rush, 1932–34," in *Historical Studies and Social Change in Western Kenya: Essays in Memory of Professor S. Gideon Were.* William R. Ochieng', ed. (Nairobi: East African Educational Publishers Ltd., 2002), 110–128.

Suttill, Keith R. "Deep Mining Challenges; South African Gold Mines at Limits of Technology Seek Ways of Mining to 4, 5000 Meters." *Engineering and Mining Journal* 191 (1990): 30–34.

Westcott, N. J. "Closer Union and the Future of East Africa, 1939–1948: A Case Study in the 'Official Mind of Imperialism.' " *Journal of Imperial and Commonwealth History* X (1981): 67–88.

Index

Abaluyha, 113; origins of the name, 134n41
Abaluhya Central Association (ACA), 113. *See also* North Kavirondo Central Association (NKCA)
Abaluyia Welfare Association (AWA), 125
Abamenya, 7
Abamilikha, 7
Adagala, Solomon, 126
Adagalas, 127
Adala, John, 104, 106, 112, 113, 127, 136n57, 137n83
Aerodrome, 181
A Failed Eldorado, 82, 220
African gold dealers, 152
African landholders, 2
Africans of Freretown, 35
Afwayi, Lazaro, 103, 113
Agoi, Chief Paul, 115, 119, 139n119, 148, 189n40
Agricultural Betterment Fund (ABF), 128, 131
Akamba, 107, 159
Aldersons, W. P., 69, 79
Alluvial claims, 62
Alluvial Concessions, 209

Alluvial Mining, 61–65, 70, 77, 78, 79, 83, 84, 90, 152, 204, 205, 207, 208, 209, 210, 222, 227
Alluvial EPLs, 207
Alluvial Miners 159. *See also* Stitt, Mrs.; Stitt, Col.
Alphega Mining Syndicate, 172, 209
Amadala, 6
Amaguru, 5, 122, 124, 126, 127, 128, 130
Amahwa, Samson, 216
Amery, Leopold S., 4, 24, 25
Andachila, Shibesi, 213
Amiani, Paul, 100
Anderson, E. L. B., 103, 144, 183, 206
Anglo-Continental Gold Mining Co., 105. *See also* Anglo-Continental Mines Company, Ltd.
Anglo-Continental Mines Company, Ltd., 70, 73–74, 77
Anglo-French Kenya Development, Ltd., 71
Anti-Slavery and Aborigines Protection Society, 201n258
Apindi, Ezekiel, 14
Aseka, Eric, 106
Ashforth, Adam, 45

257

Askari, 164, 173
Assistant Soil Conservations Officers (ASCO), 122
Atsangu, 152
Asians, the, 16, 17, 191n58, 152, 153, 155, 173, 176, 180; employees, 172, 198n200, 226; traders, 152, 161, 186
Asthma, 210
Australia, 72

Baganda, 213
Bailey, D. M., 129
Bailey, W. H., 80
Bailey, W. J., 172
Baily, G. D. M, 129
Bailward, A. N., 113
Bantu Kavirondo, 113
Bantu-Luyia, 1, 48n27
Barazas, 98, 100, 101, 103, 110, 114, 118, 123, 158, 164, 177, 200n237, 224
Baraza, Jonathan, 127
Barclays Bank, 62
Bassi, 116
Batsotos, 201n182. *See also* Butsotso; Tsotso
Beacroft Group, 74
Bewick, Moreing and Co. Ltd., 75
Berman, Bruce, 106
Blue Reefs, 70
Boma Hotel, 180, 181
Bomas, 144
Bottomley, Sir William, 32
Bowles, Dr., 210
Brandsma, Rev. Monsignor, 17, 20
British East Africa Corporation, 16
British Parliament, Petroleum Bill 1934, 9–10
Broderick Falls, 124, 181. *See also* Webuye
Brooke-Popham, 107, 136n59
Bruce, Solicitor General T. D.H., 31
Bukhayo, 182, 183
Bukura Agricultural Center, 127

Bukura Agricultural Farm, 183
Bukura Mining Company, 78–79, 80, 81, 83, 172, 175
Bukura Ridge, 71, 83
Bukusu, 114, 117, 118, 120, 121, 122, 123, 124, 126, 127, 128, 163, 183, 224
Bukusu Union, 124
Bullion refiners, 62
Bullion gold, 75, 156
Buluyia, 1, 6, 8, 39, 42, 71, 99, 101, 104, 107, 109, 111, 113, 114, 115, 116, 117, 118, 120, 121, 122, 124, 128, 129, 130, 131, 132, 143, 144, 153, 157, 162, 165, 182, 204, 205, 210, 211, 220, 223, 224, 225, 228, 229
Bunyore, 8, 64, 101, 104, 114, 115, 116, 117, 118, 120, 121, 128, 129, 131, 153, 155, 157, 165, 182. *See also* Nyore
Butere, 107, 181
Burns, Rev. Canon, 31
Butsotso, 61, 85, 103, 182, 184, 201n251
Byrne, Sir Joseph A., 28, 31, 33, 221, 227

Calderwood, Rev. R. G. M., 40
Capitalist expansion, demand for prime land, 13
Carter, Sir William Morris, 15
Carter Land Commission, 13, 44
Cassava, 125
Central Land Board, the, 146, 188n14
Central Lands Trust Board, 38, 39, 40, 41
Central Native Lands Trust Board, 27, 38, 40, 41
Central Kavirondo (CK), 202n285
Central Province, 14, 109, 124, 130, 161
Cerebro-spinal fever, 208. *See also* Cerebral-spinal meningitis
Cerebro-meningitis, 143, 208, 210, 216, 228. *See also* Cerebral-spinal meningitis

Cerebral-spinal meningitis, 207
Chambers, P. C., 128
Chaminyolo, 128
Chavakali, 120
Chinese labor, 172
Chiroko, 167, 170
Church Missionary Society (CMS), 14, 18, 19, 34, 35, 105, 106
Church of God, 101
Clive, John, H., 90, 97n125, 183, 184
Colonial capitalism, 9
Colonial Development Fund Loan, 210n260
Colonial industrial capitalism, 9, 13, 221, 224
Colonial Mining Laws, 99, 143
Colonial Office (CO), 1, 3, 4, 10, 15, 16, 83, 86, 104, 108, 109, 125, 134n51, 140n151, 160, 212, 221, 222, 225, 226; repressive colonial policies, 3, 4, 5
Colonial State, 3, 4, 10–11n2, 13, 16, 44, 45, 110, 111, 113, 114, 115, 117, 118, 122, 123, 126, 128, 130, 131,143, 144, 145, 150, 151, 152, 153, 154, 156, 163, 165, 179, 205, 207, 213, 220, 223, 224, 225, 228
Committee on the Employment of Juveniles, 166
Corkscrew Inn, 180, 181
Cornish miners, 174
Cotton, 125, 161
Class Structure, 9
Closer Union of East Africa, 14, 46n5
Chief Native Commissioner (CNC) Wade, 29, 30, 31, 33, 34, 41, 211
Combe, A. D., 59
Commissioner for Local Government, 99
Commissioner of Mines, 62, 69, 84, 99, 100, 102, 150, 154, 166, 207
Consolidated Miners Selection Trust, 74
Crown Lands Ordinance of 1915, 15, 16, 24

Crown Lands Amendment Ordinance of 1926, 24
Cumberland Hotel, 181
Cunliffe-Lister, Sir Phillip, 15, 28, 32, 33, 34, 37, 38, 43, 47n12, 102, 104, 221, 222
Coast Province, 161
Cyanide, 78, 216

Daily Express, the, 63, 64, 180
Dass, Isher, 107
Davis, Carl, 74
Davis, Sir Edmund, 74
Davies, F., 172
Defense (African Labour for Essential Undertakings) Regulations, the, 163; provisions of, 163–64
Department of Labor, 164
Deventer, J. N., 172
Dini Ya Msambwa (DYM), 123, 124, 140n145. *See also* Masinde, Elijah
Dougall, Bishop James W. C., 36
Duffy, B. S., 172
Dukas, 179

East African Chamber of Mines (EACM), 80, 89
East African Concessions, Ltd., 70, 72, 73, 111
East African Department (EAD), 26, 31
East African Goldfields, 76
East African Eldorado, 221
East African Mining Areas, Ltd., 74, 76, 208
East African Rand, 220, 223
Economy Motors, 181
Economic Depression of the 1930s, 10, 27, 38, 59–60, 88, 91n3, 205, 221
Edzaba, 61
Edzawa Ridge, 70, 78, 83, 208. *See also* Edzawa Ridge Mine; Edzawa Ridge Mining Company, Ltd.
Edzawa Ridge Mine, 80, 174
Edzawa Ridge Mining Company, Ltd., 77, 78, 147, *149*; leases, 128, 147–49

Eguiri, 61
Eldorado, 180, 220, 222
Eldoret, 66, 181
Eldoret-Kakamega Mining Venture, 69
Eldoret Mining Syndicate, 65, 66, 67, 70, 72, 73, 79
Elgon, 163
Elijah, Gari Hanson, 157
Eshahidza, Esawa, 154
Eshama sha malova, 10. *See also* North Kavirondo Central Association (NKCA)
Eshangalangwe, 6
Eshitsimi, 6
European gold mining cartel, 157, 206
European settlers, 2, 5, 9, 10, 14, 19, 27, 60, 105, 121, 122, 123, 125, 131, 148, 151, 165, 224
Exclusive Prospecting License, 62, 65, 66, 67, 69, 71, 74, 144, 148, 150, 190n49, 207, 209, 222, 223, 224

FAM, 101, 102, 106, 113
Famine of 1942–43, 164–65, 226
Farson, Negley, 18, 167
Fayle, O., 209
Fazan, S. H, 15, 109, 164
Fearn, Hugh, 60, 82, 143, 166, 167, 168–69, 174, 204, 223
Feradzi, 61
Ferguson, Corrie, 177
Feza Ltd., 69
Forest Reserves, 108
Friend, the, 181
Friends African Industrial Mission, 186

Ganda, 17, 213
Geita, 73
Gendia, 217n37
Geological and Mines Department, 116
Gibendi, 128
Gold: alluvial, 59, 61, 62, 90; dawn, 59; deep level mining, 71, 75; decline of, 223; diamond drilling, 75, 222; discovery, 1, 4, 9, 10, 13, 59, 60, 61, 90, 98, 107, 150, 171, 220, 221; economic impact, 143–87; environmental impact, 149, 175, 204–10, 216; industry, 164, 214, 216; legal African prospectors, 153, 155–57; mining, 9, 10, 13, 93n53, 122, 143, 144, 149, 153, 154, 179, 182, 186, 205, 210, 213, 221, 223, 229; mining industry, 3, 4, 9, 10, 59, 63, 65, 71, 76, 77, 78, 80, 81, 82, 84, 86, 87, 88, 90, 98, 100, 104, 105, 106, 110, 123, 126, 131, 143, 144, 150, 151, 157, 158, 161, 165, 178, 186, 207, 211, 220, 223, 224, 225, 226; mining leases 102, 106, 110–11, 145–48, 165, 222, 224, 225; mining office, Kakamega, 63; main office moved to Kisumu in 1935, 95n78; Kakamega office destroyed by fire in 1941, 197n194; organized protest, against gold mining, 104–9, 131–32; prospecting License, 62; recovery methods, 59, 61–62, 216; reef mining, 62, 64, 77, 78, 90, 155, 208, 222; rush, 10, 60–71, 84, 85, 86, 117, 211; spontaneous protest, 98–104, 131
Gold Coast, 63
Gold Corporation, Ltd., 72
Gold Mines Development Loans Ordinance, 87, 88
Gold thefts, 150, 151
Golden Hope Inn, 180
Gold production, decline, 76–89
Gori Gori River, 152
Government African School (GAS), 110, 179, 202n276
Graham, M. D., 114
Great Britain, 3, 4; House of Commons, 2, 15, 34, 38; House of Lords, 2; imperial economic design, 2
Great War, the, 108. *See also* World War I Soldier Settlement Scheme
Greeks, 212

Grigg, Edward Sir, 4, 25, 26, 28, 71, 221
Group farming, 128–29, 141n165
Guru, 105
Gusii, 116, 159
Gusiiland, 9, 116

H. S. F. Syndicate, 78, 83. *See also* Stitt, Mrs.
Habari, 105
Habwatira, Atanas, 213
Hailsham, Viscount, 37
Hall, George, 84
Harding, B., 152
Hartley, A. S., 23
Harverson, D., 153, 192n78. *See also* Maragoli Gold Mining Scheme
Harvey, Conway, 25
Health, 186, 204–10; European dispensary, 209; Native Hospital, the, 206, 209; Roman Catholic Mission Hospital, Mukumu, 209–10; Rosterman Child Welfare Clinic, 210; Seventh Day Adventist Mission Hospital at Gendia, 217n37
Hemstead, Rupert W., 15
Hilton Young Commission, 26, 107
Hislop, F. D., 113, 119, 123, 144, 163
Hitchen, Stansfield C., 73
Hobley, Charles William, 17
Hosking, E. B., 21, 160
Horne, Sir Robert, 71
Howe, Swinton, 91n18, 167
Humanitarians, British, 10
Humphrey, Norman, 124
Hunter, K. L., 130, 147, 155
Hut tax, 172
Huxley, Elspeth, 60

Idakho, 8, 34, 61, 64, 70, 72, 74, 100, 102, 103, 104, 109, 129, 145, 152 157, 158, 162, 165, 174, 182, 187, 189n42, 213, 226
Imbali, Joseph, 173, 177
Imbuye, Samuel, 184

Imperial Airways Service, 96n103, 181
Indulagi, 128
Indushi, Popai, 182
Industrial labor, 157–62, 164, 226; Chinese, 172, 226; Committee on the Employment of Juveniles, 166 (*see also* Ordinance No. XXXV, Employment of Women, Young Persons and Children (Amendment) Ordinance of 1936; Women's labor); conditions, 166–69; conscription, World War II, 163–65; ethnic composition, of, 159; food rations, 167–69, 170, 186, 226; hours, 167, 174, 226; labor camp, 110, 207, 214; labor camps, 102, 151, 164, 173, 209, 213; labor compound, 170; labor compounds, 158, 194n207; Italian POWs, 172–73, 226–27; migrant labor, 161–63, 165, 226; destinations, of, 161, 165, 194n126; mining accidents, 111, 143, 174–79, 200n229, 204; protest, 169–74; absenteeism, desertion, absconding, 169,174, 226; strikes, 169, 170, 174, 197n194, 199n212, 226; women's protest, 177–78, 227; race-relations, in, 161, 173–74, 198n208, 199n212, 226; wages, 167–68, 169, 171, 174, 186, 226; withholding, of, 171–72, 226
International Labour Conference of June 1935, 159
Inspector of Mines, 177, 179
Inter alia, 4, 70, 214
Irhanda FAM School, 101, 102, 105, 106
Isaacman, Allen F., 3, 10n1
Isikhu, 61, 64, 69
Isukha, 8, 21, 34, 61, 70, 71, 101, 102, 103, 109, 110, 129, 145, 146, 148, 157, 158, 162, 165, 174, 178, 182, 187, 189n42, 201n251, 204, 206, 213, 215, 226

262   Index

Isukha Locational Advisory Council (ILAC), 215
Italians, 212
Italian-Ethiopian War, 76, 223
Ithaka, 6

Johnson, L. A., 59, 65, 66, 77, 79; discovery of alluvial gold, 59
Jones, A. Creech, 125
Johannesburg, 64, 72
Jumba, Andrea, 104, 106, 110, 112, 119, 217n34
Jumba, Robert, 104, 106

Kabras, 8, 17, 100, 104, 117, 163, 214
Kahiya, Jonah, 101
Kaimosi, 44, 64, 66, 111, 186, 214; Settled Area, 17, 64, 106, 194n126
Kaimosi Friends African Mission (FAM), 214
Kaimosi-Maragoli-Edzawa Valley roads, 181
Kakamega 1, 2, 3, 5, 8, 9, 10, 13, 14, 20, 21, 24, 36, 37, 70, 72, 73, 74, 75, 76, 77, 79, 81, 84, 87, 88, 95n78, 98, 100, 101, 102, 104, 106, 107, 110, 111, 114, 117, 138n103, 151, 152, 153, 166, 167, 170, 174, 180, 181, 182, 183, 184, 194n126, 204, 205, 206, 211, 212, 213, 220, 221, 222, 223, 227, 228; boomtown, 179–81; beneficiaries, 184; trade, 182–85 (*see also* Luyia, petty bourgeoisie; Eldorado; East African Eldorado); gold rush, 13, 14, 26, 27, 29, 37, 59, 60, 84, 106, 107, 124, 134–35n51, 211, 229; goldfields, 34, 38, 59, 60, 61, 62, 63, 64, 66, 70, 71, 72, 74, 75, 76, 77, 78, 79, 80, 81, 83, 90, 93n51, 144, 153, 158, 159, 174, 205, 206, 223, 226, 227; child labor, 155, 159–60, 163, 166, 195n162, 226; gold mining, 59, 61, 84, 90, 98, 151, 186, 206, 221, 224; gold mining industry, 5, 9, 11–12n2, 79, 80;

Kenya European settlers, 60, 98, 151, 173 (*see also* European settlers); Land question, 37; Luyia land holders, 38, 39; prospectors, 62, 63, 106, 107, 166, 170; reef mining, 64
Kakamega-Butere Road, 181
Kakamega Forest, 43, 64, 74, 101
Kakamega Forest Reserve, 74
Kakamega-Kapsabet Road, 72–73,120, 181, 201n260
Kakamega-Kisumu Road, 181. See also Kisumu-Kakamega Road
Kakamega Township, 63, 71,113, 179, 181, 182, 194n126, 205, 206, 209, 214, 227
Kakamega Ore Reductions, 181
Kalenjin, 157
Kang'ethe Joseph, 107
Kanogo, Tabitha, 11n3, 49n49
Kariuki, Jesse, 107
Kavirondo, 18, 48n27, 96n112, 105,109
Kavirondo Gold Mines, Ltd., 70, 74–75, 77, 79, 80, 81, 154, 167, 168, 169, 173, 174, 210, 214; mining lease, 146–47, 149
Kavirondo Golf Mining Syndicate, 146; mining lease, 146–47, 149
Kavirondo Taxpayers' Welfare Association (KTWA), 19, 107, 109, 212, 136n211; grievances, 33; opposition to European invasion of Kakamega, 19; prospecting, 63; radical political association, 19; registration of land, 19
Kenealy, E. M., 26
Kentan, 67
Kentan Gold Areas, Ltd., 72, 73
Kenya, 2, 9, 99, 108, 109, 112, 122, 123, 125, 150, 155, 160, 170, 171, 210, 212, 221, 223, 225, 226, 228, 229; African land rights, 2, 22, 222, 224; African reserves, 24, 38, 42, 43; British colonial policy, 4; colonial land and mining laws, 29, 38, 99; colonial economy, 60, 81, 86, 88, 90;

European settlers, 3, 10, 23, 24, 29, 59, 60; gold discovery, 27, 28, 221, 223; Kenya Highlands, 9, 15, 19, 43, 97, 106, 108, 109, 205, 107, 108 (*see also* White Highlands); Legislative Council (LegCo), 24, 25, 26, 29, 30, 31, 32, 35, 38, 39, 40, 65, 97n120, 107, 212, 221; purchasing land outside reserves, 36; relations between Luyia and European miners, 38

Kenya African Union (KAU), 124

Kenya Consolidated Goldfield, Ltd., 70, 71, 72, 76, 79, 101, 174. *See also* Kentan

Kenya Development, Ltd., 69, 70, 71, 77, 93n52

Kenya Farmers Association, 186

Kenya Gold Mining Syndicate, 80

Kenya Land Commission (KLC), 4, 13–15, 36, 37, 38, 39, 41, 44, 45, 100, 105, 106, 107, 108, 110, 125, 145, 163, 221, 222; amending of the NLTO of 1930, 40, 106; investigation, of, 16–22, 38; Joint Select Committee of Parliament of 1931, 10, 14, 41, 46n5; recommendations, of, 42–45, 222; Luyia demands, 16; members, of, 15, 110, 221; Special African Leasehold Ares, 43; terms of reference, 14–15; on the NLTO of 1930, 24–27, 39–42; recommendations, of, 42–45

Kenya Mines Department, 65

Kenya Miners' Association, 91n18, 158, 196n170, 200n237

Kenya Missionary Council (KMC), 15, 34, 35, 40, 41

Kenya Ore Reduction, 70

Kenya Reefs, Ltd., 70, 78

Kenya-Uganda Railway (K.U.R), 56n149, 162, 180

Kenyatta, Jomo, 46n6, 106, 107, 108, 134–35n51, 160, 224

Kerebe Mines, Ltd., 73, 170

Kericho, 161, 165
Kerr-Cross, D., 170
Khaminwa, Ralph, 182, 184
Khayega Market, 184
Khwisero, 129
Kiambu, 109, 124
Kibiri, 67, 103
Kikuyu, 6, 9, 106, 109, 126, 130, 159, 224; stolen lands, 13
Kikuyu Central Association (KCA), 46n6, 105, 106, 107, 108, 109, 113, 134–35n51, 137n84, 224
Kimberley diamonds, 153, 190n53
Kimilili, 118, 124, 128, 129
Kimingini Gold Mining Company, 70, 72, 73, 76, 94n61, 145, 167, 168, 169, 170, 173, 175, 177, 182, 201n260, 226; accidents, 175; labor camp, 207; mining lease, 70, 102, 110–11, 144–45, 149; opposition, to, 145–48
Kings African Rifles, (KAR), 163, 194–95n137
Kipande, 20, 21, 22, 34, 49n49, 104, 213
Kipkarren, 17, 36, 44
Kipsigis, 9, 159, 224
Kisa, 100, 103, 129, 147, 148, 208
Kisala, Lumadede, 104, 112, 113, 119, 121, 126, 127, 217n34
Kisalas, 127
Kisii, 116, 194n126, 217n37
Kisumu, 17, 89, 95n78, 155, 180, 181, 183, 184, 185, 208, 215
Kisumu-Kaimosi, 36
Kisumu-Kakamega Road, 73, 120, 201n260
Kisumu-Londiani, 161
Kisumu town, 162, 183
Kitosh, 119
Kitson, Sir Albert, 37, 38, 63, 64, 67, 93n52
Kiswahili, 155
Koa-Milimu Gold Mining Company Limited, 94n69. *See also*

Koa-Milimu Mines; Kavirondo Gold Mines, Ltd.; Risks
Koa-Milimu Mines, 74, 77

Labour Monthly Journal, 106
Lambert, 170
Lands and Settlement, 99
Land as a factor in politics, 14
Land alienation, 22, 126, 229
Land Ownership, 6, 7
Land Ownership and Economy, 6, 7
Lake Victoria, 61, 93n51
Large Mining Companies, 61, 62, 65, 71–79, 90, 162, 169, 222
Leakey, Louis, 36
Leggett, Humphrey, 36
Liebig packing plant, 107
Ligalaba, Erasto, 104, 105
Liguru, 6, 7
Limenya, 6
Lisudza, Mmudi, 104, 112
Lisulu, 79
Litavakha, 182
LNC Ambulance, 209
Local economy based on land, 6, 7, 8
Local Land Boards, 40, 108, 148,179
Local Native Council (LNC), 19, 20, 27, 31, 101, 103, 106, 112, 114, 115, 118, 121, 123, 124, 127, 129, 130, 146, 147, 162, 165, 183, 184, 195n153, 210, 213, 214, 215, 222, 227
Local Native Councils (LNCs), 4, 27, 31, 39, 108, 116, 194n126
Locust infestation, 59, 60, 64, 221
Logoli, 8,102, 114, 115, 116, 117, 118, 119, 120, 125, 126, 128, 147, 148, 153, 158, 159, 175, 177, 183, 184, 187; migrations to Kisii and Kanyamkago, 116
Logoli, Bantu, 158, 187
London Group on African Affairs (TLGOAA), the, 34, 36, 37, 160
Lord Lugard, 35, 36
Loss of land due to gold mining, 9
Lukwaa, Amina, 213

Lumidi, Ashiono, 99
Lumwachi, Johanna, 101
Luo, 17, 18, 22, 23, 146, 158, 159, 167, 170, 175, 213
Luo, Nilotic, 14, 16, 48n26, 157, 158, 187
Lurambi, 129
Lutonyi, 61, 214
Luyia, 1, 2, 3, 4, 5, 6, 7, 8, 9, 10, 14, 15, 16, 17, 18, 19, 20, 21, 22, 23, 84, 98, 99, 100, 101, 102, 103, 104, 105, 106, 109, 112, 113, 114, 115, 116, 117, 118, 119, 120, 121, 122, 123, 124, 125, 126, 127, 128, 129, 131, 132, 143, 144, 145, 148, 149, 150, 153, 154, 156, 157, 158, 159, 161, 162, 164, 165, 167, 177, 179, 181, 184, 186, 205, 210, 211, 212, 213, 214, 215, 216, 218n40, 220, 221, 222, 223, 224, 225, 226, 227, 228, 229; agrarian economy, 9, 13, 182–86, 220; American prospectors, 63–64; appointment of Paramount Chief, 16; claim to Kaimosi and Kipkarren, 44; colonial establishment, 28; demands to KLC, 16; economy, 5–10, 31, 100, 131, 143, 160, 161, 163, 220, 226, 227; employment, 60, 143, 161, 165, 167, 169, 187, 226; European prospectors, 2, 3, 14, 28, 29, 70, 76, 86, 99, 102, 105, 144, 145, 148, 23, 221, 222, 224, 228; farming methods, 125, 127; gold mining, 3, 22, 143, 186, 211, 222, 223, 225, 227, 229; illicit gold buying, 150–51, 153, 225; illicit gold dealing, 152, 154, 225; illegal gold dealing, 143, 150–51, 225; illegal gold winning/mining, 150, 151, 225; immorality, 211; individual land ownership, 7; petty bourgeois, 126–27, 128, 129, 182–84; land grievances, 14, 16, 18–22, 224, 148; landholders, 20, 29, 30, 31, 37, 38, 98, 99, 102, 103, 106, 145, 221, 222,

223; land rights, 5, 13, 20, 22, 39, 45, 104, 221, 222, 225; land tenure, 5–8, 121, 146; loss of land to miners, 28; opposition to colonial agrarian policies, 114–31, 125, 126; opposition to registration of land on individual basis, 16, 20, 21; peasants, 21, 117, 20, 121, 122, 124, 125, 130, 160, 163, 165; repeal of Crown Land Ordinance of 1915, 16; roadside plots, 125; rural industrialization, 9, 10, 129, 131, 149, 224, 228; soil conservation, 129, 131, 229; straddlers, 125–29; surface rights, 39, 44; women, 177, 227, 228
Luyia reserve, 4, 44, 123
Lolgerian, 61, 72
Lonsdale, John, 10n2

Maasai, 9, 39, 126, 224
Macalder Mines, 158, 172
Machakos, 123, 124, 125
Makerere Technical College, 214, 218n62
Malakisi, 117, 118, 130, 163, 183
Malaria, 143, 204–8, 209, 210, 216, 227–28
Manchester Guardian, 180
Marachi, 130
Maragoli, 8, 61, 64, 66, 71, 72, 84, 84, 100, 101, 102, 103, 104, 105, 113, 114, 115, 116, 117, 118, 119, 120, 121, 126, 128, 129, 131, 138n93, 147, 154, 158, 165, 172, 175, 176, 182, 183, 185, 224
Maragoli Closed Area, 153, 225
Maragoli Gold Mining Scheme, 92n30, 225
Maragoli Society, 126, 141n154, 127; Soil Conservation Committee, 126
Marama, 61, 100, 103, 106, 130, 172, 182
Marama Locational Advisory Council, 130
Masidza, Jeremiah, 104

Marikani, 173
Marsh, C. A., 91n6, 180
Marshall, Claude, 210
Masidza, Jeremiah, 104
Masitsa, Jeremiah, 113
Matsitsa, Jeremiah, 112
Masinde, Elijah, 124, 140n145
Maxon, Robert, 9
M. de. Bord Union Miniere, 69
Mbale, 119, 120
Mbiu, Chief Koinange wa, 14
Mbwabi, Nathan, 101, 103, 104
Mbwanga, David Mulindi, 126
Medical Officer of Health (MOH), 177, 204, 206, 207, 208, 209
Messrs Risks, 71, 101, 102, 103
Metal Works, 181
Milango, 5, 122
Milimu, Chief Amaidza, 64, 71, 98, 100, 102, 109, 110, 146, 152, 174, 188n17, 190n44, 192n89, 204, 216n3
Milimu's camp, 183, 204
Mining accidents, 174–79, 199n227
Mining Ordinance of 1925, 62
Mining Ordinance of 1930, 64, 65
Mining Ordinance of 1931, 29, 37, 62; provisions, of, 64–65, 223
Mining (Amendment) Ordinance of 1932, 70
Mining Ordinance of 1933, 150
Mkaisi, Ashivenda, 178
Mnubi, 115, 118, 119
Mombasa, 165, 170
Montgomery, H. R., 22, 23, 150
Morgan, M. G., 121
Mount Elgon Forest Reserve, 43, 105, 106
Mtama, 60, 125
Mudavadi, 175, 177
Muhati, Reuben, 112
Muhanga, Moses Herbert, 104, 106, 113
Muhoroni-Koru Farms, 160
Mulama, Chief Joseph Shiundu, 106, 107, 111, 112, 113, 135n54, 137n80. *See also* Paramount Chief

Mulindis, 127
Mumias, 129
Musgraves Properties, 72, 73, 77,167, 175, 199n227
Mutsami, Luka, 101, 104
Michell Cotts, 101
Mutsembi, Chief, 201n251
Mutua, James, 14

NADAR of 1937, 166
NCA, 121
Nabwana, Pascal, 124, 127, 140n147
Nairobi, 26, 32, 39, 62, 64, 72, 84, 84, 105, 106, 107, 170, 181, 212, 225, 229
Nakuru, 165
Nambale, 129
Nandi, 9, 23, 159, 213, 214, 157
National Bank of India, 62
Native Authority Ordinance of 1937, 115, 122
Native Chamber of Commerce (NCofC), 19
Native Land Trust Bill (NLTB), 20, 25
Native Land Trust Board, 43, 62
Native Land Trust (Amendment) Ordinance (NLT(A)O) of 1932, 1–2, 4, 13, 24, 29, 31, 32–34, 35, 37, 39, 42, 43, 44, 106, 144, 145, 221, 222, 223; CO defends, NLT (A)O, 37–39; provisions, of, 144, 145, 221–23; protest against the NLT(A)O of 1932, Kenya, in, 32–34, 106; protest in England, 34–37
Native Land Trust Ordinance (NLTO) of 1930, 1, 4, 13, 16, 24–25, 26, 27, 28, 29, 31, 34, 37, 38, 39, 40, 41, 43, 56n153, 62, 106, 220; Provisions of, 221, 224, 225
Native Lands Trust (Amendment) Ordinance (NLT(A)O of 1938, 107
Native Land Trust Ordinance of 1938, 107, 108, 118
Native Tribunals, 115
Ndagalu, Habil, 184

Ndede, Leokadius, 153, 156, 157
Ndorobo, 6
Negro, 106
Nkore/Toro, 15
Ngorio, Chief, 109
North America, 81, 123
North Kavirondo (NK), 1, 2, 3, 4, 5, 8, 9, 10, 17, 18, 19, 21, 22, 24, 26, 29, 30, 31, 34, 37, 43, 53n106, 56n153, 61, 64, 66, 72, 76, 81, 82, 85, 90, 96n112, 98, 99, 101, 106, 109, 111, 113, 116, 117, 118, 121, 123, 124, 125, 127, 131, 135–36n56, 143, 145, 158, 159, 161, 162, 163, 164, 165, 170, 179, 181, 182, 202n285, 204, 209, 210, 220, 221, 222, 224; African landowners, 99; alienation of land, 143; discovery of gold, 4, 13; economy, 60; European prospectors, 99, 144, 145; gold fields, 65, 159, 166, 169, 174; gold mining industry, 59, 61, 144; Land registration, 21, 22; locust infestation, 59, 60–61; prospecting in Maragoli and Nyang'ori, 59, 129
North Kavirondo Central Association (NKCA), 98, 102, 111, 134n48, 137n77, 137, 117, 118, 119, 224; members, 104, 106, 112, 113, 127, 136n57, 137n83; protest against gold mining, 104–9; state response, to, 109–14; opposition to soil conservation, 113, 114–31, 126
North Kavirondo Chamber of Commerce (NKCofC), 182, 183
North Kavirondo District, 13, 21, 59, 60, 61, 71, 72, 76, 79, 81, 85, 90, 99, 125, 160, 161, 162, 164, 179, 188n17, 206, 210, 220, 224
North Kavirondo Local Native Council (NK LNC), 101, 118, 121, 122, 123, 124, 128, 147, 148, 165, 182, 183, 184. *See also* Local Native Council (LNC)

North Kavirondo Taxpayers' Welfare Association (NKTWA), 107–8, 136n57
North Nyanza, (NN), 129, 130, 131, 184
North Nyanza African District Council (NNADC), 142n178, 149
North Nyanza District, 86, 96n112, 128,185
North Nyanza Local Native Council, 129, 130, 141n165
Nubians, 16, 17
Nyang'ori, 59, 129
Nyani, 173, 209
Nyanza Province, 14, 15, 16, 17, 18, 23, 27, 59, 60, 61, 63, 66, 67, 77, 106, 109, 110, 115, 116, 152, 160, 162, 164, 165, 169, 175, 181, 184, 210
Nyaparas, 115, 166
Nyaribari, 116
Nyoka Nyeusi, 173
Nyore, 104, 117, 153, 182

O'Brien Property, 78, 221, 223
Obulimu, 6
Odanga, Chief William, 109
Oluanjelekha, 6
Olugongo, 6, 121, 122, 127
Omugunda, 6, 7, 8
Omulimi, 6
Omwene, 6
Omutsuru, 6
Okumu, W. Zadok, 156, 157
Ojango, H. Benjamin, 156, 157
Ordinance No. XXXV, Employment of Women, Young Persons and Children (Amendment) Ordinance of 1936, the, 159, 160
Ore exhaustion theory, 88
Osborne, P. S., 130
Osianjo, Dominikus, 146, 150
Otiende, J. D., 214
Overseas League, the, 173
Ovutsakha, 6
Owen, Walter Edwin, 18, 19, 20, 23, 33, 34, 40, 41, 48n27, 105,134n48, 160, 175, 193n111, 211, 224. *See also* Church Missionary Society (CMS)
Owino, Reverend Mathayo, 156, 157
Owombo Mining Company, 79, 81,173

Panning, 61
Pakaneusi Prospecting, 75
Paramount Chief, 16, 111, 112, 113, 137n80
Paris Green, 207
Pass Laws, 213
Pemberton, Harold, 63, 180
Penfolds Garage, 181
Piccadilly Circus, 71, 146. *See also* Sigalagala
Pitts, Arthur Pitt, 35
Plateau Transport, 69
Pneumonia, 143, 206, 207, 210, 216n1, 228
Posho, 121, 167, 168, 170, 171, 184, 185, 186
Post World War II Nyanza Province Five-Year Plan, 152
Prisoners of War (POWs), 172, 173
Prostitution, 211, 213, 214, 216, 228; colonial Nairobi, in, 212–13; Kakamega, in, 213–14; Kisumu, in, 215–16; family structure, impact of, 211, 216; social cultural tensions, 228 (*see also* Social transformation); stigmatization, 215–16, 228; response, African authorities, to, 213–15; Venereal Diseases (VDs), 213–16
Prospecting License, 62
Provincial Commissioner, (PC), 62, 110, 121, 122
Public Health Ordinance, 208
Public Works Department (PWD), 161, 169
Pugh, W. R. B., 154
Pulfrey, William, 117
Power lines, 181

Red Ridge, 114, 115, 120
Reef Mining, 65, 90

Regimental Debts Act of 1893, 44
Report of the Committee on Native Land Tenure in the North Kavirondo Reserve, 20, 11n8, 11nn10–11, 12nn15–16, 12n19, 12nn21–24, 140n141
Rhodesia-Katanga Company, 67
Riddoch, John, 89
Rift Valley, 123
Risks, 71, 74, 76, 77, 173, 174, 175, 181, 206, 207. *See also* Risks, Ltd.; Risks Mine Company
Risks, Ltd., 70, 74, 77, 94n69
Risks Mine Company, 76, 111, 206
Roberts, A. D., 166, 174, 204, 223
Robotham, H. P., 172
Rodseth, 173
Roman Catholic Mission at Mukumu, 210
Ross Mining Syndicate, 75
Ross, William McGregor, 40, 41, 56n149
Rosterman Gold Mining, Ltd., 70, 74, 75–76, 77, 79, 80, 82, 83, 85, 86, 87, 88, 103, 111, 145, 146, 147, 148, 149, 160, 162, 167. *See also* Tanami Gold Mining Company
Rosterman labor camp, 207, 168, 169, 170, 172, 173, 174, 175, 177, 178, 179, 182, 207, 209, 210, 214, 220, 226; mining lease, 145–46, 147, 148, 149, 183–84
Rulungulu, 67
Rural landless class, 9
Rutherford and Company, 181

S. Everett and Sama Syndicate, 78, 223
S. M. Syndicate, 69, 70, 71, 76
SAO, 121, 124, 147
Samia, 183
Saramanji, 81
Secretary of State for the Colonies (S of S), 4, 14, 17, 25, 83, 84, 97n120, 102, 104, 108, 109, 111, 112, 125, 137n77, 173, 212, 221, 226

Segero, Chief, 150, 214
Seychelles, 28
Seychellois, 80, 176, 191n58, 226
Seleta, Minishi, 178
Sexually Transmitted Diseases, (STDs), 215
Shamba(s), 115, 117, 119, 120, 144, 148, 160
Shenzi, 173
Shiberenge, James, 104
Shibonje, Museve, 178
Shields, W. P., 159
Shiels, Drummond, 36; criticism of NLT(A)O of 1932, 36
Shikokho, 61, 67
Shikomera, 184
Shikomera, Benjamin, 182
Shikonde, 74
Shilisia, Chimoi, 213
Shitambasi, Etolondo, 178
Shitanda, Mulupi, 100, 214
Shitoli, 61
Shivachi, Chief William, 189n41
Shivoga, Mutangale, 213
Sierra Leone, 35
Sigalagala, 74, 146, 154, 175, 215
Silicosis, 210
Simekha, Joseph, 155, 191n75
Simsim, 125, 170
Singh, Curchuran, 153
Singh, Sauta, 153
Slatter, E. P. C., 211, 212
Sluicing, 61–62
Small men, 70
Smallwood, 78, 223
Smallwood and O'Brien Property, 78, 223
Socialist Fabian Bureau, 50n75
Social transformation, 204–16
Soda Ash, 86
Soil Conservation, 5, 9; Kenya Land Commission, recommendations, on, 107; Luyia opposition, to, 98, 117, 118, 126, 130–31, 224; state response, to, 128–9; Luyia women,

and, 131; Maize cultivation, and soil erosion, 117–24, 128, 129, 130, 163
Sore, Chief, 21, 100, 102, 109, 174
South Africa, 2, 72, 75, 80, 123, 126, 153
South Kavirondo, 9, 66, 72, 79, 116, 158, 162, 164, 199n212, 217n37
Southern Rhodesian Land Commission of 1925, 15
Spectator, the, 63
Standard Bank of South Africa, 62
Stanley, D. H., 209
Stitts, Col., 79, 152, 191n61, 223
Stitts, Mrs., 78, 80, 83, 152, 191n60, 209, 223
Stone Company, 181
Sunman, W. O, 185
Swahili, the, 16, 17
Sweatman, E. A., 161

Taita, 107
Tanami Gold Mining Company, 69, 70, 71, 75
Tanami Gold Mining Syndicate, 69
Tanganyika Concessions Limited, 66, 67–9, 70, 72, 73, 77, 79, 93n40, 101, 110, 111
Tanganyika territory, 59, 84, 159, 162
Tax, 160
Taxation, 158, 226. *See also* Hut Tax
Taxes, 172, 225
Telegraph line, 181
Thika, 165
Thompson, C. B., 21
Tintax, 147, 148
Tiriki, 61, 79, 100, 103, 104, 110, 111, 114, 115, 120, 121, 129, 130, 165, 174
Tisdall, E. G. St. C., 154
Tooley, H., 172
Trade Union, 111
Timber Concessions, 101
Timber Licence, 74

Trading in Unwrought Precious Metals Ordinance of 1933, 151
Trading in Unwrought Precious Metals (Amendments) Ordinance (TUPM(A)O) of 1940, 151, 154, 225
Trading in Unwrought Precious Metals (Amendment) Ordinance (TUM(A)O) of 1952, 154, 225
Trans Nzoia, 17, 23, 106, 123, 161, 162, 163, 165
Trent, A. I., 170
Trusteeship, 2, 4, 29, 55n137, 222
Tsotso, 201n251

U. G. Tank, 181
Uasin Ngishu, 17, 104, 161, 162
Uganda, 15, 159, 213, 214
Ukamba Members Association, 107
Ukambani, 107, 126

Vagrancy Laws, 213
Ventures, Ltd., 71
Vihembo, Adriano, 177
Vihiga, 119
Vidal, M. R. R., 110

Wagner, Gunter, 5
Wakukha, Imbande, 213
Wanakacha, Andrea, 104
Warden of Mines, 62, 73, 99, 100, 102, 104, 110, 163, 172, 179
Watende Mines, Ltd., 199n212
Watetezi, 105
Watkins, Oscar, 40, 56n143
Watson, T. Y., 114
Watt, Lynne, 60
Watu, 152
Webb Sydney (Lord Passfield), 4, 24, 25, 26, 27, 28, 36, 50n75, 221; "Black Papers", 27; "Passfield pledge", 27; White Paper on Native Policy of 1930, 27, 106
Webuye, 124
West African Gold Coast, 150

Westerberg, P., 153
Western Civilization, 127
Western Kenya, 1, 3, 4, 5, 8, 9, 10, 45, 84, 88, 98, 204, 207, 208, 220, 228; African land rights, 4, 108; discovery of gold, 98
White Highlands, 27, 44, 56n143, 105, 107, 108, 123
White, Luise, 212, 224
Whitehouse, L. E., 122
Williams, C. H., 127
Williams, D. K., 117
Wilson, Frank O'Brien, 15, 25, 47n13
Wimbi, 60, 125
Women's *barazas*, 131
Women's labor, 163, 226
Workers' Educational Association (WEA), 107, 134–35n51
Workmen's Compensation Scheme, 179, 227
World Agriculture Census, 130
World War I Soldier Settlement Scheme, 17, 108
World War II, 76, 78, 79–85, 89, 90, 117, 123, 151, 152, 160, 162, 163, 165, 194–95n137, 220, 223, 225, 226
World War II Soldiers Settlement Scheme, 123

Yala, 61, 64, 67, 69, 70, 72, 74, 194n126, 208, 209

Zambesia Exploring Company, 67
Zawadi, 81
Zeleza, Tiyambe, 49n49, 213